Playing with Fire

Playing with Fire

Histories of the Lightning Rod

Edited by

Peter Heering,
Oliver Hochadel,
and
David J. Rhees

American Philosophical Society
Philadelphia   2009

ISBN: 978-1-60618-995-5
US ISSN: 0065-9746

Library of Congress Cataloging-in-Publication Data

Playing with fire : histories of the lightning rod / edited by Peter Heering, Oliver Hochadel, and David J. Rhees.
    p. cm. — (Transactions of the American Philosophical
  Society held at Philadelphia for promoting useful knowledge ; v. 99,   pt. 5)
  Includes bibliographical references and index.
  ISBN 978-1-60618-995-5
 1. Lightning rods—History. 2. Lightning rods—Social aspects—History.
3. Lightning rods—Philosophy—History. 4. Lightning—Research—History.
5. Lightning—Social aspects–History. 6. Lightning—Philosophy–History.
I. Heering, Peter. II. Hochadel, Oliver, 1968– III. Rhees, David J.
  TH9061.P58 2009
  693.8'98—dc22

                                                          2009033624

TRANSACTIONS
of the
AMERICAN PHILOSOPHICAL SOCIETY
Held at Philadelphia
For Promoting Useful Knowledge
Volume 99, Part 5

Contents

**Part 3: The Lightning Rod as Commodity in Nineteenth
Century America**

**Part 4: Looking at Lightning: Models, Instruments, and
Modern Research**

Acknowledgments

The editors would like to express their sincere gratitude to Mr. Thomas F. Peterson, Jr. for financial support that made possible our 2002 conference on the history of the lightning rod at the Bakken Museum.

We also thank Mr. Earl Bakken, inventor of the first transistorized cardiac pacemaker, for inventing the Bakken Museum itself. The Bakken's remarkable exhibits and collection of books and artifacts documenting the history of electricity and magnetism provided an appropriate (and sometimes literally shocking) venue for our discussions. Thanks also are due to the Bakken's staff, particularly Elizabeth Ihrig, Librarian, who assisted many of the contributors to this book in conducting their research.

We thank the anonymous reviewers for their helpful and insightful comments, and also BookComp, Inc. for their excellent copy editing and book design.

We especially wish to thank our editor, Mary McDonald, Director of Publications at the American Philosophical Society, and the Society itself. We feel honored that our book has been published by an institution founded by Benjamin Franklin himself, in 1743, which makes it even older than the lightning rod!

Illustrations

Introduction
Revisiting an Invisible Technology

Oliver Hochadel and Peter Heering

DOES THE LIGHTNING ROD HAVE A HISTORY? A modern encyclopedia suggests that Benjamin Franklin invented the lightning rod in 1752, and it has been in use ever since, protecting human beings and their property from heavenly destruction.[1] End of story. Indeed, the lightning rod's construction seems so simple and its function so obvious that even a child can understand it. What more could possibly be said?

If the story were really so pat, this book might end here. However, the matter is far from being so simple. In our own research on eighteenth century electricians and research on electricity, we each encountered these odd-looking metal constructions time and again. For one thing, Benjamin Franklin's invention of the lightning rod is one of the classic, emblematic stories of the Enlightenment.

The most famous episode is Franklin's kite experiment, which showed both artificial and natural electricity to be electrical and which was related in these stories to the invention of the lightning rod. The demonstration of lightning as an electrical phenomenon was hailed as a momentous breakthrough in science, credited by some with the breakdown of an old system of religious beliefs and heralded by others with the very invention of America. The second best-known episode is probably the Purfleet controversy in which the proper design of the tip of the lightning rod was hotly debated in Britain in the 1770s. This was epitomized by the anecdote in which the president of the Royal Society (in favor of points) told King George III (in favor of knobs) that it was not in his power to alter the laws of nature.[2] And the third cause célèbre is the trial of St. Omer in Northern France in the early 1780s concerning the right of Charles Dominique de Vissery to install a rod on his roof. This spectacular trial was followed throughout France (and beyond) and provided the first public stage for a young lawyer named Maximilien de Robespierre. The future revolutionary scolded the people of St. Omer for their backwardness and their failure to embrace enlightened values.[3]

1

On one level, then, the lightning rod has played a historical, if stereotypical role as the stuff of breakthroughs, founding moments, and anecdotes that celebrate the triumph of reason and science over superstition and dogma. In our own research, lightning rods have played less exalted but perhaps more interesting parts in the story of Charles Augustin Coulomb's classic investigations that led to the formulation of the inverse square law of electrical charge (Heering) and in the humbler story of itinerant lecturers struggling to scratch a living from public demonstrations of electrical marvels and the installation of lightning rods (Hochadel).[4] These tales were less about triumph and more about conflict, both epistemological and economic. Our stories dealt with the contested use of new instruments and questions of authority: Who was to decide the proper design of a lightning rod? Could these quarrels simply signify the difficult birth of a new technology that, once established, stirred little interest or controversy after 1800? Or could there be more?

Our curiosity was aroused and made us probe deeper, well beyond our initial topics. Surveying the publications in our field, we quickly learned that research on lightning and the lightning rod remained controversial throughout the nineteenth century. For example, the august *Académie des Sciences* in Paris devoted several committees to the subject, arguing for decades about design issues.[5] Beginning in the 1820s the British electrician William Snow Harris promoted his new lightning protection system for warships. In his numerous publications he tried to alert the Royal Navy to the fact that several warships and many men had been lost due to insufficient lightning rods.[6] In the late nineteenth century the British physicist Oliver Lodge tried to re-create lightning in the laboratory to improve the construction of lightning rods.[7] And during World War I, Nikola Tesla, the flamboyant inventor well known for his high-voltage discharge experiments, obtained a patent for an improved lightning rod.[8]

In the early twentieth century German scientists attempted to virtually catch lightning in order to perform high-energy experiments attempting to use the lightning bolt as a source of energy.[9] Lightning rods turned out to be relevant even in mining: lightning could strike aboveground buildings and be conducted into the mine.[10] We learned that Swiss lightning rods in the 1950s contained radioactive substances in order to improve their efficacy.

Some researchers are now proposing high-energy laser beams as a new method of securing airports and other buildings with sensitive electronics against lightning.[11] In recent decades electrical engineers have been arguing about lightning rod tips and fighting rearguard battles against purveyors of "high tech" electronic lightning rods. At the beginning of the twenty-first century, the "war of the lightning rods" is raging again.[12] Franklin would be amazed.

Clearly there were more twists and turns to the story of the lightning rod than we expected. So does the lightning rod have a history? Yes, we concluded, but—

to put it bluntly (or might we say pointedly?!) — it has not been written yet. In 1984 Rudolf Stichweh pointed out: "In retrospective, the lightning rod is in danger of being underestimated. It appears as an instrument of such simple matter of course — based on almost trivial knowledge."[13] Because the lightning rod seems to be so common and self-evident in its utility, it has not attracted a great deal of attention from scholars. The lightning rod became invisible because it seems so easy to understand and purportedly remained technologically stable through time and across cultures.

The Historiography of the Lightning Rod

Taking a closer look at the history of this seemingly ahistorical technological object, we learned that there have been a few attempts to write a "history" of the lightning rod. With respect to the nineteenth century, two monographs stand out: Richard Anderson's *Lightning Conductors, Their History, Nature, and Mode of Application* (first ed., 1879) and Heinrich Meidinger's *Geschichte des Blitzableiters* (1888).[14] They were labeled "history," and they did indeed gather an enormous amount of source material that is still of great value today. Yet they were both predominantly concerned with practical questions of design. History served merely to provide examples of "how to" and "how not to." Meidinger's motivation, for example, was to update his knowledge of the subject when he was asked to give advice to the administration of the duchy of Baden in southwestern Germany. In that sense, they were not disinterested histories but continuations of the Enlightenment enthusiasm for the lightning rod.

The next attempt was not published until 1950. Yet neither is Basil Schonland's *The Flight of Thunderbolts* a history of the lightning rod as such.[15] It is best described as a popular science book on lightning as a natural phenomenon with a brief historical introduction. Typical of the historiography of the 1950s, the invention of the lightning rod serves as an example of the transition from "myth" to "science." This triumphal Enlightenment theme has often been echoed by historians who cite the lightning rod as one of the key achievements of eighteenth century research.[16]

This has changed, of course, in the past fifty years. The history of the "great men" and historical accounts that smack of teleology or tell a history of linear progress have gone out of fashion, at least within the discipline of history of science itself. The word "superstition" is nowadays only used in quotation marks to avoid simply replicating the Enlightenment worldview. Yet these shifts in the history of science have not led to a significant change in the perception of the lightning rod. Instead of being glorified as a scientific achievement, it is now simply omitted, as in two recent publications that are in many other respects excellent:

the encyclopaedic *Eighteenth-Century Science* (942 pages) gives the lightning rod just two brief nods; in the collected volume *The Sciences in Enlightened Europe* (566 pages), the lightning rod is not even indexed (though lightning is).[17]

Due to the 250th anniversary of his kite experiment in 2002 and the 2006 tricentennial of his birth, public interest in Franklin received an immense boost.[18] In the wake of this booming "Franklin industry," three books were published that mention lightning or the lightning rod in their titles.[19] Yet Thomas Tucker's *Bolt of Fate* focuses more on the intriguing, if somewhat tangential issue of "did he or didn't he" actually fly that famous kite. Michael Schiffer's *Draw the Lightning Down* deals with "electrical technology in the age of enlightenment," as the subtitle of his book specifies. Only one chapter is dedicated to the lightning rod. However, he is not offering a study of the lightning rod but, due to his broad approach, remains very much on the surface of the historical background of these technologies. And Philip Dray's *Stealing God's Thunder* is a popular biography of Benjamin Franklin that focuses on the scientific aspects of his life and relies on existing scholarship. None of the volumes looks far beyond the actual invention of the lightning rod.

The lightning rod has received some attention in the narrative of the "disenchantment of the world," specifically with respect to the shift in mentalities that took place in the eighteenth century, at least among the upper classes.[20] The Almighty is not a punishing God anymore but a loving father. In former times, a thunderstorm was supposed to remind the sinner of his bad conscience; now it is only the unenlightened projecting their fear onto thunder and lightning. The German physicist Georg Christoph Lichtenberg called the lightning rod a "Furchtableiter," a rod that drains off not only electrical charges but also fear. This mental shift in turn enables the aesthetic appreciation of natural events such as thunderstorms. In Immanuel Kant's view, instead of being at the mercy of the elements, we experience the sublime. Yet the Enlightenment also causes a metaphysical "deficit." If being struck down by lightning is not God's punishment, it is arbitrary and, in a moral sense, "scandalous."[21]

All in all, the scholarship on the history of lightning rod is still very patchy. Quite a few fine articles have been published, but they remain isolated. There has not been a major scholarly monograph on the history of the lightning rod for more than fifty years, and there is certainly none that lives up to modern standards of scholarship.

It is nevertheless somewhat surprising that the lightning rod has not attracted more scholarly attention, in particular taking into account the very fruitful trend in cultural history to research the often intriguing and revealing stories of simple artifacts. As Henry Petroski has shown, one can learn an immense amount about the interplay between materials, modes of productions, and the needs of customers by looking at everyday objects such as the paper clip, zipper, and pencil.[22] It was rather a longwinded story until the paper clip had reached its final form, which was both functional and aesthetically appealing. There are countless examples of cultural histories of "*x*." Take, for example, the history of salt, a seemingly "natural" and therefore ahistorical substance. Two recent monographs have shown how salt was shaped by human culture and how the "white gold" has been all kinds of things: a substance that was mined and traded to preserve and to spice food. At times, salt was an economical and political resource, a commodity, and an object of scientific research. In earlier times, salt was fraught with symbolical meanings, in particular in alchemy, before it was "banalized" by industrial production.[23]

In the history of science and technology, this turn toward objects is sometimes labeled "material culture." A shift from "disembodied" theories toward experimental practices has proven very fruitful.[24] With respect to the production of scientific knowledge, the role ascribed to instruments has changed from mere accessories to fundamental catalysts, comparable to theories in their importance. Moreover, as the use of apparatus has gained attention, a narrative in which actors must develop specialized skills in order to use their devices appropriately has replaced the notion of impersonal experimental set-ups and protocols. Recent studies have shown that the production of scientific knowledge is highly localized and culturally embedded. Innovation depends on the availability of expertise and material, and such technological resources are unevenly distributed.[25]

At the same time, the natural sciences are more and more seen as a part of human culture. This shift has led to a sharper perception of political, social, and religious contexts. Natural philosophers are bearers of particular attitudes and mentalities embedded in specific traditions. Knowledge is transformed in being communicated and diffused, and therefore social interactions and ways of communicating scientific knowledge have gained attention.

Applying these new approaches to the lightning rod, a host of questions comes to mind: Who were the people doing research on the lightning rod? Who manufactured, marketed, and installed it? How did their designs differ? Were there local appropriations or some kind of universal standard? What kind of instruments were used or developed in this context? How did the new technology spread and at what speed? Who promoted the lightning rod and with what kind of rhetoric? What resistance did it meet and why? These questions suggest that there is not just one but multiple histories of the lightning rod. It is an object of scientific inquiry and technological development as well as a commodity, a materialization of ideas, and a metaphor.

We admit: we were and still are astounded that this simple iron bar could cause such reverberations. This fascination was the starting point of a project whose end is marked by this publication. We organized a small conference on "The History and Cultural Meaning of the Lightning Rod." The ideal venue for such a conference was The Bakken, in Minneapolis, Minnesota, where we both had been visiting research fellows. David Rhees, the director of The Bakken, quickly became as enthusiastic as we were about the idea of hosting a conference on lightning rods. And given that all these discussions took place at the beginning of this millennium, the conference simply had to be held in 2002, celebrating the 250th anniversary of the Marly experiment, the "official" birth date of the lightning rod.

Toward the end of the conference, David Rhees told us about one of the reactions he received while preparing for the event: "Hmm, the lightning rod—interesting, but isn't it a bit limited as a topic?" The twenty or more participants of the conference could not help smiling. Our call for papers had generated more than a dozen contributions from various disciplines, including history of science and technology, engineering, history, history of art, and several branches of literature. We were amazed to see how varied the historical and cultural contexts were in which lightning rods played an important role. And in spite of the great variety of the papers, we were at the same time surprised to see how many common themes emerged. At the end of the conference, we all agreed that it would be more than worthwhile to publish a volume devoted to an object that cuts so broad a swath through time and space as well as through disciplines. The papers given at The Bakken have gone through numerous revisions and have benefited from our discussions during and after the conference as well as from the review process.

So what kind of book is this collected volume? It is not "a" or even "the" history of the lightning rod. Even if we had attempted to write or edit a "complete" or "definitive" history, it is difficult to imagine how such a book could cover more than 250 years of a convoluted history. In the epilogue of this volume, we sketch a few of those aspects that we did not touch upon (or barely did). Far from being comprehensive, the essays in this volume tell eleven very different episodes in the history of the lightning rod.[26] Thus the diversity of the material presented here turns out to be an asset. As a whole, they give an idea of the breadth (and the depth) of the topic, pointing in very different directions. Yet these essays are not isolated; they intend to show the potential of interdisciplinary discussions. In the remainder of this introduction, we elucidate how the essays are interconnected in various ways. In doing so, we also show how they address broader issues that have been recently addressed in the history of science, in Enlightenment studies, and in other fields.

Lightning Rods as Symbols of the Enlightenment

The utility of natural philosophy was a commonplace of the eighteenth century. There is hardly a treatise, preface, or book review that does not extol the practical benefits mankind will reap from the investigation of the natural world. Such statements generally make historians suspicious, so it is not surprising when Martina Lorenz identifies them as "nearly stereotypical phrases." "Promoting the Publick Good" (Robert Hooke) became part of the enduring rhetoric of the rise of science.[27] If we look closer at the actual achievements in the eighteenth century, however, we realize that natural philosophy at that time had little to show for itself. True, experimentalists were able to produce effects that appeared to their contemporaries as wondrous, most notably in electricity. New epistemological approaches provided much more convincing frameworks to understand, for example, the way the heavens worked. The foundation of scientific academies and other learned associations mark a crucial stage in the institutionalization of science. Yet in terms of applicable inventions or useful discoveries, the results to that point were rather unimpressive. Or, as Geoffrey Sutton remarked, "science announced its achievements before it achieved any notable successes." Recently Michael Lynn pointed out that "natural philosophy, in particular, was useful because it could be used for social reform and the overall rationalization of society rather than because of its more concrete applications."[28] In other words, the public barrage of stressing the utility of natural philosophy was largely more rhetoric than reality, and existed more in print than in practice. Natural philosophy was only about to establish itself as something that is worth pursuing and worth funding. Particularly in the first half of the eighteenth century, natural philosophy still had to fight notions of being something utterly useless, pedantic, or outright esoteric ("weighing air"!).[29]

This ironic situation helps to explain the extraordinary symbolic meaning that was attributed to the lightning rod. It qualified as one of the few tangible achievements of natural philosophy that did indeed promote "the Publick Good." And what a spectacular achievement it was, controlling the most powerful fire coming down from above. Thunder and lightning had in many cultures been interpreted as emanations of God's power. Research into the workings of nature had yielded a result that did benefit mankind in a very concrete sense. Historians have described the lightning rod as a "materialization of Enlightenment," "a popular metaphor for the late Enlightenment," or "a telling symbol of New World Enlightenment."[30]

Is the lightning rod "a thing that talks," to use Lorraine Daston's phrase? Is it one of those objects that "press their message on attentive auditors"?[31] Not anymore. The lightning rod has grown more silent in the past 250 years. Yet in the second half of the eighteenth century, the lightning rod spoke out loud. In 1784 Georg Christoph Lichtenberg gave the *siècle des lumières* also a voice so it could enumerate all its achievements: "I have learnt to resist thunder" was second on its list.[32]

Indeed, the existence of the lightning rod was a clear indicator that Enlightenment had arrived. When Gottlieb Wilhelm in 1791 wanted to defend the imperial city of Augsburg against recurring criticism of its backwardness, he pointed out "that lightning rods were erected on several buildings," a fact that many would "not expect from such a sinister town."[33] Lightning rods became a measuring stick of how far Enlightenment had spread.

In the *Histoire de l'Académie Royale des Sciences* from 1777 we find complaints that although in England and the United States the utility of the lightning rod had been accepted, and even in Italy, and the "Maison d'Autriche" lightning rods had been erected, the French were still hesitant (Heering). In 1783 the Italian scholar Sebastiano Canterzani admitted "with regret and confusion that in Bologna and surrounding areas" no building was equipped with lightning rods (Bertucci). José Antonio Alzate y Ramírez nearly despaired when in 1793 the lightning rod he had erected twenty-two years prior was still the only one in Mexico City (Clark).

In the nineteenth century the symbolic meaning of the lightning rod was still very present. Although these protection systems had been extremely rare in rural areas in the eighteenth century, lightning rod men began peddling them from roughly the 1840s onward. Lightning rods affixed on farmhouses and barns were spearheading the "village enlightenment" (Mohun and Cavicchi).

It is only in the twentieth century that lightning rods have become both ubiquitous and invisible. Nowadays it takes a trained eye to spot the small upper termination of the rod, barely higher than the roof itself. As a topic, lightning rods have long left the public arena of discussion.

New Meanings, New Images

Lightning can acquire positive meanings as soon as its nature is understood and its force controlled. With a lightning rod on one's roof, the "beauty" of a thunderstorm can more readily be appreciated. Yet the new coding of lightning goes further as Christian Fuhrmeister shows in his analysis of political iconography of the French Revolution. On numerous engravings and sketches, lightning is depicted as a liberating force striking down the *ancien régime*. In the traditional iconography, paintings and sculptures show God Almighty casting down bolts of lightning. It is certainly no coincidence that the last works of art of this kind were made in the 1780s. The revolutionaries seized God's instrument, and soon afterward lightning as an icon became a metaphoric weapon of other political powers too. Lightning symbolized the sheer force to destroy enemies, a metaphor that lends itself particularly well to all variants of fighting for freedom. Without the taming power of the lightning rod, this shift in iconography would have been impossible.

Slow to Spread

The lightning rod was a powerful symbol of the Enlightenment to control the forces of nature. Yet to what extent was the celebrated invention a reality? Even, or particularly, in the case of the lightning rod, we should be careful not to fall for Enlightenment rhetoric. The "invention" of a new technology often blurs its initial ambiguities and the complex circumstances. So it is important to note that in its early phase the lightning rod meant two different things that were not always easy to distinguish: an instrument of research into atmospheric electricity and a protection mechanism for buildings. These meanings were situated somewhere in between science and technology. The first rod erected in Marly near Paris on May 10, 1752, by the French translator of Franklin's work, Thomas François Dalibard; the kite flown by Benjamin Franklin in or around June 1752 at Philadelphia; and the electrical apparatus used by Giuseppe Veratti on the observatory tower of Palazzo Poggi on July 27, 1752 (Bertucci), all belong to the former category. The sources on these devices are not rich in technical details, but it seems that some of them were insulated and not grounded. After all, their purpose was to prove the electrical nature of the atmosphere and of lightning. Therefore they were very dangerous. On August 6, 1753, Georg Wilhelm Richmann, a member of the St. Petersburg Academy of Science, was killed during a thunderstorm by lightning that he had drawn down into his laboratory.

The ambiguities in this early phase were multifold. What exactly was a lightning rod? Was the multipointed "machina meteorologica" erected by the Moravian priest Prokop Diviš on June 15, 1754, in his garden a real lightning rod? Natural philosophers of the time and historians of science have had their disagreements on this point. After all, this was a challenge to Franklin's prominent place as the sole inventor of the rod.[34]

And one last ambiguity: how does the lightning rod actually work? Does it avert lightning by dissipating atmospheric electricity or by attracting lightning? Benjamin Franklin was equivocal on this point.[35] The idea that the rod might actually increase the likelihood of a building being struck by lightning was the key argument of the lightning rod's early opponents such as the Abbé Nollet (Heering).

In British America, numerous lightning rods were already erected in the 1750s. Yet in Europe it took three decades for this allegedly marvelous device to spread to at least some degree.[36] The first lightning rod in England was mounted in 1760 on a lighthouse south of Plymouth; in Italy, in 1766 at the campanile of San Marco in Venice; in the German Empire, in 1770 at the bell tower of the Jacobi church in Hamburg; in Austria-Hapsburg, in 1770 at a church in Penzing near Vienna; in Switzerland, in 1771 on the house of Horace Bénédicte de Saussure, a natural philosopher in Geneva; and in France, in 1773 at the Academy of Sciences in Dijon.[37] Yet these lightning rods were lonely forerunners. There were still relatively few rods around by the end of the 1770s.

There was certainly a tipping point around 1780. After that lightning rods proliferated rapidly, at least compared to the previous pace. This, of course, raises more questions. Why around 1780? Why not earlier—or even later? The new technology needed time to "stabilize" as an artifact and to win the recognition of more than just natural philosophers. By 1780 the clergy had been won over, the enlightened rulers had become staunch supporters of the lightning rod (not least because they wanted to protect their gunpowder magazines), and the press had waged a crusade in favor of the new technology. Even the weather helped. The "dry fog" of the summer of 1783 brought with it an unprecedented number of thunderstorms and arguably accelerated the spread of the rod, at least in the German Empire (Hochadel). But even with the vast increase in the number of lightning rods in the 1780s, they still remained an elite phenomenon throughout the eighteenth century, both because of their high cost and because they were closely connected to Enlightenment culture.

Superstition

Superstitious beliefs in alternative practices of lightning protection such as ringing the church bells were central in the pro–lightning rod crusade of the Enlightenment in the eighteenth and even the nineteenth century. These practices provided a strategic foil to the rule of reason, the backward past that had to be overcome: Without superstition, there could be no victory of the Enlightenment. It is no surprise that in Andrew Dickson White's *A History of the Warfare of Science with Theology in Christendom* (first edition, 1896) the invention of the lightning rod plays a central role. White put forward the "conflict thesis," that is, that science and religious doctrine were fundamentally incompatible and that in the course of history reason eventually triumphed over dogma. What could illustrate this thesis better than contrasting a safe protection device with an utterly useless and even dangerous religious practice such as the ringing and blessing of bells inscribed with pious sayings? White warmed to his task with evident relish: the "old sacred theory" of demonic agency in thunderstorms "received its death-blow," he wrote. "In 1752 Franklin made his experiments with the kite on the banks of the Schuylkill; and, at the moment when he drew the electric spark from the cloud, the whole tremendous fabric of theological meteorology reared by the fathers, the popes, the medieval doctors, and the long line of great theologians, Catholic and Protestant, collapsed."[38]

The case of the lightning rod is an excellent example for the discursive construction of this dichotomy, a dichotomy that was asymmetrical from the beginning; the "superstitious" had no voice of their own. Therefore any account of "resistance" against the erection of lightning rods should be interpreted with extreme care. The "common people" were equated with ignorance and superstition while the clergy was often accused of deliberate deceit to control the people through fear of divine punishment (Hochadel, Bertucci). The scolding of opponents or critics of the lightning rod helped the natural philosophers to gloss over their own inconsistencies and knowledge gaps. Did the lightning rod neutralize or attract lightning? If it attracted thunderstorms, did that endanger buildings that were not equipped with protective devices? Such objections (not entirely "unscientific" by the standards of the time and reminiscent of Nollet's warnings) were by the 1780s simply brushed aside by advocates of the lightning rod such as the German Johann Jacob Hemmer as being ignorant and unworthy of further discussion.[39] For Robespierre, the people of St. Omer who brought Vissery to court to get rid of his lightning rod also brought shame to the entire kingdom of France. However, as Jessica Riskin points out, their objections were "practical rather than superstitious and express no general hostility toward modern science, but instead a dubious attitude toward Vissery's engineering credentials."[40]

The Public and the Media

"The public" has long been a central category of Enlightenment studies. The essays in this volume dealing with this period make clear that "the public" can mean different things. The public can be the uneducated "common man"—the crowd that is easily frightened, unreceptive to reason, and that might even turn into a mob. For example, in July 1752 the whole town of Bologna watched Giuseppe Veratti's experiments, most of them with disapproval. Carrying out experiments and measurements in such public places demanded careful negotiations with the civil authorities. In the case of the observatory tower of Palazzo Poggi in Bologna, the electrical experiments were stopped by the magistrate due to complaints of nearby residents (Bertucci). During this period, reports abound of incidents in which lightning rods had been destroyed or their erection had been obstructed by furious or fearful neighbors. The following episode might have served as a master narrative: in March 1760 enraged peasants allegedly tore down the lightning rod of Prokop Diviš in Moravia because they blamed it for the severe drought.[41]

Yet "the public" of the Enlightenment can also be identified with the media of the time. The journals and newspapers, the publications of the scientific academies, and the writings of natural philosophers were instrumental in promoting the lightning rod. They published reports of incidents in which lightning struck and caused damage and claimed lives as well as reports in which a lightning rod prevented harm. These highly detailed reports tried to persuade the reader by placing him or her in the position of a virtual witness (Heering).

In the summer of 1783, with its unusually high number of severe thunderstorms all over western and central Europe, newspapers practically went on a pro–lightning rod crusade. In the German Empire this campaign resulted in the erection of hundreds of new rods and led to a ban on the ringing of church bells (Hochadel). José Antonio Alzate pursued a similar strategy, but the situation in Spanish Mexico was quite different. With his *Gazeta de Literatura de México*, Alzate tried to disseminate scientific knowledge and convince his fellow citizens of the merits of the lightning rod by providing examples and informing his readers about the European debates. Yet he was very much on his own, and how far he succeeded is difficult to say (Clark). Nevertheless, Alzate's campaign illustrates the role of the media as the forum in which the battle for "public opinion" took place.

Strategies of Persuasion

Enlightenment reformers were quite inventive when it came to convincing the allegedly unenlightened public of the protection that the lightning rod could provide. In addition to skillful use of print media, the most persuasive propaganda was, of course, the actual installation and testing of successful lightning rods in what often became a public spectacle. The acid test was of course the next thunderstorm when the lightning rod would be struck, leaving the building unharmed. Perhaps the most celebrated of these "collective experiments" was conducted on Piazza del Campo in Siena. On April 18, 1777, a large and initially skeptical crowd watched the successful channeling of a lightning strike into the ground (Bertucci). The incident was widely reported throughout Europe.

Johann Jacob Hemmer, one of the chief advocates of the lightning rod in the German Empire, even designed a removable lightning rod terminal consisting of five spikes and a horizontal cross, to enhance its propaganda value. Once the lightning had struck and "injured" the terminal—for example, melted one of the spikes—it was removed and placed on public display in the cabinet of the ruler.[42] Hemmer hoped "to convince the common man who would then understand that lightning had actually struck but left the house unharmed."[43]

Recent research in the history of science has identified three-dimensional models as crucial experimental and conceptual tools in many areas of the natural sciences. Models were "central . . . to projects of Enlightenment."[44] In this vein, Willem Hackmann deals with eighteenth century model experiments such as the very popular and widespread thunder house experiment, still performed today at The Bakken Museum and other venues; spellbound audiences would watch a model house being hit by a huge stroke without harm (or "exploding" if the lightning rod had been "turned off"). In the Enlightenment, the thunder house was vital to the promotion and popularization of the lightning rod, showing *en minia-ture* its beneficial uses (and the disastrous consequences of not having one).

At the same, natural philosophers and engineers debated the possible pitfalls of scale models and conceptual models, some arguing that models of ships or of engines could not be trusted to represent the natural world, challenging "the legitimacy of projecting from microscopic to macroscopic systems."[45]

The British electrician Benjamin Wilson tapped into this tradition of using models in a public display when he staged his famous experiments in London's Pantheon in 1777. Drums coated with tinfoil represented the charged clouds. Wilson wanted to convince the audience that the tip of the lightning rod should be a knob. He thought it far safer than the pointed rod favored by Franklin and his allies because it was less "attractive."

The Rise of the Expert Scientist

R. W. Home sheds new light on this cause célèbre in the field of lightning rod research: the Purfleet controversy on "knobs versus points." Home argues, contrary to previous accounts, that the rejection of Wilson's experiments at the Pantheon in London cannot be explained in terms of different political backgrounds, that is, British royalists versus American revolutionaries. Rather, Wilson's opponents were able to discredit his reputation by describing his experiments as fraudulent. Ultimately, what decided the matter were neither the paying audiences of the Pantheon performances nor the King himself but the vote of Wilson's peers. In this case, "going public" backfired, and Wilson was cast as a dubious showman.

The research on lightning rods exemplifies the change electrical studies underwent toward the close of the century. Coulomb claimed that he carried out high-precision experiments with the torsion balance. This experimental style broke with the Enlightenment tradition and can be explained as an attempt of Royalist researchers to establish a scientific practice, based on the acceptance of the authority of experts, that could serve as a model for society (Heering).

The role of the public has rightly been stressed in recent accounts of Enlightenment science. Yet in the last decades of the eighteenth century we witness the rise of the expert scientist. Governments and other bodies relied increasingly on committees recruited from academies of sciences and other learned associations. On the one hand, Enlightenment science was "public" and seemingly open to everybody with an interest in the matter; on the other hand, it was elitist and involved constant boundary work to distinguish experts from non-experts. This Janus face of Enlightenment science revolves continuously when it comes to the question of how lightning rods should be built and installed. On the one hand, Georg

Christoph Lichtenberg claimed that it is as easy to erect a lightning rod as it is to make an umbrella.[46] On the other hand, he worked hard to discredit itinerant lecturers and warned local administrators not to buy their dubious lightning rods. As the market for lightning rods emerged around 1780, the competition between itinerant lecturers, local instrument makers and natural philosophers was fierce. This competition took place on two levels—epistemological and economic—over the authority to decide how a lightning rod is properly installed and who will get commissioned to do so.[47] In virtually all cases, the professor was the final arbiter. In the 1780s, Lichtenberg was constantly asked by magistrates in the German states to judge the quality of lightning rod designs proposed to them.[48] The reputation of the roving salesmen did not improve in the next one hundred years, if we follow Anderson's verdict: "The tramping 'lightning-rod men' of the United States have been notorious for extortion and ignorance."[49]

On the Italian peninsula, Giambattista Beccaria (in the kingdom of Piedmont) and Giuseppe Toaldo (in the republic of Venice) were able to pursue careers with their research in electricity and become lightning rod experts, sought after by their respective governments (Bertucci). These examples from Britain, France, Germany, and Italy all testify to the increasing reliance of governments on scientific academies or university professors for scientific expertise.

For Alzate, the lightning rod was to some degree an instrument in a metaphoric sense. In the virtual absence of real lightning rods in Mexico City, he instructed the readers of his *Gazeta* about European achievements in this field. The rational method of acquiring knowledge about the natural world was to be emulated. At the same time, he stressed that the situation in Mexico was different from that in Europe, both culturally and meteorologically. Calling for specific adaptations, Alzate tried to form a national consciousness (Clark).

These examples show that the rise of the expert scientist is in no way teleological. Local contingencies played a vital role, a point particularly stressed by Bertucci in her comparison of three Italian settings. The success or failure of natural philosophers was critically dependent upon the availability of local resources, whether political, financial, or cultural, as well as their ability to access and manipulate them. In the Bolognese case, as we shall see, research on atmospheric electricity had to be aborted altogether.

Unintended Consequences:
Trigger Effects and the Uses of Instruments

Research done on the lightning rod was never an isolated activity; it is interesting to see how one development triggered another. Following Franklin's lead, Giambattista Beccaria's experiments on atmospheric electricity led to a new foundation of the ancient science of meteorology as an empirical study of weather that would benefit agriculture and society as a whole. The weather, Beccaria and many other natural philosophers at the time came to believe, was predictable—at least in principle.

Bertucci also points out an interesting case of electrical research being forced in a different direction. When in 1752 the Bolognese authorities forbade further research on atmospheric electricity, the local electricians had to focus on medical electricity. Even though the findings of Veratti and Pivati were rejected by their contemporaries, they mark an important step. Medical electricity became the other central topic in discussing and demonstrating the utility of electricity during the second half of the eighteenth century.[50] One outcome of this shift in the research agenda of Bolognese electricians is well known: Galvani's research on "animal electricity" that led to the origins of electrochemistry. It would surely overstate the case to claim a causal connection between the ban on lightning rod experiments and the Voltaic pile. But the ban might be considered as one of many predisposing circumstances.

The history of science is replete with chance inventions and unintended discoveries.[51] The history of lightning rod research, too, illustrates very well that research is a messy business, full of surprises and unexpected twists and turns.

Charles Augustin Coulomb's electrical experiments in the 1780s, starting with the formulation of the inverse square law in electrostatics, have been interpreted by some historians as "basic research." Yet these experiments, including those with the torsion balance, were actually part of a research program on the design of lightning rods for military installations (Heering). It was the fear of gunpowder magazines being struck by lightning that led to the formation of a commission by the French Academy of Sciences in 1784. Coulomb was included on the committee because of his knowledge of the design of military fortifications. Nevertheless, torsion balances became a standard for very sensitive measurements throughout the nineteenth century.[52]

The rapid spread of telegraphy in the United States in the second half of the nineteenth century led to serious problems with respect to protecting the wires and poles during electrical storms. Different sorts of "lightning arresters" were installed to protect the transmission of signals from high voltage disturbances. These devices were invented by the telegraphers themselves, not by scientists or engineers. Cavicchi describes these interactions between the telegraphers and atmospheric electricity as a genuine learning process.

In continuation of this line, Paolo Brenni shows that there was a fruitful link between research on lightning rods and wireless telegraphy at the end of the nineteenth century. Physicists up until the 1930s even tried to tap the seemingly boundless energy of lightning as a source of terrestrial power. Despite their failure to do so, the instrumental devices they developed were quite ingenious.

Brenni focuses on instruments that were employed in conducting research on lightning rods and atmospheric electricity. In his analysis, he shows that eighteenth century research was carried out with instruments that were based on standard devices of natural philosophers working on electricity. The way to do research changed substantially in the nineteenth century, when highly specialized apparatus was employed that could not be used outside its designated area.

In Business: The Lightning Rod as a Commodity

Already in the second half of the eighteenth century, the lightning rod had become a commodity, an object that was marketed but not yet mass produced. One might think that a seemingly simple object consisting of metal bars and wires should have been rather cheap. The opposite was the case. To give but one example: Giambattista Beccaria paid 1,600 lires for the installation of three rods, and his initial annual salary as a professor was 1,200 lire (Bertucci). This was one of the reasons why the lightning rod remained largely an elite phenomenon in the eighteenth century.[53] It was not enough to be enlightened; one also had to be wealthy to afford the protection devices.

The fact that Alexandre Lapostolle tried to promote lightning rods made of straw in France in the 1820s indicates that the cost of the material was still a major obstacle for its general implementation, particularly in the countryside.[54] This only changed gradually. Lightning rods were still relatively costly in the nineteenth century. Yet the rods evolved more from custom-made products into commodities of mass production, "patented, manufactured in factories, and sold by specialized companies" (Mohun). This story of engineering businessmen and lightning rod salesmen walking the hinterland of the United States is told by Mohun and Cavicchi. Squabbles about the best way to construct and install the device continued into the nineteenth century. It seems obvious that the differences between lightning rods in material and form were often due to commercial interests rather than to technological advantages. Nowadays these often very decorative designs, among them fanciful trident tops and multiple-pointed "crowns," have become collectors' items.

The final essay by Charles B. Moore, Graydon D. Aulich, and William Rison brings us to the present. The authors—physicists actively engaged in lightning research—focus on the problem of lightning rod design. In discussing some of the experiments carried out during recent decades and describing some of the continuing controversies over the proper design of lightning rods, their essay offers another opportunity to reflect on the eighteenth century case studies presented earlier in the book. From their essay it becomes clear that lightning rods are still the subject of research even today, offering another indication that the lightning rod is by no means an ahistorical object that, after coming into the world, remained unchanged forever. Moreover, in some sense, their essay brings the reader full circle: according to their modern research, knobs are actually better than points! Or, to go back to Home's study, Benjamin Wilson had been right, in spite of the discrediting of his reputation.

Today lightning rods are mostly invisible to us. We rarely notice them, save at sixty miles per hour on a country road, when we might occasionally note a few vestigial rods adorning the roof of a barn or farmhouse. And if we do see them, we rarely give them much thought, except perhaps to remember their puckish inventor, Benjamin Franklin. The essays that follow, however, shed new light on the lightning rod. These unimposing metal bars do have a fascinating history replete with contention and attributed meanings, both in the scientific and the public arenas. The lightning rod has indeed attracted more than just lightning over the past two and a half centuries, and even today it remains much more than just a simple metal rod.

Notes

1. To give but a few examples from current encyclopedias: "Les études faites au milieu du XVIIIe s. . . . ont abouti au dispositif dit 'de Franklin' encore en usage"; "Paratonnerre," *Grand Larousse Universel*, vol. 11 (Paris: Librairie Larousse, 1997), 7818–19. The German *Brockhaus* incorrectly claims that Franklin was the first to erect a lightning rod: "Blitzschutz," in *Brockhaus. Die Enzyklopädie*, 20th ed., vol. 3 (Leipzig: Brockhaus, 1996), 436. The respective article in the *Encyclopædia Britannica* is ahistorical: "Lightning Rod," *Encyclopædia Britannica Online*, February 1, 2007, *http://www.britannica .com/eb/article-9048229*. The canonical story of Franklin's invention is told in the article "Kite," *Encyclopædia Britannica Online*, February 1, 2007, *http://www.britannica .com/eb/article-215106*.

2. See R. W. Home's chapter in this volume, including further references.

3. Charles Vellay, "Robespierre et le procès paratonnerre (1780–1784)," *Annales Révolutionnaires* 2 (1909): 201–19; and Jessica Riskin, *Science in the Age of Sensibility* (Chicago: The University of Chicago Press, 2002), ch. 5: "The Lawyer and the Lightning Rod." An earlier version of this chapter was published in *Science in Context* 12, no. 1 (1999): 61–99.

4. Peter Heering, *Das Grundgesetz der Elektrostatik: Experimentelle Replikation und wissenschaftshistorische Analyse* (Wiesbaden: DUV, 1998); Oliver Hochadel, *Öffentliche Wissenschaft: Elektrizität in der deutschen Aufklärung* (Göttingen: Wallstein, 2003).

5. The reports of these committees were published in 1823, 1854, 1867, and 1903.

6. William Snow Harris, *The Meteorology of Thunderstorms: With a History of the Effects of Lightning on 210 Ships of the British Navy, as Recorded in the Official Journals of the Respective Ships* (London: Hunt, 1844).

7. Nani Clow, "Lightning in the Laboratory," in *Instrument—Experiment: Historische Studien*, ed. Christoph Meinel (Berlin: Diepholz, Verlag für Geschichte der Naturwissenschaften und Technik, 2000), 376–85; Ido Yavetz, "A Victorian Thunderstorm. Lightning Protection and Technological Pessimism in the Nineteenth Century," in *Technology, Pessimism, and Postmodernism*, ed. Yaron Ezrahi et al. (Dordrecht: Kluwer, 1994), 53–75; and Ido Yavetz, "Between High Science and Practical Engineering: Two Studies of Lightning by Simulation," *Physis* 33 (1996): 221–58.

8. U.S. Patent 1266175, issued May 14, 1918.

9. Burghard Weiss, "Blitze für Kernphysik und Strahlentherapie: Die Stoßspannungsexperimente von Brasch und Lange am Monte Generoso und bei der AEG in Berlin 1925–1935," *Technikgeschichte* 66, no. 3 (1999): 173–203.

10. Jens Kugler, *Das Donnerwetter kann in Gruben schlagen: Ein Beitrag zu Blitzschlägen und Blitzschutzeinrichtungen an bergbaulichen Anlagen in Sachsen* (Kleinvoigtsberg: Jens-Kugler-Verlag, 2000). We thank N. Pohl, who drew our attention to this study.

11. R. Ackermann et al., "Triggering and Guiding of Megavolt Discharges by Laser-Induced Filaments under Rain Conditions," *Appl. Phys. Lett.* 85 (2004): 5781–83.

12. Friedhelm Noack, "Early Streamer Emission Devices—Verbesserung des Blitzschutzes?" *Etz. Elektrotechnik + Automation; Organ des VDE Verband der Elektrotechnik Elektronik Informationstechnik e.V. und der Energietechnischen Gesellschaft im VDE (ETG)* 123, nos. 3/4 (2002): 22–24; and Abdul M. Mousa, "War of the Lightning Rods," *Electricity Today* (2004): 45–47.

13. Rudolf Stichweh, *Zur Entstehung des modernen Systems wissenschaftlicher Disziplinen: Physik in Deutschland 1740–1890* (Frankfurt a.M.: Suhrkamp, 1984), 279. Even though Stichweh made this acute observation, he did not do any further research on the history of lightning rods.

14. Richard Anderson, *Lightning Conductors: Their History, Nature, and Mode of Application* (London: E. & F. N. Spon, 1879). Heinrich Meidinger, *Geschichte des Blitzableiters* (Karlsruhe: G. Braun, 1888). See also Felix Sestier, *De la foudre: de ses formes et de ses effets sur l'homme, les animaux, les végétaux et les corps bruts de moyens de s'en préserver et des paratonnerres*, 2 vols. (Paris: J. B. Baillière et Fils, 1866).

15. Basil F. J. Schonland, *The Flight of Thunderbolts* (Oxford: Clarendon, 1950).

16. I. Bernard Cohen, "Prejudice against the Introduction of Lightning Rods," *Journal of the Franklin Institute* 253 (1952): 393–440. Another example is Hans Prinz, "Gewitterelektrizität (Nach dem nachgelassenen Manuskript bearbeitet von Hans Steinbigler)," *Deutsches Museum: Abhandlungen und Berichte* 47, no. 1 (1979): 5–74.

17. Roy Porter, ed., *Eighteenth-Century Science. The Cambridge History of Science*, vol. 4 (Cambridge: Cambridge University Press, 2003), 370, 754–755; and Jan V. Golinski, William Clark, and Simon Schaffer, eds., *The Sciences in Enlightened Europe* (Chicago: The University of Chicago Press, 1999). Alan Charles Kors, ed., *Encyclopedia of the Enlightenment*, 4 vols. (Oxford: Oxford University Press, 2003) also has no entry on the lightning rod.

18. See, for example, H. W. Brands, *The First American: The Life and Times of Benjamin Franklin* (New York: Anchor Books, 2002); Walter Isaacson, *Benjamin Franklin:*

An American Life (New York: Simon & Schuster, 2003); Edmund S. Morgan, *Benjamin Franklin* (New Haven, CT: Yale University Press, 2003); and Gordon S. Wood, *The Americanization of Benjamin Franklin* (New York: Penguin Books, 2004). A tercentennial exhibit titled *Benjamin Franklin: In Search of a Better World* toured in the United States and Europe in 2005–7 and is said to be the largest collection of his materials ever brought together in one place.

19. Michael B. Schiffer, *Draw the Lightning Down: Benjamin Franklin and Electrical Technology in the Age of Enlightenment* (Berkeley: University of California Press, 2003); Tom Tucker, *Bolt of Fate: Benjamin Franklin and His Electric Kite Hoax* (New York: Public Affairs, 2003); and Philip Dray, *Stealing God's Thunder: Benjamin Franklin's Lightning Rod and the Invention of America* (New York: Random House, 2005).

20. Christian Begemann, *Furcht und Angst im Prozeß der Aufklärung. Zu Literatur und Bewußtseinsgeschichte des 18. Jahrhunderts* (Frankfurt a.M.: Athenäum, 1987), ch. 3, in particular, pp. 90–95; Heinz Dieter Kittsteiner, "Das Gewissen im Gewitter," *Jahrbuch für Volkskunde* 10 (1987 NF): 7–26, esp. pp. 18–23; Heinz Dieter Kittsteiner, *Die Entstehung des modernen Gewissens* (Frankfurt a.M.: Suhrkamp, 1995), 79–93; Engelhard Weigl, "Entzauberung der Natur durch Wissenschaft—dargestellt am Beispiel der Erfindung des Blitzableiters," *Jahrbuch der Jean-Paul-Gesellschaft* 22 (1987): 7–39; and Engelhard Weigl, *Instrumente der Neuzeit: Die Entdeckung der modernen Wirklichkeit* (Stuttgart: Metzler, 1990), 174–200.

21. Olaf Briese, *Die Macht der Metaphern. Blitz, Erdbeben und Kometen im Gefüge der Aufklärung* (Stuttgart/Weimar: J. B. Metzler, 1998), 26.

22. Henry Petroski, *The Evolution of Useful Things. How Everyday Artifacts—From Forks and Pins to Paper Clips & Zippers—Came to Be as They Are* (New York: Alfred A. Knopf, 1992).

23. Pierre Laszlo, *Salt: Grain of Life. Arts and Traditions of the Table* (New York: Columbia University Press, 2001); and Jakob Vogel, *Ein schillerndes Kristall. Eine Wissensgeschichte des Salzes zwischen Früher Neuzeit und Moderne* (Köln: Böhlau, 2007).

24. David Gooding, Trevor Pinch, and Simon Schaffer, eds., *The Uses of Experiment: Studies in the Natural Sciences* (Cambridge: Cambridge University Press, 1989).

25. Harry M. Collins, *Changing Order: Replication and Induction in Scientific Practice*, 2nd ed. (Chicago: The University of Chicago Press, 1992); and Jan V. Golinski, *Making Natural Knowledge: Constructivism and the History of Science* (Cambridge and New York: Cambridge University Press, 1998).

26. Two of the participants published their papers separately; see E. Philip Krider, "Benjamin Franklin and Lightning Rods," *Physics Today*, 2006: 42–48; and James Delbourgo, *A Most Amazing Scene of Wonders: Electricity and Enlightenment in Early America* (Cambridge/Mass.: Harvard University Press, 2006), ch. 2.

27. Martina Lorenz, "Präsentation und Legitimation von Naturkunde im deutschen Sprachraum der Aufklärungszeit," in *Transactions of the Ninth International Congress on the Enlightenment*, vol. 3 (Oxford: Voltaire Foundation, 1996), 1069–73, at 1071; and Larry Stewart, *The Rise of Public Science. Rhetoric, Technology and Natural Philos-*

ophy in Newtonian Britain, 1660–1750 (Cambridge: Cambridge University Press, 1992), 16.

28. Geoffrey V. Sutton, *Science for a Polite Society: Gender, Culture, and the Demonstration of Enlightenment* (Boulder: Westview Press, 1995), 284; and Michael Lynn, *Popular Science and Public Opinion in Eighteenth-Century France* (Manchester, England: Manchester University Press, 2006), 34. This is certainly the case for France, to which Sutton and Lynn refer, and for Germany, to which Lorenz refers. Natural philosophers and popularizers of science in Britain were more geared toward solving concrete technical problems. Larry Stewart, "Public Lectures and Private Patronage in Newtonian England," *Isis* 77 (1986): 47–58, at 55.

29. For the precarious status of "the man of science" in the Enlightenment, see Steven Shapin, "The Image of the Man of Science," in *Eighteenth-Century Science*, vol. 4, ed. Roy Porter (Cambridge: Cambridge University Press, 2003), 159–83, esp. 171–73.

30. Schiffer, *Draw the Lightning Down* (cit. n. 19), 204; Dray, *Stealing God's Thunder* (cit. n. 19), p. xvi; and Delbourgo, *Most Amazing Scene* (cit. n. 26), 75.

31. Lorraine Daston, ed. *Things That Talk: Object Lessons from Art and Science* (New York: Zone Books, 2004), 12.

32. Georg Christoph Lichtenberg, "Vermischte Gedanken über die aerostatischen Maschinen," *Göttingisches Magazin der Wissenschaften und Literatur* 3, no. 6 (1784): 930–53, at 930.

33. Gottlieb Tobias Wilhelm, *Ueber Augsburg. Gegen die unwahre Darstellung dieser Reichsstadt in dem Geographisch Statistisch Topographischen Lexikon von Schwaben* (Augsburg, 1791), 120n.

34. Prokop Diviš, *Längst verlangte Theorie von der meteorologischen Electricite, welche Er selbst Magiam naturalem benahmet: Samt einem Anhang vom Gebrauch der electrischen Gründe zur Chemie*, ed. Friedrich Christoph Oetinger (Tübingen/Frankfurt a.M., 1765/1768); Karel Hujer, "Father Procopius Divis—The European Franklin," *Isis* 43 (1952): 351–57; and I. Bernard Cohen and Robert Schofield, "Did Divis Erect the First European Protective Lightning Rod, and Was His Invention Independent?" *Isis* 43 (1952): 358–64.

35. Schonland, *Flight of Thunderbolts* (cit. n. 15), 24.

36. A rhetoric that praises potential achievements of technological developments that are not real but just potential is not only found in case of the lightning rod. Another example would be the praise of the steam engine in German-speaking areas during the first part of the nineteenth century, when few machines existed but the device was seen as a means to liberate men from nature; see Elizabeth Neswald, *Thermodynamik als kultureller Kampfplatz: Zur Faszinationsgeschichte der Entropie 1850–1915* (Freiburg i.Br.: Rombach, 2006), 79f.

37. We partly follow the dates given in Meidinger, *Geschichte des Blitzableiters* (cit. n. 14), 35–44; for Germany and Italy we found different dates. Technically speaking, Geneva became part of Switzerland only in 1815.

38. Andrew Dickson White, *A History of the Warfare of Science with Theology in Christendom*, vol. 1 (New York: Braziller, 1955; first ed. 1896), 364. For the ringing of church bells, see 344–50.

39. Fritz Dross, "Gottes elektrischer Wille? Zum Düsseldorfer 'Blitzableiter-Aufruhr' 1782/83," in *Landes- und Reichsgeschichte: Festschrift für Hansgeorg Molitor zum 65. Geburtstag*, ed. Jörg Engelbrecht and Stephan Laux (Bielefeld: Verlag für Regionalgeschichte, 2004), 281–302, at 300.

40. Riskin, *Science in the Age of Sensibility* (cit. n. 3), 150.

41. Hujer, "Father Procopius Divis" (cit. n. 34), 356.

42. See the reports from Mannheim and Karlsruhe, *Augsburgische Stats- und Gelehrte Zeitung*, July 24, 1779, and *Augsburger Ordinari Postzeitung*, October 7, 1784.

43. Johann Konrad Gütle, *Unterricht vom Blitz und den Blitz- und Wetter-Ableitern zur Belehrung und Beruhigung sonderlich der Ungelehrten und des gemeinen Mannes von Johann Friedrich Luz, neu bearbeitet von Johann Konrad Gütle* (Nürnberg, 1804), 201.

44. Soraya de Chadarevian and Nick Hopwood, eds., *Models: The Third Dimension of Science* (Stanford: Stanford University Press, 2004), 1.

45. Simon Schaffer, "Fish and Ships: Models in the Age of Reason," in *Models: The Third Dimension of Science*, ed. Soraya de Chadarevian and Nick Hopwood (Stanford: Stanford University Press, 2004), 71–105, p. 71.

46. Georg Christoph Lichtenberg, "Neueste Geschichte der Blitzableiter (im Jahr 1779)," in *Physikalische und mathematische Schriften*, vol. 6, *Vermischte Schriften*, ed. Ludwig Christoph Lichtenberg and F. Kries (Göttingen, 1803), 210–20, at 211; Lichtenberg to Ramberg, April 8, 1782, Georg Christoph Lichtenberg, *Briefwechsel*, vol. 2, ed. Ulrich Joost and Albrecht Schöne (Munich: Beck, 1985), 303.

47. Oliver Hochadel, "Martinus Electrophorus Berschütz: Georg Christoph Lichtenberg und die wissenschaftlichen Schausteller seiner Zeit," *Lichtenberg-Jahrbuch*, 1998: 155–75.

48. Lichtenberg to Hollenberg, February 18, 1788, Lichtenberg, *Briefwechsel*. vol. 3 (cit. n. 46), 482. In 1782 Lichtenberg had to cast his judgment on two differing schemes for a lightning rod on the church of Mandelsloh. He found both lacking; Georg Christoph Lichtenberg, "Gutachten über den Blitzableiter zu Mandelsloh," ed. Ulrich Joost, *Lichtenberg-Jahrbuch*, 1994: 72–80; and Ulfrid Müller, "Der Bau des Wetter-Ableiters auf der St. Osdag-Kirche in Neustadt-Mandelsloh," *Lichtenberg-Jahrbuch*, 1994: 81–92.

49. Anderson, *Lightning Conductors* (cit. n. 14), 133.

50. Utility had been one major aspect with respect to discussions of electricity in this period. Several authors pointed out that medical electricity and the lightning rod were the two beneficial uses of electricity, see, e.g., Tiberius Cavallo, A *Complete Treatise of Electricity in Theory and Practice; with Original Experiments* (London, 1777); and Karl Gottlob Kühn, *Geschichte der medizinischen und physikalischen Elektricität und der neuesten Versuche, die in dieser nützlichen Wissenschaft gemacht worden sind*, vol. 1 (Leipzig, 1783). For a discussion of medical electricity, see Paola Bertucci and Giuliano Pancaldi, eds., *Electric Bodies: Episodes in the History of Medical Electricity. Bologna Studies in the History of Science*, vol. 9 (Bologna: CIS, University of Bologna, 2001).

51. For a popular account of such discoveries by chance, see Martin Schneider, *Teflon, Post-It und Viagra. Große Entdeckungen durch kleine Zufälle* (Weinheim: Wiley-VCH, 2002). For an interesting discussion of the role of contingency in science, see Giu-

liano Pancaldi, *Volta. Science and Culture in the Age of Enlightenment* (Princeton, N.J., and Oxford: Princeton University Press, 2003), ch. 9. However, as the case of Ørsted shows, the attribution of chance to a discovery can depend very much on the interpreter, see, e.g., Nahum Kipnis, "Chance in Science: The Discovery of Electromagnetism by H.C. Oersted," *Science & Education* 14, no. 1 (2005): 1–28.

52. Matthias Dörries, "La standardisation de la balance de torsion dans les projets européens sur le magnétisme terrestre," in Christine Blondel and Matthias Dörries, eds., *Restaging Coulomb: Usages, controverses et réplications autour de la balance de torsion* (Florence: Leo S. Olschki, 1994), 121–50.

53. Schiffer, *Draw the Lightning Down* (cit. n. 19), p. 203; and Delbourgo, *Most Amazing Scene* (cit. n. 26), 74f.

54. Alexandre Ferdinand Léonce Lapostolle, *Traité des Parafoudres et des Paragrêles en Cordes de Paille* (Amiens: Caron-Vitet, 1820). Some ten years later the lightning protection for the cathedral in Strasbourg had an estimated cost of 15,000 francs; see Antoine Fargeaud, *Établissement d'un Paratonnerre sur la flèche de la Cathédrale de Strasbourg* (Strasbourg: F.-G. Levrault, 1833).

Part 1
ERECTING THE ROD—
SPREADING ENLIGHTENMENT

Enlightening Towers

Public Opinion, Local Authorities, and the Reformation of Meteorology in Eighteenth Century Italy

Paola Bertucci

In the spring of 1777 Piazza del Campo at Siena became the site of a collective experiment. One year earlier the Grand Duke of Tuscany, Pietro Leopoldo, ordered that a lightning rod be affixed to the tower of the town hall, the very heart and symbol of the city. His decision prompted the heated reaction of a Siena nobleman, the marquis Alessandro Chigi, who published an attack against the theory that lightning was an electrical phenomenon, claiming that metallic conductors would be ineffective in preventing damages to building and people during thunderstorms.[1] Chigi's opposition to lightning rods found quite a few supporters, and when, on April 18, 1777, black clouds darkened the sky above Piazza del Campo, a large crowd gathered in the square to observe the effects of the conductor erected on top of the tower. They saw a bolt of lightning strike the tower and be conducted safely into the ground, channeled by the metallic rod. The professor of physics at the University of Siena, Domenico Bartaloni, examined the tower and the conductor after the storm. Although "the incredulous" expected "a completely different result, almost wishing to see the tower flashing, so as to expose to ridicule the holy laws of physics," Bartaloni declared the complete success of the conductor in protecting the tower. His official report was published in the transactions of the Academy of Siena as well as in the local newspaper. The collective witnessing of the experiment sanctioned the success of lightning rods in the public sphere.[2]

The Siena episode highlights typical elements that characterized eighteenth century debates on the effectiveness of lightning rods: the involvement of public opinion, the role of local authorities, the experts' engagement in the popularization of their views, and the spectacularly visible setting of the experiments. Towers were main protagonists of the early history of lightning rods. Highly tangible symbols of political, religious, or financial power, towers had always been frequent targets for the fiery meteor of lightning. From the mid-eighteenth century, they became favorite sites for experimenting with lightning conductors. Not only in Siena but also in Florence, Pisa, Milan, Turin, Venice, Genoa, Bologna as well as in smaller towns south of the Alps, natural philosophers affixed metallic conductors on top of the towers of churches, city halls, castles, and palazzi.[3] In the philosophers' opinion, the pointed conductors would slowly draw the electric fire from thunderclouds and channel it into the ground, thereby preventing huge discharges that would damage buildings. Or, they would attract lightning, forcing it to pass through the metal, with the same result. Yet because of their visibility and symbolic significance in the everyday life of Italian cities, towers also became highly debated experimental sites that attracted the inhabitants' attention and made the debate over lightning rods a public concern.

This essay shows that before lightning rods became marketable commodities, they were experimental devices used to substantiate or criticize Franklin's theory of electricity, which held that the matter of lightning and that of artificially produced electric sparks were one and the same. The study of the nature of lightning contributed to the reformation of Aristotelian meteorology in terms of the new science of electricity: each flash of lightning that struck a metallic conductor created the experimental setting for electricians to study the behavior of such a disruptive natural "meteor." The reports of their observations made up a sort of transnational repository of experimental results on which lightning rods advocates relied to support their arguments. Because of their unusually visible setting, however, such experiments acquired a public dimension that obliged electrical experimenters to confront public opinion and local authorities. In some cases, this confrontation brought electricians to engage in campaigns of popularization of electrical science, which aimed at highlighting the public benefits deriving from the installation of lightning rods and from the study of electrical meteorology. Yet the electricians' attitude to public opinion was not unanimous. Marsilio Landriani, for example, who in 1784 published a complete list of lightning conductors affixed on private houses, powder magazines, and public buildings in Italy and the rest of Europe, in his *On the Usefulness of Electrical Conductors* declared that he was "convinced that examples and authority have more effects than reasons on people's dispositions."[4] Nonetheless, if Pietro Leopoldo's decision was clearly an authoritative example, it did not seem to suffice to sedate controversies that characterized debates over lightning rods throughout the century.[5]

This essay explores the multifaceted interactions between local authorities and lightning rods advocates in three different experimental sites where three experts of electricity experimented with lightning conductors, each with different fortune. The fragmentation of the Italian peninsula into several states made of each site a unique combination of the electricians' aspirations with the pressures exerted by local decision makers. Yet active interest in the practical applications of electricity linked together the Italian experimenters who worked with lightning conductors. Each of the three cases highlights the dynamics that affected the fortunes of lightning rods at the local level; taken together, they offer an analysis of the emergence of atmospheric electricity as a new branch of experimental philosophy that redefined the ancient science of meteorology, implementing the Enlightenment rhetoric on the usefulness of science. Pointing upward to the sky, lightning rods provided new means, both theoretical and practical, to interpret the nature of the meteors that fell from the heavens onto the ground. In doing so, they contributed to taming unruly atmospherical phenomena with the Enlightenment ideal of a law-obeying universe.

The Tower of the Istituto delle Scienze in Bologna: Giuseppe Veratti, the Pope, and the Fear of Lightning Rods

In 1752 at Marly, France, the French electrician Thomas François Dalibard performed a crucial experiment. Following Franklin's directions, he erected a pointed, long metallic rod toward a group of thunderclouds. Approaching the rod with another conductor near the ground, he managed to extract sparks that appeared identical to those produced by artifice by means of electrical instruments. "Natural" and "artificial" electricity, he concluded, were one and the same thing. When news of the experiment reached Bologna, the town's leading electrician, Giuseppe Veratti, hurried to replicate it. Veratti was a lecturer of anatomy at the university and a member of the Bologna Institute of Sciences. A few years earlier, he was one of the main actors in a controversy of international resonance over the healing properties of electricity; his work, *Physico-Medical Observations on Electricity*, published in Bologna in 1748 was translated into French and was well known in the republic of letters. With Laura Bassi, his more famous wife, he set up a laboratory of experimental philosophy in their house where they both offered demonstrations to students and visitors. Electricity figured prominently in their experimental activity.[6]

Veratti performed the experiment with lightning conductors on top of the observatory tower of Palazzo Poggi, the building that hosted the Istituto delle Scienze. Founded at the beginning of the century by Gen. Luigi Marsili, who envisaged his institution as a new "House of Solomon," the Istituto complemented the range of lectures offered by the university. Contrary to other contemporary scientific academies, such as the *Royal Society* and the *Académie des Sciences*, it was intended primarily as a site for experimental research: located at walking distance from the city center, in the elegant Palazzo Poggi, its rooms were all equipped with the necessary instruments for the members to lecture and carry out their research. It also included a library and an observatory (fig. 2.1). Each of the rooms for experiments was dedicated to one branch of natural philosophy and was directed by a member. Veratti was in charge of the physics room.[7] The Institute of Sciences was Bologna's most prestigious scientific institution; it hosted large collections of instruments, wax models, and natural specimens, and it was a destination that learned travelers would not miss. In 1740, when the Bolognese cardinal Prospero Lambertini became pope Benedict XIV, the Istituto could boast of a very powerful patron who added expensive items to its scientific collections and supported its research activities.

Veratti's choice to carry out the experiment on top of the observatory tower of Palazzo Poggi was an obvious one. The observatory was a tall building conveniently located just a few floors above the physics room where Veratti kept his electrical instruments. He sought the collaboration of two fellow members of the Institute of Sciences, the astronomer Petronio Matteucci and his assistant Tommaso Marini. Together they erected a metallic rod on top of the observatory tower. The rod was sixteen Parisian feet high (about five meters) and half a digit wide, and it was connected to a metallic chain that allowed them to test the electric state of the rod without climbing the tower.[8]

When thunderclouds appeared in the sky Veratti repeated Dalibard's experiment as it was described in the gazettes: he approached the chain connected to the conductor with a metallic key and observed strong sparks that issued from it. In drawing conclusions on the nature of such sparks he relied on his experience as an electrical experimenter, making use of his senses and of his own body. The color of the sparks and the "disagreeable sensation" that they produced on the body did not leave any doubt as to their similarity with those produced by artifice with common electrical instruments, even though the strength with which they shocked the experimenters' arms and legs indicated that the quantity of electricity involved was exceedingly greater. The physiological effects of the shock did not differ from those produced by the Leyden jar, an instrument that enjoyed international popularity because of the strong shock that it could provoke when discharged through one's body. In his practice as a medical electrician, Veratti, as well as his fellow colleagues in the rest of Europe, used the Leyden jar also as a medical remedy against paralysis and other ailments. The electric phenomena revealed by the lightning conductor disappeared after a few minutes, when rain started to fall, just as electrical experiments lost their vigor in humid weather. The experimenters then climbed the tower to examine the rod, convinced that they would not observe anything interesting for the rest of the day. However, when a bolt of lightning struck nearby, Matteucci and Marini, who were touching the iron bar, experienced a terrible shock, as if they had been struck by lightning. They recovered completely after a few hours of convalescence and even resumed the experiments, but their accident did not go unnoticed.[9]

Figure 2.1.   The Bologna Institute of Sciences and the Observatory Tower.
De Bononienis Scientiarum et Artium Instituto atque Academia Commentarii. Tomus Septimus
(Bologna, 1791)

The local gazette praised Veratti, who had equaled experimenters in Philadelphia and Paris and confirmed the "hypothesis of those who believe that by this
means lightning can be pushed away from those places, where iron bars are
erected above resinous or sulphurous substances." With a view toward highlighting the expertise of the experimenters, the anonymous author of the article
referred his readers to Veratti's work on medical electricity and informed them
of his recent success in the treatment of a paralytic arm that was restored to its
normal state after seven minutes during which the physician extracted sparks
from it.[10] Electric sparks, the author seemed to imply, were not dangerous if properly managed, as Veratti was able to do. However, the experiments at the observatory tower did not elicit unanimous approval. Upon hearing of the terrible
shock experienced by Matteucci and Marini, the people who lived near the Institute of Sciences protested that the experiment demonstrated that the metallic
rod attracted the electric matter from the clouds and was therefore dangerous.
Their vociferous complaints reached the Assunteria d'Istituto, the board of magistrates (*assunti*) who legislated on matters related to the Institute of Sciences. Although the magistrates were surprised by the hostility that such experiments
engendered among the Bolognese and did not intend to comply with what they
regarded as unreasonable fears, as a cautionary measure they decided to suspend

Veratti's work at the observatory tower, ordering him and his colleagues to re-move the apparatus from the tower.[11] The secretary of the Institute of Sciences, Francesco Zanotti, was distressed about the resolution to put an end to the "won-derful electrization" and bitterly commented that vulgar fears had infected even the learned. From his point of view, the Istituto delle Scienze would miss the op-portunity of making important discoveries in a field of research that, being rela-tively new, promised to yield interesting results.[12] The ensuing tension between a number of members of the Institute of Sciences and its magistrates prompted the assunti to inquire about the opinion of Benedict XIV, patron of the Istituto. They explained to the pope that in light of popular hostility against erecting metallic conductors on top of the observatory tower, they had resolved to wait and "see the results of the experiments that will be done elsewhere, without jeop-ardizing a piece of art as valuable as the Institute."[13] The pope declared that al-though he was "little inclined to believe the dissipation of thunderstorms" he "praised the experiments" and disapproved the assunti's endorsement of "popu-lar fears, especially after the example of the experiments [on atmospheric elec-tricity] carried out by the French at the presence of Their King." The pope believed that the French would not have exposed the King's "life to any dan-ger."[14] However, Benedict XIV's criticism had little to do with the advocacy of lightning rods. In fact, the pope was not enthusiastic about the electrical craze of the age that had already exposed the Istituto to international ridicule and wanted to express to the assunti his disappointment about his latest donations to the Istituto (a collection of statues and anatomical tables, plus Campani's opti-cal instruments) that had not yet been displayed in Palazzo Poggi. The pope did not take any measure to invite the assunti to withdraw the ban on lighting rods because he worried that other experiments with metallic conductors would fur-ther delay the proper collocation of his donations.[15]

Although Tommaso Marini erected a metallic conductor on the roof of his house and carried out experiments there while Veratti discussed Franklin's *Ex-periments and Observations* with the members of the Istituto,[16] de facto the as-sunti's decision prevented the installation of lightning rods in Bologna until the turn of the nineteenth century.[17] In 1770 Laura Bassi told the musician Charles Burney, who was touring Italy, that since her husband replicated the Marly ex-periment on the observatory tower, demonstrating soon after Franklin the iden-tity between lightning and the electric spark, "no lightning rods have been erected in this town."[18] In 1783 Marsilio Landriani, who was compiling an in-ventory of all the lightning rods erected in Europe, asked Sebastiano Canterzani, physics professor and new secretary of the Institute of Sciences, to provide him with a list of the conductors that had been affixed in Bologna and surrounding areas, recalling the priority of the institute's members in replicating the Marly ex-periment. Canterzani had to admit with regret and confusion . . . that in Bologna and surrounding areas, as far as I know, there is no building equipped with electric conductors. This will probably do little honor to Bologna, and to the Bolognese. But this is the spirit of the nation: we are as easy in picking up whimsical foreign clothing fashions and behavior, as we are slow at implementing the new methods, useful to citizens' health and safety. I believe that the reason of this inconvenience lies in the coincidence, that new methods promptly find many opponents, and also many scoffers, who feed on disputes and controversies, whereas those few who perceive their reasonability and value, are peacelovers, and they do not want to become the subject of the debate, and maybe also of the mockery of the many.[19]

Canterzani's complaint about local indolence toward new scientific ventures was partly informed by the Enlightenment cliché attacking the opponents of the new science as vain people, but in part it expressed the Institute's frustrated ambition of being protagonist of an important discovery in the field of electricity. To Landriani's surprised request of further explanations about the absence of lightning rods in Bologna, Canterzani mentioned the suspension of Veratti's experiments in 1752, and commented that probably the Senate's decision "has then taken away from physicists the courage to propose the use of conductors in Bologna."[20]

If Bologna remained without lightning rods for the remaining of the eighteenth century, the Institute of Sciences and its members maintained an active interest in the study of the electricity of the atmosphere. Only two years after Veratti's experiments, the Institute supported the experimental activity of Giambattista Beccaria, professor of physics at the University of Turin and author of *Natural and Artificial Electricity* (Turin, 1753). Beccaria traveled to Bologna and performed experiments on the electricity of the atmosphere in the countryside together with a group of members of the Institute of Sciences who encouraged his research and offered to publish his next work in Bologna. Indeed, his letters on natural electricity, addressed to Jacopo Bartolomeo Beccari, former president of the Institute, were published in Bologna in 1758.[21]

The suspension of experiments with metallic conductors separated the study of the electricity of the atmosphere from the history of lighting rods in Bologna. Given the Institute's long-lasting interest in electricity, the suspension also played the crucial role of diverting the members' attention from public safety to public health, with physicians and electricians focusing on the role of electricity in the animal frame. It is a singular coincidence that in the 1780s Luigi Galvani, a student of Veratti and a reader of Beccaria, performed some of his celebrated experiments by testing the effects of the electricity of the atmosphere on a dissected frog. At a time when the identity of artificial and natural electricity was an undisputed result, for the Bolognese it was still a matter of experimentation. Galvani's

experimental set-up was curiously similar to Veratti's apparatus on the observatory tower.[22]

The "Electric Observatory" in Turin:
Giambattista Beccaria and the Formulation of the New Meteorology

The relationship between Beccaria and the Institute of Sciences of Bologna was
to last a long time. Even after the publication of his *Letters to Beccari*, Beccaria
continued to work with Bassi, Veratti, and Beccari in Bologna. In 1758 he decided to address the letters forming his *On Atmospheric Electricity* to the president of the Institute of Sciences because, as he explained to Franklin, "the others,
and mostly this part of experimental physics, are cultivated there."[23]

Beccaria's interest in maintaining good relations with electricians outside Piedmont was no exception for a member of the republic of letters. Yet the circumstances of his appointment as the physics professor of the University of Turin
played a crucial role in directing his research interests toward electricity and in his
keeping an extensive network of foreign correspondents. In 1748, when the chair
of physics was offered to him, the kingdom of Piedmont-Sardinia had entered a
period of peace after several decades of war. The king, Carlo Emanuele III, concerned with building up the kingdom's economy, encouraged his magistrates to
plan a new system of education that would link the centers of learning to the state.
To implement the project, the king was aware that it was necessary to render the
university independent from the control of the Minims, the clerical order that had
colonized the university during the previous decades.[24] Beccaria's appointment
was one of the steps that the reformers undertook in this respect. Beccaria, still a
cleric, did not belong to the order of the Minims. Hence, he worked in a hostile
environment in which his colleagues, looking with suspicion at the attempts of
the reformers to limit their power, tried to discredit his work.[25] Aware of these hos-
tilities, one of the reformers, the marquis Giuseppe Morozzo, upon reading in the
newspapers of the Marly experiment, advised Beccaria to replicate it and to investigate the subject further.[26] The ambitious professor realized that the experiment could lead to a reformulation of the ancient science of meteorology: "I
cannot tell how much enthusiasm of the most cheerful [giocosissima] confidence
took my heart when I read of the greatest experiment. Here, I said to the most hon-
orable Mr. Marquis, has been opened a new and very wide field for the investigation of nature's most wonderful effects; here is the way from which to proceed
to the very important study . . . of meteorology."[27]

The connection between electrical research and meteorology informed Beccaria's experimental work on "natural" electricity. Soon after Morozzo's communication, he set up the apparatus on the roof of his house, and on July 2, 1752,

he was the first Italian to extract sparks from a conductor pointed to the sky. In the course of the following months he collected experimental results, studied Franklin's experiments and observations, and published his first work, *Natural and Artificial Electricity*, a work in two volumes dedicated to the king.[28] The first systematic exposition of Franklin's theory of electricity, the text gained Beccaria fame and reputation in the international community of electricians. He was elected a member of the Bologna Institute of Sciences and a fellow of the Royal Society.[29] The king was so pleased with Beccaria's achievements that he awarded him an increase in salary, the first of a series of rewards.[30] He also required Beccaria to affix a lightning rod upon the Royal Palace.

Whereas in his *Natural and Artificial Electricity* Beccaria only conjectured that the electric fluid could play a role in the operations of nature, in his *Of Atmospheric Electricity: Letters to Beccari*, published in 1758, he proposed a comprehensive theory of the role of the electric fluid in the natural world. Not only lightning, but also earthquakes, whirlpools, whirlwinds, auroras borealis and falling stars, Beccaria argued, could be explained in terms of the motion of the electric fluid. His program linked the new science of electricity to meteorology, reshaping the Aristotelian science in terms of contemporary experimental philosophy. To carry out his electrometeorological observations, Beccaria conceived of an "electric observatory": a place where he could measure, on a daily basis, the electric state of the air. The experimental apparatus was a simple one: he placed a pointed metallic conductor on the roof and connected it by means of a long insulated metallic wire (which he called "exploratory wire") to an electrometer placed in a room. Extra wires, which he called "safety wires" (*fili di salute*), grounded the apparatus to avoid the risk of being struck by lightning. The electric observatory was a mobile experimental site where Beccaria would "live day and night" recording measurements: it could be the tower of a church, castle, or private house "in which the deferent wire brings electricity to me, and subjects it for me to a delicate electrometer."[31] In the course of two decades he set up electric observatories on the tower of the Valentino castle, at his house in Mondovì, and on the hills of Superga and Garzegna, near Turin. In his electrical observatory, he studied the electricity of the atmosphere in different meteorological situations: clear sky, rain, fog, dew, hail, and thunder.[32] With this apparatus he replaced Franklin's kites, which he noted could not be used in the absence of wind. By recording humidity, pressure, and temperature of the air together with its "quantity of electricity" he investigated whether there was any relation between the electric state of the atmosphere and the weather, becoming among the first in Europe to weave together electricity and meteorology.[33] It was a self-conscious program, which he intended as a new foundation of meteorology based on electrical experiments. If in his laboratory practice he was able to "verify the artificial circulation of the new and very active element," then by means of his electric observatory he was able to "discover, and draw the right lines of its natural circulation." He proudly claimed that he was one of the first to "find by experiment and to demonstrate that such an element [electricity] is present also in the air," and that he could show that "the exterminatory instantaneous fire is the main cause of the various watery meteors." In sum, he boldly reminded his patron, "I discovered, and made plain, the true principle of a very important science: meteorology."[34]

Beccaria's reference to meteorology contributed to a new understanding of electrical science, which began in the late 1740s with the first attempts to apply the electric fire to medical therapies, in terms of one of the Enlightenment themes: the usefulness of science. At a time in which meteorologists claimed that knowledge of the weather could have useful applications to agriculture and medicine, its connection with meteorology gave electrical research new relevance, and it made electricians fit the new image of natural philosophers as experts in the service of the state.[35] Beccaria had acted in this function soon after his appointment, when he worked on the standardization of weights and measures to be used for commerce in Piedmont, and on the design of canals that would channel water from the river Po to the fields inland.[36] His work with lightning rods was part of this program. He was involved in the construction and design of lightning conductors to be affixed on powder magazines and private buildings in various sites in Piedmont, and he was asked to supervise the lightning rod to be affixed on the Duomo in Milan.[37]

To respond to the objections against the usefulness of lightning conductors, Beccaria focused on the nature of lightning and on the examination of the path that it followed when it struck houses, towers, or other buildings. His empirical observations shifted his attention from the shape of the conductors' terminations to the best method to make the electric fire disperse into the ground. He noticed that a bad junction between the conductor's elements could cause sparks that, issuing from the conductor, could set fire to inflammable substances. On the basis of Henry Cavendish's experiments of the conductivity of water (that demonstrated that water conducted less than iron), he observed that when the electric fire reached the water, it could gush back to the ground with the same, disruptive results. Hence, he suggested that no lightning rods should be affixed on powder magazines, but that they should rather be erected in front of them, at a safe distance.[38]

His experimental observations of lightning strokes informed his recommendations about conductors' shape, material, and insulation, and his work attracted the attention of younger electricians involved in the lightning rod campaign. Felice Fontana from Tuscany and Giuseppe Toaldo from the Republic of Venice both corresponded with Beccaria when their governments requested them to supervise the construction of lightning rods to be affixed upon public buildings.[39] From the 1770s onward, enlightened governments as well as individuals invested consistent sums into the display of their faith in scientific progress: lightning rods became symbolic icons of that faith. In 1784, when Landriani published his catalogue, there were more than eighty lightning rods on top of private houses south of the Alps, and three dozen distributed evenly between public edifices, churches, and powder magazines.[40] The business of making lightning conductors proved profitable. For the construction of three lightning rods to protect the powder magazine at the Tortona castle near Turin, Beccaria and his assistants estimated expenses of 1,600 *lire*, which at the time very few people could afford. For a comparison: Beccaria's annual salary, when he was first appointed, was 1,200 lire. But lightning rods could be tailored to different pockets, as Beccaria himself noticed while giving advice on which materials to choose to make lightning rods: "Gold would be a very performing conducting material, but because of its price it has no other place than the premises of the greatest kings, otherwise copper, which resists better to the weather, but if it is too expensive, iron is good too, also because, not costing too much, one can easily make a thick conductor."[41]

The price of lightning rods was one of the issues that Giuseppe Toaldo, the most famous advocate of lightning rods south of the Alps, addressed in his numerous pamphlets in favor of lightning rods.

The Padua Observatory Tower:
Giuseppe Toaldo, the Expert and the People

When he became involved in the lightning rod campaign, Toaldo was professor of "Astronomy, Geography and Meteors" at the University of Padua. In the long list of objections and replies that made up his *Of the Use of Conductors: New Apology* (1774), he argued that costs could be consistently reduced if decorations were left out and, in any case, no expense could be regarded as excessively dear for the safety of the people. He advocated the necessity of lightning conductors to protect public buildings such as theatres, and reminded his readers that a conductor would cost twenty or thirty *scudi*, which, compared to the hundreds of thousands of *scudi* required to build a theatre, was as inexpensive as it could be.[42]

Contrary to Beccaria's relatively private work on lightning conductors, Toaldo made the "information of the people" his mission. Aware of popular resistance against lightning rods, he envisaged in the popularization of electrical science and of the numerous cases in which conductors had preserved buildings from disasters the path toward a better reception of lightning rods. Arguing that "authority is worth nothing when it comes to physics," his view on how to gain popular consensus was different from Landriani's insistence on authoritative examples as more effective means to forge the people's opinions. He was aware that local authorities needed to be educated just as ordinary people. The case of Bologna demonstrated that even magistrates could concede to popular fears:[43]

"In order to extend the use of conductors, we need to educate Magistrates and people of the administration, we should not talk only to the learned, we need to enlighten the people of the world, dissolve their prejudices, and reassure them about their fears."[44]

Toaldo, who was a Catholic priest, subscribed to the French amateur electrician Barbier de Tinan's view on the popularizing mission of the enlightened philosopher; his texts were conceived as "information to the people" that, while spreading natural knowledge, attempted also to combine enlightened natural philosophy and Catholicism. His works in favor of lightning rods were distinctively marked by the intention to spread knowledge about natural electricity and its role in meteorological events. When he was archpriest in Montegalda, a small village near Padua, Toaldo realized that multiple observations in time and space were necessary to find the "causes" of meteorological changes.[45] His work on meteorology was imbued with the conviction that better knowledge of the weather would result in improvements in agriculture and medicine, and Beccaria's theory of natural electricity fit perfectly with his idea of researching the natural causes of meteorological events. With the intention to sedate popular fears about lightning rods by offering a rational understanding of the nature of lightning, in his *Meteorological Essay* (Padua, 1770) he embraced Beccaria's view of the electric fire as "the great instrument of nature, the principle of evaporation, winds and thunderstorms, earthquakes, aurorae borealis and, above all, *lightning*."[46] All these, together with snow, fog, hail, and rain, were "meteors" whose motions could be explained in terms of the natural tendency of the electric fire to reach balance; the climatic features of geographically different places resulted from the interactions between such meteors. Thanks to the new meteorology, the weather was no longer to be seen as unpredictable or resistant to rational understanding. If the place and time where lightning would strike were still beyond exact predictions, meteorologists could foretell the likeliness of events based on meteorological records, observations, and measurements. All meteors behaved according to a pattern that meteorologists could decode; thereby their work would help people not to be passive victims of the weather. "Toaldo gives me rain on the 4th, and on the 4th it rained; I did the same on the 13th, 20th and 26th and rain fell. Everyone is amazed and stood bewitched, and they all shout; bravo Toaldo."[47]

Padua, just like Siena, Toaldo argued, was particularly exposed to lightning. Its towers were notoriously favorite targets for bolts of lightning; therefore conductors were highly recommended.[48] In the course of the 1770s, when he looked after the construction of a new, well-equipped observatory for the University of Padua, Toaldo designed a lightning rod to be affixed on top of the tower. The new observatory was completed thanks to the pressures that the professors of the university exerted on the Venetian Senate. The university where Galileo once taught was in visible decline, and the new observatory, to become the most mod-

Figure 2.2.   The Padua observatory tower with its lightning rod. Giuseppe Toaldo, "Dell'uso dei conduttori" (Padua, 1774). Franklin Collection, Yale University Library.

ern south of the Alps, was presented by Toaldo as the essential step toward the renovation of the university.[49] A lightning rod on its top, the first conductor in the Veneto affixed on a public building, would crown — symbolically and physically — the progressive inclinations of the university and of its institutional supporter, the Venetian Senate. When a bolt of lightning struck the observatory tower in 1772, it caused damages of 500 Venetian lire. One year later, on September 28, 1773, a lightning rod appeared on top of the observatory.[50] It cost a little less than 200 *ducati* (1,600 lire).[51]

The lightning rod on the observatory tower was the first of many others that Toaldo would design. Wealthy individuals asked him to design lightning rods for their own palaces and, given the number of people who died struck by lightning every year, the senators of the Republic became sensitive to Toaldo's appeal to public safety. Toaldo could count on firsthand information on the damages caused by lightning in the Venetian countryside. He had arranged a network of meteorological observers spread in the domain who sent him barometric and thermometric measurements together with records about occasional yet meaningful events. Deaths by lightning and the damages caused by severe weather were among them, and in 1787 his correspondents urged him to petition the Senate for a ban against bell-ringing during thunderstorms.[52]

In Toaldo's advocacy, expertise was essential in the making of conductors. The death of the physics professor Georg Richman at St. Petersburg (1753) demon-

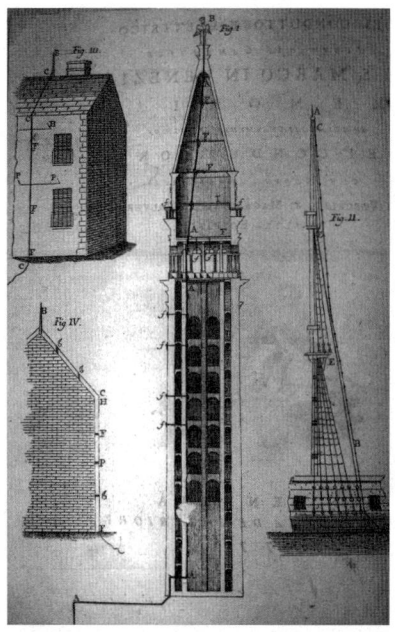

Figure 2.3.   The lightning rod that Toaldo designed for the church of San Marco in Venice. GiuseppeToaldo, "Del conduttore elettrico posto nel campanile" (Padua, 1776). Franklin Collection, Yale University Library.

strated the risks of inexpert management of the electric fire. Toaldo argued that "conductors are God's gifts to preserve human life, but, like medical remedies, they have to be well prepared."[53] It was on the basis of their observations of the marks left by bolts of lightning upon conductors or other materials that electricians, such as Beccaria and Toaldo, had drawn their conclusions on how to make lightning conductors. Expertise derived from practical knowledge and familiarity with facts, not from abstract theory: "We can well ignore the intimate nature of this fire: it is not necessary to know the nature of materials to make use of them."[54] Indeed, theory could be revised on the basis of the evidence that new facts offered. Each bolt of lightning that struck towers with or without conductors left signs that electrical experts could decode and use to improve the design of lightning rods. The public, by witnessing the results of such observations, could partake of expert knowledge. Toaldo published the reports of the effects of lightning upon conductors to educate "the class of people that is not much dedicated to studies, and that is less versed in physical erudition."[55]

For those who could not go to Padua and see with their own eyes, Toaldo's *Of the Use of Conductors: New Apology* offered a detailed description of the lightning rod on top of the observatory, together with some notions on the electric nature of lightning. Similarly, after his supervision of the lightning rod that was affixed on the bell tower of San Marco in Venice, he felt it his "duty" to account for his work to the senators who ordered it, and to the public.[56] Aware that lightning rods engendered controversies on both sides of the Alps and the Atlantic among "the people not yet philosophical,"[57] Toaldo's strategy was to popularize the recent discoveries in the science of electricity so as persuade these people that preserving edifices from the effects of lightning was now possible, just as it was possible to prevent floods by means of embankments.[58] The government responded sympathetically to his activity. On May 9, 1778, the Venetian Senate ordered that lightning rods should be affixed on all the powder magazines of the Republic, and eight years later, in 1786, it extended the order to all bell towers.[59]

Conclusion: Lightning and the Enlightened Philosopher

Brandished like sharp swords against threatening thunderclouds and wrapped in a scabbard of controversies, lightning rods symbolized the enlightened philosopher's victorious campaign against superstitious beliefs about the nature of lightning. "What else is left to say about the ancient opinion that lightning froze wine, which, while melting, provoked death, or furor? And what about the modern one, that lightning dissipates wine without offending its barrel? If not that, the malice of drunken servants has contrived them and has imposed them upon ignorant masters, who, after exaggerating the real effects, have then been abused by all these kinds of oddities to deify lightning. Thus they have reciprocally deceived the deceitful people and oppressed them under the yoke of blind superstition."[60]

Popular culture had interacted with lightning since antiquity. It had produced notions on the nature of lightning and of its relation to divine agency, and it had devised methods to prevent lightning's disruptive power. If the habits of ringing bells and firing cannons during thunderstorms were relatively recent, that of burning laurel dated back to ancient times. Laurel and olive trees were commonly held to be immune from lightning strikes, and people used to place laurel leaves or branches of olive trees on top of their houses, around their fields, or on their beds.[61] Enlightened philosophers like Toaldo did not confront such traditions with sarcasm. On the contrary, Toaldo engaged in the reformulation of popular beliefs in terms of Franklinian electrical science. Confused and uncertain, he argued, popular opinions were nonetheless rooted in observation and practical knowledge, and they could be explained by experimental physicists on more authoritative grounds.[62] Because of the resinous nature of plants such as laurel or olive trees, he explained, lightning abhorred them, and indeed, experience showed that they were only rarely struck.[63]

Throughout history, lightning terrified not only because of its disruptive nature but also because of its relation with the godhead. In monotheistic religions as well as in pagan traditions, it was perceived as the manifestation of the most awe-inspiring form of divine power: wrath. Enlightened philosophers who belonged to a clerical order, such as Beccaria and Toaldo, engaged in the task of read-dressing popular beliefs and containing people's fears, showing the compatibility between Catholicism and faith in scientific progress. According to Scriptures, neither evil spirits nor sorcerers have in themselves the evil power to rouse lightning, thunder, hail, or storming winds; all such effects are produced by the powers of nature, preordained by God with infinite knowledge and benevolence.[64]

By his own admission, Beccaria was not the first who stated that no human being could produce lightning: he quoted extensively from the bishop of Lyon, Agobard, who in the eleventh century wrote against "the silly opinion of the people about hail and thunder." Beccaria invited those who were in charge of "educating the clergy" to read Agobard's works. A properly educated priest would reassure the people and would make them "admire the immensity of nature's operations and its Creator's omnipotence" instead of letting them be prisoners of their superstitious fear.[65] The demonstration that lightning was one and the same with artificial sparks and that the electric fire was responsible for the "unusual operations of nature" left no room for supernatural powers, if not those of a benevolent, omnipotent being.

The appeal to education as a means through which to emancipate the people from ignorance and superstition in the name of enlightened Catholicism was also a refrain in Toaldo's campaign in favor of lightning rods. He was familiar with traditional beliefs about the powers of certain people to attract lightning. Pliny recorded that Numa Pompilio, the Roman emperor, mastered the art of evoking lightning. Tullio Ostilio, his successor, died struck by lightning while trying to steal the secret. The Etruscans were also acquainted with the art of attracting lightning. A traditional folk tale told the story of a monster that, after periodically ravishing the peoples who lived around the lake of Bolsena, was eventually killed by a flash of lightning called on for that purpose.[66]

Toaldo did not intend to support such beliefs, yet he wanted his readers to appreciate that the eccentric behavior of lightning could be explained in terms of the main tenets of Franklin's electrical philosophy: the electric fire's natural tendency toward balance and its predilection for metals.[67] With this in mind, it was possible that humans could artificially attract lightning. By linking the earth with the sky, metallic conductors attracted the electric fire and forced it through themselves, preventing it from setting fire elsewhere. However, the modern Prometheuses who stole fire from the sky acknowledged that lightning rods were God's gifts: if they snatched lightning from Jupiter's hands, it was to remit it in God's. In Toaldo's work, modern meteorology was to be seen as a human attempt blessed by divine benevolence and not as a new, presumptuous Icarus flight.

Debates on lightning rods prompted educational campaigns whose goal was to advocate a new alliance between Catholic faith, popular beliefs, and enlightened natural knowledge. The new meteorology advocated by Beccaria and Toaldo domesticated the terrifying meteor of lightning within a philosophical system in which the apparently unpredictable did not undermine the ideal of a

God-ordained, law-obeying universe. With a lightning rod on top, towers now symbolized the power of the new science of meteorology to tame the unpredictable forces of nature. On their part, towering above ancient beliefs, lightning rods pointed to the real direction of Enlightenment.

Notes

1. Alessandro Chigi, *Dell'elettricità terrestre-atmosferica* (Siena, 1777).

2. Domenico Bartaloni, "Relazione dello stato della Torre colpita da fulmine," *Giornale Letterario di Siena* 3 (1777): 267–68.

3. For an analysis of the debates on lightning rods in the English context see Trent A. Mitchell, "The Politics of Experiments in the Eighteenth Century: The Pursuit of Audience and the Manipulation of Consensus in the Debate over Lightning Rods," *Eighteenth-Century Studies* 31 (1998): 307–31; for the French context see Jessica Riskin, *Science in the Age of Sensibility: The Sentimental Empiricists of the French Enlightenment* (Chicago: University of Chicago Press, 2002), ch. 5; for the American, James Delbourgo, "A Most Amazing Scene of Wonders. Electricity and Enlightenment in North America" (Cambridge, MA: Harvard University Press, 2006), ch. 4; for the Italian, Ferdinando Abbri, "La 'spranga elettrica': Frisi e l'elettricità," in *Ideologia e scienza nell'opera di Paolo Frisi (1728–1784)*, vol. 1, ed. Gennaro Barbarisi (Milano: Franco Angeli, 1987), 161–99; Stefano Casati, "Storie di folgori: il dibattito italiano sui conduttori elettrici nel Settecento," *Nuncius* 13 (1998): 493–512; and Antonio Pace, *Benjamin Franklin and Italy* (Philadelphia: The American Philosophical Society, 1958), ch. 2.

4. Marsilio Landriani, *Dell'utilità dei conduttori elettrici* (Milano, 1784), unpaginated dedication.

5. Pace, *Franklin and Italy* (cit. n. 3), 28–30.

6. See Paola Bertucci, "Sparking Controversy: Jean Antoine Nollet and Medical Electricity South of the Alps," *Nuncius* 20 (2005): 153–87; and *Viaggio nel paese delle meraviglie. Scienza e curiosità nell'Italia del Settecento* (Turin: Bollati Boringhieri, 2007). On Laura Bassi there is an extensive literature that is impossibile to list here: see at least Marta Cavazza, "Laura Bassi e il suo Gabinetto di Fisica Sperimentale: realtà e mito," *Nuncius* 10 (1995): 715–53; and Paula Findlen, "Science as a Career in Enlightenment Italy. The Strategies of Laura Bassi," *Isis* 84 (1993): 440–69.

7. On the history of the Institute of Sciences, see Annarita Angelini, ed., *Anatomie Accademiche*, vol. 3: *L'Istituto delle Scienze e l'Accademia* (Bologna: Il Mulino, 1987); and Marta Cavazza, *Settecento Inquieto* (Bologna: Il Mulino, 1990).

8. Giuseppe Veratti, *Osservazione fatta in Bologna l'anno MDCCLII dei fenomeni elettrici nuovamente scoperti in America, e confermati a Parigi* (Bologna, 1752).

9. Ibid., 2. On the most popular electrical experiments in the eighteenth century see Paola Bertucci "Sparks in the Dark: The Attraction of Electricity in the Eighteenth Century," *Endeavour* 31 (2007), 88–93; for an overview of eighteenth century electricity, John L. Heilbron, *Electricity in the 17th & 18th Centuries. A Study of Early Modern Physics* (Berkeley and Los Angeles: University of California Press, 1979).

10. [anon.; untitled article] *Bologna* (August 1752): 31, p. 1. On Veratti's involvement in medical electricity, see Bertucci, "Sparking Controversy" (cit. n. 6). On electrical experiments on the human body, see Paola Bertucci, "The Electrical Body of Knowledge," in *Electric Bodies: Episodes in the History of Medical Electricity*, ed. Paola Bertucci and Giuliano Pancaldi (Bologna: CIS, University of Bologna, 2001), 43–68.

11. Archivio di Stato, Bologna (hereafter ASB), *Assunteria d'Istituto, Lettere dell'Istituto* 3 (Institute to the Ambassador, August 2, 1752).

12. Gino Rocchi, ed., *Carteggio tra Giambattista Morgagni e Francesco M. Zanotti* (Bologna: Zanichelli, 1885), 404 (Zanotti to Morgagni, August 4, 1752).

13. ASB, *Assunteria d'Istituto, Lettere dell'Istituto* 3 (Institute to the Ambassador, August 19, 1752). On the pope's donations to the Institute, see Angelini, *Anatomie Accademiche* (cit. n. 7), 215–38.

14. ASB, *Assunteria d'Istituto, Lettere all'Istituto* 4 (Fulvio Bentivoglio to the Institute, August 9, 1752 and August 12, 1752). .

15. On the pope's attitude to electricity see my "Sparking Controversy" and *Viaggio nel paese delle meraviglie* (cit. n. 6).

16. Tommaso Marini, *Esperienze sulla Elettricità che chiamano celeste* (Bologna, 1748). See also Giorgio Tabarroni, "La torre dell'Università di Bologna e l'elettricità atmosferica," *Coelum* 34 (1966): 1–14.

17. On Bassi's and Veratti's physics laboratory, see Cavazza, "Laura Bassi" (cit. n. 6).

18. Charles Burney, *The Present State of Music in France and Italy, or The Journal of a Tour through Those Countries* (London, 1771), 75.

19. Biblioteca Universitaria, Bologna (hereafter BUB), *Lettere di Landriani a Canterzani*, Cod. 2096, busta V: draft of a reply by Canterzani to Landriani dated September 15, 1783.

20. Ibid., draft of a reply by Canterzani to Landriani dated June 15, 1784.

21. Giambattista Beccaria, *Dell'elettricismo: Lettere di Giambattista Beccaria dirette a Giacomo Bartolomeo Beccari* (Bologna, 1758).

22. On Galvani's laboratory, see Marco Bresadola, "Exploring Galvani's room for experiments," in *Luigi Galvani International Workshop* (Bologna: CIS, University of Bologna, 1999), 65–82; and Marco Piccolino and Marco Bresadola, *Rane, torpedini e scintille: Galvani, Volta e l'elettricità animale* (Turin: Bollati Boringhieri, 2004).

23. American Philosophical Society Library (hereafter APS), Beccaria Papers, B B385, No. 12 undated, although it was written after Beccari's death (1766) and in the course of Franklin's supervision of the English translation of Beccaria's *Artificial Electricity* (London, 1773). Beccaria's book referred to in the quote was printed twice in 1758. The two editions differ only in the title: the first is *Dell'elettricismo: Lettere di Giambattista Beccaria dirette a Giacomo Bartolomeo Beccari* (Bologna, 1758), and the second edition is *Elettricismo atmosferico: Lettere di Giambattista Beccaria* (Bologna, 1758).

24. See Tommaso Vallauri, *Storia delle Università degli studi del Piemonte*, 3 vols. (Torino: Stamperia Reale, 1845). Also Marco Ciardi, "Medicina, tecnologia civile e militare, filosofia naturale. L'insegnamento della fisica nel Regno di Sardegna," *Studi Settecenteschi* 18:217–47.

25. See Heilbron, *Electricity in the 17th & 18th Centuries* (cit. n. 9), 362–72; Pace, *Franklin and Italy* (cit. n. 3), 50f.

26. APS, Beccaria Papers, B B385, No. 36, f. 1 (Osservazioni fatte in Torino nell'anno 1752).

27. Ibid., No. 36, f. 1.

28. Giambattista Beccaria, *Dell'elettricismo artificiale e naturale. Libri due* (Torino, 1753).

29. On the Italian fellows of the Royal Society in the eighteenth century, see Marta Cavazza, "The Institute of Science of Bologna and the Royal Society in the Eighteenth Century," *Notes and Records of the Royal Society* 56 (2002): 3–25.

30. Antonio Maria Vassalli-Eandi, *Memorie Istoriche intorno agli studi del Padre Giambattista Beccaria delle Scuole Pie* (Torino, 1783), 37.

31. Giambattista Beccaria, *Dell'elettricità terrestre atmosferica a ciel sereno* (Turin, 1775), p. 12.

32. APS, Beccaria Papers, B B385, Nos. 15, 19.

33. Joseph Priestley gave ample credit to Beccaria for his observations on the electricity of the atmosphere and on the "unusual appearances" in nature, remarking that many of his experiments were performed prior to those of other electricians in other countries. See Priestley, *History and Present State of Electricity* (London, 1767), 366–97.

34. Giambattista Beccaria, *Elettricismo atmosferico* (Bologna, 1758), unpaginated dedication.

35. On the new role of natural philosophers as civil servants, see Giuliano Pancaldi, *Volta: Science and Culture in the Age of Enlightenment* (Princeton, N.J.: Princeton University Press, 2003), ch. 2. For the case of Piedmont, see Vincenzo Ferrone, *La Nuova Atlandide e i Lumi: Scienza e politica nel Piemonte di Vittorio Amedeo III* (Torino: Meynier, 1988), especially ch. 1. Studies on eighteenth-century meteorology are still sparse; for the British context, see Jan Golinski, *British Weather and the Climate of Enlightenment* (Chicago: University of Chicago Press, 2006).

36. Vassalli-Eandi, *Memorie Istoriche* (cit. n. 30), 12f.

37. APS, Beccaria Papers, B B385, No. 31.

38. Ibid., No. 33. This and the following examples demonstrate that debates on lightning rods in the Italian states did not verge on the controversy about pointed vs. blunt terminations analyzed by Mitchell (see note 3) and R. W. Home in this volume. Indeed, the Italian electricians themselves acknowledged that the debate was confined to England.

39. Ibid., Nos. 31, 35.

40. Landriani, *Dell'utilità* (cit. n. 4), 285–93.

41. APS, Beccaria Papers, B B385, No. 35, f. 1v.

42. Giuseppe Toaldo, *Dei conduttori per preservare gli edifizj da' fulmini*, ed. Stefano Casati (Firenze: Giunti, 2001; 1st ed. Venice, 1778), 207. This work was a collection of all his previous works on the subject.

43. Giuseppe Toaldo, "Dell'uso dei conduttori: Nuova apologia" in Toaldo, *Dei conduttori* (cit. n. 42), 115.

44. Barbier de Tinan, *Nuove considerazioni sopra i conduttori* in Toaldo, *Dei conduttori* (cit. n. 42), 162.

45. On Toaldo's work as a metereologist, see Giampiero Bozzolato, *Giuseppe Toaldo: Uno scienziato europeo nel Settecento veneto* (Padova: Brugine, 1984); also Luisa Pigatto, ed., *Giuseppe Toaldo e il suo tempo nel bicentenario della morte. Scienze e lumi tra Veneto e Europa* (Padova: Cittadella Bertoncello artigrafiche, 2000).

46. Giuseppe Toaldo, *La Meteorologia applicata all'Agricoltura* (Venezia, 1775), 21.

47. Tragin della Bastia to Toaldo (Brescia, September 5, 1784), quoted in Bozzolato, *Giuseppe Toaldo* (cit. n. 45), 108.

48. Giuseppe Toaldo, "Del conduttore elettrico posto nel campanile di San Marco in Venezia" (Padova, 1776), in Toaldo, *Dei conduttori* (cit. n. 42), 154.

49. On the attempts to renovate the University of Padua, which included a plan to set up an autonomous press, see Bozzolato, *Giuseppe Toaldo* (cit. n. 45). In 1744 Toaldo published the Padua edition of Galileo's works, which included the first authorized reprint of the *Dialogue on the Two Chief World Systems* (in the Index of Forbidden Books since 1633).

50. Archivio Antico Osservatorio Astronomico, Padova (hereafter AOP), Cod. I, *Osservazioni meteorologiche. Padova 1766–1804* (September 17, 1772; September 28, 1773). Also Toaldo, *Dei conduttori* (cit. n. 42), 100f.

51. Toaldo, *Dei conduttori* (cit. n. 42), 114f.

52. AOP, *Lettere a Toaldo*, fasc. 43, 47.

53. Giuseppe Toaldo, "De' conduttori, o parafulmini," in Toaldo, *Dei conduttori* (cit. n. 42), 211.

54. Toaldo, *Dei conduttori* (cit. n. 42), 96.

55. Giuseppe Toaldo, *Tre lettere sul conduttore elettrico*, in Toaldo, *Dei conduttori* (cit. n. 42), 198.

56. Toaldo, "Del conduttore elettrico posto nel campanile" (cit. n. 48), 125–58.

57. Ibid., 128.

58. Giuseppe Toaldo, *Informazione al Popolo*, in Toaldo, *Dei conduttori* (cit. n. 42), 74.

59. AOP, Cod. I., Toaldo, *Osservazioni meteorologiche: Padova 1766–1804* (July 1786); Toaldo, *Dei conduttori* (cit. n. 42), 67.

60. APS, Beccaria Papers, B B385, No. 20, Libro VI (addressed to Priestley).

61. Toaldo, *Dei conduttori* (cit. n. 42), 70.

62. See Giuseppe Toaldo's prize-winning *Meteorologia applicata all'agricoltura, memoria che ha riportato il premio della società reale di Montpellier* (Venezia, 1775).

63. Toaldo, *Dei conduttori* (cit. n. 42), 70.

64. APS, Beccaria Papers, No. 20, Libro III (to the Bishop of Mondovì).

65. Ibid.

66. Toaldo, *Dei conduttori* (cit. n. 42), 69.

67. Jessica Riskin interprets this attitude toward the explanation of natural phenomena as "sentimental empiricism"; see Riskin, *Science in the Age of Sensibility* (cit. n. 3), especially chs. 3 and 5.

"In nebula nebulorum"

The Dry Fog of the Summer of 1783 and the Introduction of Lightning Rods in the German Empire

Oliver Hochadel

THE YEAR 1783 HAS A FIRM PLACE IN HISTORY BOOKS and chronologies of human achievements. It was June 4, 1783, when the brothers Montgolfier let their first hot air balloon ascend into the sky over Annonay in the southwest of France. The news spread fast, and in the following months, people all over Europe tried to launch their own balloons. Ballooning was the talk of the day, and the media bursting with new accounts of fortunate and less fortunate attempts to fly.[1]

Yet if we had a chance to ask the people of the time what was most memorable and peculiar about that summer of 1783, they would probably give a different answer. Yes, there was something in the air, but the people in many parts of Europe might rather have referred to the strange weather that started roughly in mid-June, two weeks after the ascent of the first montgolfière balloon. It was a summer marked by a most strange "dry fog," a white veil in the air that had no moisture in it like ordinary fog. The sunlight was clouded to such an extent that people were able to look right into the sun. Visibility was poor, and in some places the air smelled of sulfur. At sunrise and at sunset the sun often turned crimson or cherry-red and heightened the frightening appearance of the elements. The fog lasted for about five weeks, disappearing around July 20, only to return with lower intensity in August until the beginning of September (with considerable local variation). The fog was joined by very high temperatures. According to recent reconstructions of climate historians, the summer of 1783 was one of the three hottest summers in western Europe in the last three centuries.[2]

As if that were not enough, the summer of 1783 was also marked by a frequency and ferocity of thunderstorms without parallel in the memory of the living. In 1788 Anton Pilgram, an early climate historian and statistician from Vienna, calls 1783 the "strongest thunderyear I ever experienced."[3] Others went further, claiming that human memory had not recorded a summer with such

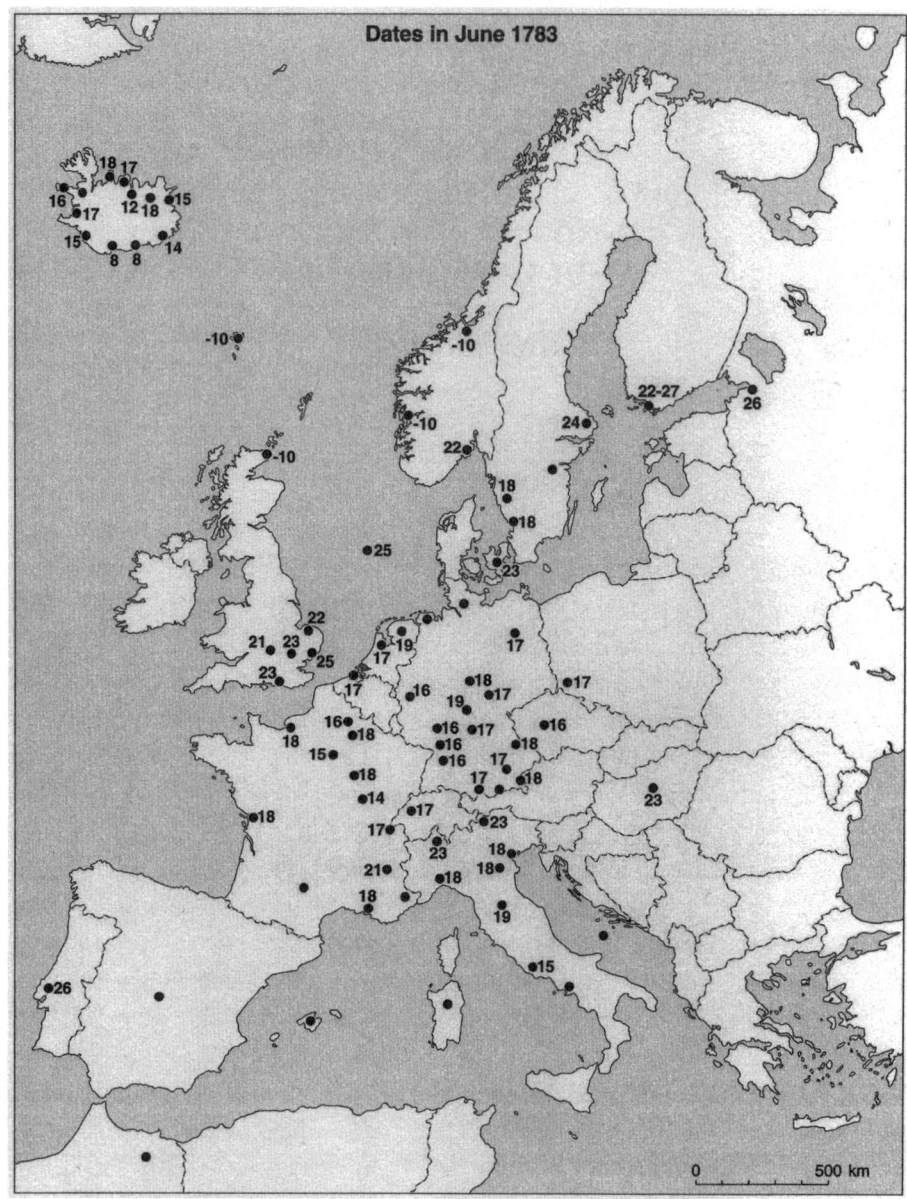

Figure 3.1.   The dates show the first appearance of the dry fog in June 1783. I thank John Grattan for the permission to reproduce this map. Grattan and Brayshay, "An Amazing and Portentous Summer," *The Geographical Journal* 161, no. 2 (1995): 646.

heavy thunderstorms.[4] With respect to England, the *Gentleman's Magazine* "for July reported a country-wide increase in death by lightning, a phenomenon 'more fatal, during the course of the present month, than has been known for many years'."[5] In July and August 1783 hardly a newspaper issue was without reports describing the frequent lightning and the immense damage it caused to buildings, deploring the lives it took when it struck people in fields, houses, or while ringing church bells.[6]

In this essay, I will concentrate on the German Empire, but the phenomenon as such—the dry fog, and the huge number of thunderstorms—was a European one.[7] The reading public of the time was very well aware of the European dimension of the strange weather.[8] The dry fog was a weather phenomenon as much as it was a media phenomenon. The people remembered the simultaneous appearance of the fog and the heavy lightning. And they remembered having read about it. A decade later, the Swabian minister and amateur scientist Gottlieb Christoph Bohnenberger still recalled: "Everybody knows, how dry the air was at the time of the high smoke in the summer of 1783, but also how numerous the thunderstorms were, and how heavy their outbreaks were nearly everywhere. All the time one would hear about the unlucky ones who were killed by lightning, and in all newspapers one would read about the terrible thunderstorms and the destruction and the havoc they caused."[9]

Pauper and Professor

How did the people react in the face of these most unusual phenomena? Due to the limited sources it is hard to tell what the "common man," the farmer in the countryside or the artisan in the city, thought about the fog. Mostly we can only retrieve some indirect evidence. Some of the newspapers reported growing public fear and apprehension. These articles tell us, of course, only what the authors thought the people were thinking. These writers often draw on the stereotype of the common man who resorts to superstition when confronted with unusual phenomena instead of reason and natural philosophy.

A few sources allow us to get closer to the experience of "ordinary people" without looking through the lens of the Enlightenment discourse. One such source is Ulrich Bräker (1735–98), the famous "poor man from Toggenburg" in the northeast of Switzerland. Bräker, a small farmer and the son of a day laborer, was forced to fight as a mercenary in the Prussian army at the beginning of the Seven Years' War, but he deserted. His autobiography, including accounts of his harrowing wartime experiences, was published in 1789 and turned him into a kind of celebrity. He also kept a diary from around 1768 until his death.

Bräker lived in a world that was imbued with God's active presence. He was engaged in a constant dialogue with his Creator. Bräker put all his trust in God, castigated himself daily for his sins and carnal weaknesses, and exhorted himself to lead a moral life. He commented on the strange weather of 1783 accordingly. The recurring theme of his entries between June 20 and September 5 describes the numerous thunderstorms as being the voice of God in the clouds.[10] Yet this did not exclude a sober and rather rational approach to life and to the strange dry fog he experienced. He was a close observer of the forever changing weather, with one eye on the sky and the other on his barometer. He also noticed that his fellows tended to get worried. Some of them even claimed that there had been a similar haze before the last plague. Bräker himself could not subscribe to this view. He did not interpret the strange phenomena as signs of an impending evil and was much more concerned with the temptations of the flesh he had to fight off every day.[11]

The thunderstorm on June 27, 1783, was so ferocious and long lasting that not even the very oldest citizens were able to remember anything like it.[12] Bräker read the papers and learned about the victims the lightning strikes claimed. On one of the extremely hot days, a newspaper reporting on the heavy storms was read aloud to the people of the village, most of them would have been illiterate. This prompted a discussion about lightning rods. "Some of the peasants became very agitated because such things tempted God."[13] Bräker was not very explicit on what he himself thought about the lightning rod, but he was certainly no staunch supporter of it. Yet he did not side with the peasants either; perhaps his attitude is best described as indifferent.

Thanks to Bräker, we know that the strange fog was also mentioned in sermons along with other extraordinary natural phenomena that marked 1783, such as earthquakes and floods. "Such sermons of God in the clouds will work more on people than one thousand sermons of pastors." Bräker bitingly commented that even "the most able natural philosophers are not able to make sense of it and that God alone knows the things that are bound to happen." Bräker often wondered whether thunderstorms and hail were natural phenomena or instruments of God punishing sinners. He was inclined to believe the latter but at the same time thought that people speculated too much.[14] The "poor man from Toggenburg" neither matches the stereotype of the superstitious and credulous common man nor does he subscribe to an enlightened view of nature according to which natural philosophy is able to explain strange phenomena. He did not need the kind of reassurance "natural explanations" were supposed to offer because of his belief and trust in God.

How did the natural philosophers experience the fog? A well-documented example is the reaction of Georg Christoph Lichtenberg, professor of experimental physics at the University of Göttingen. One of his main fields of research was electricity, and he became a sought-after expert in lightning rod matters. In particular the proper installation of lightning rods is a recurring theme in his letters in the 1780s. Both the dry fog and the intensity and frequency of the thunderstorms were widely reported and discussed in the letters to and from Lichtenberg between the end of June and the end of August 1783. On July 3, 1783, he wrote: "The dry fog extends . . . far beyond Strasburg, in a northern direction beyond Hanover and in a southern direction beyond Gotha; I received letter upon letter about it from all places."

At first he contented himself with capturing and measuring the strange phenomena with his instruments. He checked the quality of the air with his eudiometer and found it harmless—seemingly harmless, Lichtenberg specified, because the air had to pass through water and might therefore have lost some of its content.[15] In late July Lichtenberg measured in Göttingen about 45 degrees Celsius or 114 degrees Fahrenheit, the hottest day he could remember. With respect to the reason for the dry fog, Lichtenberg was not too inquisitive at first. He rather wished that the dry fog and "the sun with its red face" had stayed in Calabria because his study desk was flooded with all kinds of letters asking for explanations. Annoyed by the amount of requests, he poked fun at the weather by signing one of his letters "in nebula nebulorum," in the fog of fogs, an allusion to "in saecula saeculorum," the Christian formula for eternity.[16] Eventually he wanted to find out more. In September 1783 he read Franz von Beroldingen's "Ideas on the Persisting Extraordinary Fog," one of the numerous ready-made publications that claimed to explain the reason for the dry fog.[17] Lichtenberg followed the ensuing debate closely, as we shall see later.

Contemporary Explanations of the Strange Weather

So what was the reason for the dry fog? The contemporaries came up with all sorts of explanations. Franz von Beroldingen, a proponent of the Catholic Enlightenment, thought the fog to be "inflammable air," echoing the current debates on phlogiston theory. Johann Ludwig Christ, a minister from Rodheim, believed the dry fog was caused by subtle evaporations from the ground. Christ thought that, due to the solar irradiation, electrical particles were also raised, which were in turn responsible for thunderstorms.[18] An anonymous author linked the fog to the earthquakes that had shaken the south of the Italian peninsula and Sicily in February and March 1783. A subterraneous fire that found no outlet through volcanic eruptions produced fiery parts that turned into the dry fog. That explained the sulfurous smell and the lightning.[19]

Readers of German magazines and newspapers also learned about explanations from the European community of natural philosophers. Italian scholars such as Giuseppe Toaldo also linked it to the earthquakes that had hit Messina and parts of Calabria particularly hard. The French physicist Jérôme de Lalande held a combination of heavy rain and extreme heat responsible.[20]

Connections with earthquakes and preceding rain were widespread, but there were also "electrical" explanations put forward, most notably by the French natural philosopher Pierre Bertholon. He saw a clear connection between the dry fog and the high frequency of thunderstorms. Bertholon held a disturbance of the "electrical equilibrium" responsible and thereby offered a kind of holistic explanation for the whole set of strange phenomena of the summer of 1783, including the appearance of a new island off the Icelandic coast.[21]

There is one more "electrical" explanation of the dry fog that held the lightning rods responsible. An anonymous author wrote: "If one considers, how much the use of lightning rods in England and the Netherlands has gotten out of hand in the past couple of years, and how much it has nearly turned into abuse, then naturally one has to come up with the idea that this invention may have contributed a lot to the present vaporization of the fog. One tries to protect the houses from lightning, but in general the damage caused thereby is disproportionately greater."[22] The skeptic argued that the lightning rods drew the electricity from the air into the ground. Consequently, the sulfurous vaporization in the air was no longer bound, so it caused the dry fog.

The skeptic elicited several replies, refuting the argument quickly and even ridiculing it. Writers who opposed the introduction of the lightning rod or questioned its use had a hard time in the public arena of the day. In the newspapers and journals, virtually all of the authors were very much in favor of the lightning rod.[23] In the case cited, the editor of the newspaper did not intend to represent an opinion worth discussing. He printed the skeptic's arguments as an example of ill-founded hypothesizing in matters of natural philosophy. The skeptics usually have no name, while the champions of the lightning rod mostly do. Why the skeptics remained anonymous one can only speculate. Maybe they wanted to avoid being cast as backward, or they might even have been invented. A skeptic could be scolded to propagate the lightning rod and the enlightened values that went with it. Potential disbelief was thereby anticipated and refuted.

This does not mean that there were no reservations about the lightning rod. The scolding of the skeptic shows that the natural philosophers of the Enlightenment felt that they still had to reiterate their case over and over again. Regarding the potential harm of lightning rods, the dry fog of 1783 kept the pressure on. In early July 1783 an article containing four questions from another anonymous lightning rod skeptic was published in different papers. The skeptic wondered, for example, whether the fertility of the fields might suffer from the lack of electricity withheld by the lightning rods. He also suggested (somehow contradicting himself) that lightning rods might be responsible for earthquakes because they conducted more electricity into the ground. With his objections the skeptic was by no means outside the discourse of natural philosophy of the day. It was widely believed at the time that atmospheric electricity had a strong influence on all kinds of living beings, including plants. A possible connection between lightning rods and earthquakes was also discussed. The skeptic's questions were taken seriously and elicited a host of replies. The ones I found were all in favor of the lightning rod, declaring it innocent. Some defenders published their replies in newspapers; others, like the Viennese scholar L. P. von Weiskirch, published a thirty-page brochure to "answer" the questions. Weiskirch actually tried to calm the public. The Augsburg instrument-maker Jakob Langenbucher pointed out that the cities recently struck by earthquakes such as Lisbon and Messina lacked lightning rods. Johann Jacob Hemmer, a natural philosopher from Mannheim and leading meteorologist of his time, went even further, claiming that lightning rods would make cities earthquake proof.[24]

Again there was no real debate, but a quasi-unanimous consensus existed among the authors responding to these questions that the lightning rod would do no harm whatsoever. To the contrary, they pointed out additional benefits of the lightning rod, such as the fostering of the growth of plants or the protection of cities from earthquakes. In hindsight, these arguments are hardly more "scientific" than the objections of the skeptic. Yet public opinion, or should we say publicized opinion, had equated the lightning rod with the Enlightenment itself, and there was no going back.

The State against the Bells

For centuries people had resorted to a different way of protecting their dwellings against lightning. They rang the consecrated church bells to warn the villagers in the face of an approaching storm but also to ask for divine protection. This traditional practice was very elaborate and often included the ringing of particular bells that bore special engravings. A certain person was assigned to do this job and would receive a fixed due. This person was usually the bell ringer or sometimes the priest himself. This practice was very widespread in Catholic territories while Protestant communities usually resorted to prayers only.[25]

If the lightning rod is a prototypical invention of the Enlightenment, then the ringing of the church bells is the superstitious practice par excellence.[26] This practice had come under severe criticism. It was not only considered entirely useless but also extremely dangerous because bell towers were particularly prone to being hit by lightning. The thousands of flashes of lightning in the summer of 1783 highlighted this danger. Numerous reports on the destruction caused by lightning deplored that yet again somebody was killed while ringing the church bells in an attempt to keep away the black clouds. In early July 1783 several newspapers published statistics compiled by Johann Nepomuk Fischer. The former professor of mathematics at the Bavarian University of Ingolstadt counted 386 church towers hit and 103 people killed in the previous thirty-three years.[27]

The ruling authorities of numerous German states felt it was time to take countermeasures. The territories in question were either Catholic or had a substantial Catholic population (such as Prussia) and were largely in the south and southwest of the German Empire. Bavaria prohibited the ringing of the bells in early August 1783 by order of the elector Karl Theodor: "After which it is proven through bitter experience by the numerous reports coming in and the newspapers that the common practice of ringing the bells does more harm than good." Prussia followed in September, and Austria in late November 1783.[28] These were among the largest "secular" territories in the German Empire. But smaller territories such as Pfalz-Zweibrücken also took measures and prohibited the ringing of the bells in late August 1783.[29] The territories ruled by bishops did not lag behind. Already in July 1783 the ecclesiastic territories of Mainz and Trier forbade the ringing of the church bells.[30] The diocese of Augsburg followed in May 1784, and the diocese of Würzburg in the fall of that year.[31] After some initial measures taken in Salzburg in 1783 and 1784, a proper order was passed on February 1, 1785, including a catalogue of penal measures against the ringing of the bells and shooting during thunderstorms. The fine was 12 Reichstaler; in case of recurrence, 24 Reichstaler or confinement in a fortress. On May 16, 1789, several peasants broke into a church tower to ring the bells. They were stopped by soldiers and sentenced to severe imprisonment.[32] In Bavaria, too, the decrees got harsher. Offenders were threatened with incarceration.[33]

Obviously, it is quite a different issue as to how far these bans were actually effective. Ringing the bells was a well-established practice at the time and could not simply be abolished by a piece of paper. Who, for example, was going to compensate the person who was paid to ring the bells? Allegedly, popular resistance was widespread. Sometimes the people forced the priest to hand over the keys to the church tower so they could pray and ring the bells when a thunderstorm neared.[34] In most regions the edicts had to be reissued or even modified. In many regions such as Franconia, Vorarlberg, Swabia, and Tyrol, this tradition continued well into the nineteenth century.[35]

Yet it is remarkable that so many different territories prohibited the ringing of the bells within a relatively short period starting in the summer of 1783. It indicates a direct influence of the debate going on in the newspapers and journals of the time, which was no proper debate but a public propaganda crusade against ringing the bells and for the lightning rod, trying to refute and sometimes even to ridicule possible objections.

To be sure, the dry fog only accelerated a legislative process that had already been on the way. A few territories had already abolished the ringing of the church bells before. The imperial city of Nuremberg did so in May 1775; in the territory of Jülich-Berg, a decree was issued in November 1782.[36] Other territories did not react immediately to the incidents of the summer of 1783. This was partly due to the respective confession: in Protestant territories the ringing of the bells was far less common than in Catholic ones. But clearly the concentration of laws passed in 1783 and shortly thereafter is causally related to the dry fog of the summer and the high number of thunderstorms.

The Introduction of Lightning Rods

In 1783 we find several reports of lightning rods either being destroyed or their installation being obstructed or at least delayed. In the summer of that year, the Leipzig mathematician Christian Ludwig suggested erecting a lightning rod on a church in the estate of Gotha. The local administration consented, but according to a newspaper report, the peasants turned rebellious, pointing to the fact that their ancestors had known nothing about devices such as the lightning rod. The peasants considered it an intrusion into God's omnipotence and obstructed the installation. The incident was reported to Gotha, and the ruling duke ordered the lightning rod to be put up under the protection of military force.[37] How far this newspaper report is accurate or simply follows a master narrative of the successful battle of the Enlightenment against superstition remains an open question.

A similar event is reported from Düsseldorf. The lightning rods that had been installed on the castle and on the powder magazine in 1782 were blamed for attracting the heavy lightning in the summer of 1783. The people, led by superstition and ignorant of the rod's purpose, wanted to tear down the rods and had to be stopped by the soldiers of the garrison. This "lightning rod uproar" quickly became a celebrated episode in the history of the city to be told over and over again. Yet upon close inspection, as Fritz Dross has shown, there is no evidence in the sources of the time that there was any kind of insurgence at all. Consequently, these episodes tell us far more about the prejudices of the philosophes about the common people than about popular superstition itself.[38]

In any case, the pro–lightning rod crusade worked. Of course, lightning rods had been introduced to German-speaking territories some years before 1783. The first one was set up in Hamburg in 1770, and there were a few stray ones in the late 1770s.[39] Yet the dry fog of 1783 immensely accelerated the introduction of lightning rods in German-speaking territories. A Viennese paper tells us in late August 1783 that in these terrible months full of thunderstorms, more than six hundred new lightning rods were installed in Germany—"to the honor of philosophy." In Hamburg there was such a high demand for lightning rods in the summer of 1783 that the smiths could not satisfy the immense number of orders.

In many other German towns such as Augsburg and Stuttgart, the summer of 1783 marked the erection of the first lightning rod in town.[40] In some of the small German principalities such as Salzburg and Anhalt-Dessau, 1783 was the year when lightning rods were introduced.[41]

In the summer of 1783 Lichtenberg was busy writing up an expert opinion on a proposed lightning rod installation on the church of Mandelsloh, a village near Hanover. At the same time he exchanged letters with the Göttingen theologian Johann David Michaelis. The two professors were wondering why the temple of Solomon was never struck by lightning and whether the golden spikes on top of the roof served as conductors.[42]

Johann Jacob Hemmer, chief advocate of the lightning rod in southern Germany, traveled Swabia that summer overseeing the introduction of lightning rods and being called on as an expert by the authorities. He was invited to Tübingen where the university boasted it would erect the first lightning rod in town. The mechanic Franz Joseph Schropp from Obergünzburg near Kempten had to interrupt his research on atmospheric electricity in August because he was asked to erect lightning rods on different buildings and churches in the Neckar and Danube area.[43] Some of the numbers given earlier might be exaggerated, but 1783 was certainly not a "normal" year in the course of the introduction of the lightning rod in Germany where their installation had been sporadic in previous years.

This essay focuses on the German-speaking territories. There are no studies of the effects of the dry fog with respect to the introduction of lightning rods and the anti-bell-ringing legislation for other countries. Because the dry fog and the density of thunderstorms were both European phenomena, it would be most intriguing to compare in particular the Catholic territories.

An excellent point of departure for an international comparison is Marsilio Landriani's *On the Usefulness of Electrical Conductors*, published in 1784. The Italian natural philosopher was asked by the Austrian government in northern Italy to compile information on conductors affixed to the top of powder magazines, public buildings, and private houses.[44] The idea behind this endeavor was to sway readers by the power of numerous examples. Landriani used his European-wide network of correspondents and came up with a list of nearly four hundred buildings. As he conceded himself, the compilation was incomplete and contained little information on the north and virtually no information on the east of Europe.[45]

Unfortunately Landriani named only a few cases when the lightning rods were actually erected. In other words, it is hard to tell which ones were installed in 1783 and which ones had been installed prior to that. Nevertheless the dry fog of 1783, or rather the summer of thunderstorms, is mentioned several times. Quite a few lightning rods had been installed before the summer of 1783, for example, in the towns of Genoa (fourteen), Lucca (twenty-three), and Zurich (seventeen). In these cases, the unprecedented lightning did not so much trigger the installation of new lightning rods but was referred to as strong proof for the usefulness of electrical conductors, to cite Landriani's title once more. Landriani's correspondents from these three cities pointed out that even in this period of most ferocious thunderstorms, no damage was caused by strokes of lightning. In other cases, such as the little margraviate of Ansbach, the number of lightning rods increased. In that territory, today in northern Bavaria, fifty-four lightning rods were installed between October 1783 and November 1784.[46] The margrave of Ansbach even sent his court mechanic to Mannheim to learn the art of installing a lightning rod properly from Johann Jacob Hemmer.[47]

Further research might shed interesting light on the patterns of how lightning rods spread across Europe. It seems to have been patchy at the start, depending on the initiative of private citizens or local scientific societies. Soon thereafter quite a lot of the authorities followed suit who were mainly worried about their powder magazines but were also eager to be hailed as enlightened rulers, caring for the well-being of their subjects. This first wave of lightning rod installations occurred largely between the late 1770s and the mid 1780s.[48] There was certainly a peak in 1783 in the German territories.

The situation in France needs more research. It is a coincidence but nevertheless fits in nicely with our story that the famous trial of St. Omer ended in mid-June 1783, just as the dry fog set in. Vissery re-erected his lightning rod on July 31 that year. In January 1784 the French minister of war Ségur asked the *Académie des Sciences* for an expert report on how to protect the planned powder magazine in Marseille from lightning.[49] It seems that the first orders prohibiting the ringing of consecrated church bells in France were issued in July 1784.[50] In Italy, Giuseppe Toaldo was asked to approach the Venetian Senate to effect such a ban in early 1787.[51]

The Lakagígar Eruption in Iceland

This interesting link between a most strange weather phenomenon and a crucial phase in the history of the lightning rod leaves us with one intriguing question that contemporaries were at pains to answer: What did cause the dry fog of the summer of 1783? As the summer came to an end and the fog had already more or less disappeared, news first reached Copenhagen on September 1, 1783 about the eruption of a volcano in Iceland. It took nearly two months for this information to spread to the rest of Europe.[52] To be more precise, there was a sixteen-mile- or twenty-five-kilometer-long fissure eruption at the side of the Laki volcano in the southeast of the island. The Laki, or Lakagígar, as it is called in Iceland, fissure erupted in midmorning of June 8, 1783. It was Whit Sunday, eight to ten days before the dry fog was first detected in Germany, France, and other places. The eruption continued until February 1784 with varying intensity.

Climate historians are quite unanimous about the extraordinary character of one of the greatest volcano eruptions in modern times, which initiated "the largest terrestrial lava flow of the present millennium. . . . A total of 15 km³ of magma was erupted." About 122 million tons of sulfur dioxide were emitted into the atmosphere "and maintained a sulphuric acid (H_2SO_4) aerosol veil that hung over the Northern Hemisphere for at least five months." Nowadays climate historians are able to trace the spread of the dry fog far beyond Europe, as far as northern Africa and the Near East, and there is even strong evidence for Labrador and some for western China. Therefore the dry fog had a "truly hemispheric character."[53]

For Iceland itself, the eruption brought disaster. The grass was poisoned by the acid rain, and the number of cattle was halved. About ten thousand people, about 24 percent of the island's population, died in the ensuing "famine of the mist" which was primarily caused by the eruption. Adverse effects on agriculture due to aerial pollution and acid deposit have been identified for Great Britain, Scandinavia, and the area of the Baltic Sea.[54] Recent studies suggest that the air pollution through sulfur dioxide in combination with the high temperatures was responsible for a significant increase in the mortality rates in England and France in the late summer of 1783 and in the extraordinary severe winter of 1783–84. Claire Witham and Clive Oppenheimer reckon that for England the "combined mortality for August–September [1783] is 40% higher than the mean [between 1759 and 1808] and that for January–February [1784] 23% higher. Nationally this would have been equivalent to approximately 11,500 and 8,200 additional deaths in these periods, respectively." Grattan and colleagues come up with similar numbers for 53 parishes in France: They found that mortality had increased by 38 percent between August and October 1783 and by 25 percent for the period August 1783 to May 1784 if compared with the general mortality between January 1782 and December 1784.[55] Regarding the German Empire and central Europe, there are as of yet no studies of possible adverse effects of the dry fog. At the time, there were even reports about an "unknown fertility" in agriculture. Ulrich Bräker mentioned the plenitude of the fields several times. In particular, the grape harvest of 1783 seems to have been abundant in some regions.[56]

The eruption of the Laki fissure was clearly the cause of the dry fog. Yet was it also responsible for the high frequency of thunderstorms all over Europe? The Belgian climatologist Gaston Demarée thinks that a large-scale perturbation of the atmospheric conditions remains possible. But in his view, such anomalous

Figure 3.2.   Cinder cones along the Laki fissure in Iceland.

weather patterns are not necessarily linked to the Icelandic volcano eruptions. The American climatologist Richard Stothers is also cautious. He concedes that an increase in thunderstorm activity could well have been related to the dry fog, which seems to have changed local atmospheric circulation and precipitation patterns. Yet the link cannot be made definite because regional climate changes are much harder to compute from global circulation models than are global changes, and because the Laki eruption's effects on the atmosphere have just begun to be studied theoretically. The British climatologist John Grattan is more assertive. He is convinced that the summer of lightning was caused by the high temperatures and the concentration of aerosol and microscopic particulate material in the lower atmosphere. Volcanic eruption clouds are well known for their lightning shows. And he quotes from an English newspaper: "As the storm came on the stench of sulfur was greater than before, where there was no rainfall the stench of sulfur remained and the temperature became unbearable." A causal connection between the dry fog and the high frequency of thunderstorms seems at least possible if not likely.[57]

Let us now return to the contemporaries who were at pains to try to explain the great dry fog of that summer. As mentioned before, due to the delay in communication, there was at first no connection established between the strange weather and the Laki fissure.

It is an intriguing question to ask, who was the first to suggest that the volcanic eruption in Iceland caused the dry fog in 1783? For a long time Benjamin Franklin was credited as the one who first established the link to Iceland.[58] Christian Gottlieb Kratzenstein, a professor of physics at the University of Copenhagen, may have identified the cause of the fog even earlier, in 1784.[59] Yet recently the little-known French naturalist (and later politician) Jacques Antoine Mourgue de Mont-Redon (1734–1818) is said to have been the first.[60] His paper was published in 1784 but was apparently read at one or several meetings of the Société Royale des Sciences in Montpellier in 1783. But a serious contender is the Swiss meteorologist J. R. Salis-Marschlins. He is the only one who published this idea as early as 1783 in a local Swiss weekly paper.[61]

Whoever actually was the first person to suspect the causal link between the volcano eruption in Iceland and the dry fog of the summer of 1783, the four persons mentioned are representative of a very heterogeneous but also growing community of scholars turning to meteorology. As Richard Hamblyn put it: "an entire generation of European and American researchers, whose primary scientific interests had hitherto been earthbound, renewed their attentions to the actions of the air." And for young Luke Howard, who was born in 1772 and would later classify the forms of clouds, the strange events of 1783 marked "the turning point, the fulcrum of his engagement with the evolving science of weather and climate."[62] In the second half of the 1780s natural philosophers were still in dispute about which explanation was to be preferred, but eventually scholars such as Lichtenberg accepted the theory attributed to Franklin.[63]

Climate historians always had an interest in the strange weather of 1783. In the past fifteen years, this interest has yielded a substantial body of research dealing with the events and consequences of the Laki fissure eruption. Many of these studies do mention the high frequency of thunderstorms during that summer, but they do not establish the connection with the lightning rod. Neither have scholars of the Enlightenment or historians of science picked up on the interaction between the atmosphere and events on the ground. The trigger effect of the dry fog of 1783 for the introduction of the lightning rod and the abolition of the ringing of the church bells as well as the efforts of the natural philosophers of the time to come to grips with the strange phenomena have so far gone unnoticed.

Yet phenomena such as the dry fog might prove to be very promising objects of study because they are situated right at the intersection of many fields and disciplines, including environmental and climate history, history of science and technology, as well as history of mentalities studying attitudes to nature and the ongoing conflict between "enlightened" and "superstitious" ways of dealing with natural phenomena.

Conclusion: An Incomplete Victory

For the celebrated French writer and journalist Louis-Sébastien Mercier, 1783 was "l'année des merveilles." The greatest marvel was, of course, the conquest of the sky. He enumerates many more discoveries and inventions from natural philosophy but also the return of Franz Mesmer to Paris—in other words everything "scientific" that had a large popular appeal. The "faits extraordinaires de l'année" were made even numerous by the Italian earthquakes and "les volcans d'Islande."[64]

Prompted by the spectacular air balloon launches, the philosophes were becoming ecstatic. The Enlightenment finally seemed to live up to its promises. In his *Miscellaneous Thoughts about Aeronautical Machines*, published in early 1784, Georg Christoph Lichtenberg let the eighteenth century triumphantly speak for itself. Most of the roughly two dozen hallmarks of progress enumerated belong to the sphere of natural philosophy. The second achievement mentioned by Lichtenberg was the lightning rod: "I have learnt to resist thunder."[65] In the spring of 1784, Lichtenberg's brother Ludwig Christian oversaw the installation of a lightning rod on the tower of St. Margaret's church in Gotha. The inscription attached to the top of tower was addressed "To posterity" and it boasted: "We have prescribed the lightning its path." In 1783 Robespierre called the lightning rod "the boldest and most surprising idea mankind ever came up with."[66]

Why were the Lichtenberg brothers, a future revolutionary, and many others so proud of this simple bar of metal? As the German philosopher Hans Blumenberg observed, the lightning rod is one of the prototypical achievements of the Enlightenment. What had been out of reach and a reason for fear became cognizable and to some degree even controllable.[67] Probably more than any other achievement, the lightning rod materialized Enlightenment in its purest form. The practical use of natural philosophy that the Enlightenment had always claimed had rather been a promise than an actual reality.[68] Yet the lightning rod seemed to prove in a spectacular way that, through the experimental study of nature, man created knowledge that could be put into practice and would serve the common good. Therefore, the lightning rod could attain symbolic status for the Enlightenment itself. It was hailed as a "sign of triumph over superstition," particularly if it was erected on the houses of farmers.[69]

The Enlightenment had tried "naturalizing" peculiar or disastrous phenomena such as comets and earthquakes.[70] In 1783 this capacity was put to the test once more. People were afraid of the dry fog, the red sun, and the occasional sulfurous smell in the air, not to mention the heavy and frequent thunderstorms. In some parts of Europe there were also earthquakes, floods, and epidemics. Were these to be taken as signs? Was there a deeper meaning behind all of these most peculiar occurrences? The answer given in the printed sources is overwhelmingly no. In the summer of 1783 natural philosophers constantly addressed the public to explain the reasons for the fog. The explanation given by Lalande, mentioned earlier, was not published in an academic periodical but in the widely read *Journal de Paris* and was quickly translated into English and German. Lalande, a member of the Académie des Sciences, pointed out similar events in the past to dispute the alleged uniqueness of the strange fog. That was meant to calm the public and reassure the people that they only witnessed a natural and recurring phenomenon.[71]

Grattan and Brayshay analyzed English newspaper reports from the summer of 1783 singling out the adjectives describing the weather. The word "portentous" crops up only once as opposed to "violent" (46), "tremendous" (27) and "dreadful" (23), therefore making "portentous" less than 1 percent. The total count of adjectives is 183.[72] Virtually all articles in the German press tried to demystify the fog. An example is a letter published in the *Augsburgische Stats- und Gelehrte Zeitung* addressed to a "Madame." She had been wondering where the fog that had hung over Germany for the past few weeks was coming from. The anonymous author reassured her: "You will find just how few people have died in Stuttgart these days. Yet as soon as one thinks the fog comes from Italy, one feels one's chest tightening, one sees fiery balls falling from the sky and divine earthquakes. Then here, as in a thousand other cases, the ghost is in our heads."[73]

All these attempts to calm the public indicate that there was a lot of apprehension in face of the strange weather. The *Gentleman's Magazine* reported that in England "the churches and saints are more respectfully attended than usual," and that there was a "fear of impending calamities." The Protestant Lichtenberg reported that in a village near Fulda, special "fog prayers" were held and the priests (former Jesuits) were scolded for agitating their flock.[74] In Lausanne, the dry fog was interpreted by some as a sign from the Apocalypse of John (9:2), namely smoke ascending from the underworld. In other parts of Switzerland there were ceremonies of repentance and fasting to prepare the people for the approaching end of the world. The reactions of the "enlightened" Swiss observers such as F. Verdeil or N. E. Tscharner varied between compassion for the easily frightened people and ridicule and contempt for their credulity.[75]

From a historical point of view, the strange weather of 1783 provides a very revealing insight on how the late Enlightenment dealt with extraordinary natural phenomena. In a sense, the dry fog covered Europe at the right time. The "Enlightenment machinery" propagating the lightning rod was already well oiled at the time. In this essay, I mainly focused on newspaper and journals, which I think played a pivotal role with respect to the public (including the political authorities). Yet this Enlightenment machinery consisted of more parts: the natural philosophers, able to naturalize the phenomena and going public with it; the authorities, both local and "national", pushing the new technology with sermons, laws, and sometimes even military force; and the instrument-makers and itinerant lecturers installing lightning rods and filling their pockets.[76]

Before 1783, the introduction of the lightning rod did not exactly proceed at great speed. The strange weather and, in particular, the large number of thunderstorms provided vocal supporters of the lightning rod with a stepping stone to push the new technology a decisive move forward, both regarding its actual installation and convincing the public of its protective bliss. This was not a new strategy as such.[77] Such success-stories had been previously reported in the press. Yet the summer months of 1783 are without parallel. The papers were full of reports describing the damage that was done by lightning to property and people. The journals and newspapers recounted at the same time numerous examples of castles, churches, gunpowder-magazines, houses, and barns being saved from lightning thanks to previously erected conductors.[78]

To give but one example, in early August 1783 a weekly journal published in Augsburg ran a special feature on lightning rods. The article explained the way the rod worked and stressed that lightning struck the rod not the building. "This has already been proven by experience, and newer examples taken from the frequent thunderstorms of last and this month have confirmed this." The article enumerates several incidents from all over Germany and Austria.[79] One of the examples mentioned is Düsseldorf, a case that was turned into the Enlightenment tale of the "lightning rod uproar" mentioned earlier. In August 1782 Hemmer had equipped the gunpowder magazines of Düsseldorf with lightning rods, and everything seemed to go according to plan. When the magazine was struck on June 28, 1783, the rod did its job. Witnesses (guards and officers) were questioned in a "special interrogation," and the good news was published in several German papers. This well-documented event was to "calm the clamor of the people from all estates against the machines which was fed by ignorance" one author hoped.[80]

Far from it. The people of Düsseldorf were not impressed. They blamed the lightning rods for the heavy thunderstorms getting "stuck" over the city instead of moving on. The members of the magistrate of the town therefore asked their ruler Karl Theodor to tear the lightning rods down again! They did not want to judge the technology as such—it was "all too new"—but they obviously felt the need to communicate the concerns of the "public here." One can imagine Hemmer's indignation when he was asked for an expert opinion on this matter. If two things occurred at the same time (the erection of the rods and the persistent thunderstorms), one was not necessarily the cause of the other, he retorted. He explained the resistance of the people by their lack of Enlightenment. The petition of the magistrate he considered as coming close to lèse-majesté. How dare they question the measures taken by a wise and enlightened ruler? All in all, Hemmer resorted to authority rather than to scientific argument.[81]

The people of Düsseldorf did not put forward any religious or superstitious argument. Instead they used a rational point made by Nollet earlier in the century that lightning rods attracted lightning, thereby increasing the danger instead of averting it. And in the summer of 1783 the people of Düsseldorf could claim with some justification that their objections were empirically grounded, too. The appeal to experience could cut both ways.

As Jessica Riskin has pointed out with respect to the trial of St. Omer, "many since have accused early opponents of lightning rods of 'prejudice' and called their arguments 'pseudoscientific.' Perhaps they were, but if so, they were no worse than the arguments in favor of lightning rods, whose proponents were altogether unable to explain how their devices were meant to work."[82]

In the same vein, one might say, if only with hindsight, that at first the explanations for the dry fog given by natural philosophers all over Europe only differed in their degree of imaginative power. And it is somewhat ironic that—contrary to the reassuring words quoted above—more people did die in the summer of 1783 due to the dry fog, as we now know.

In the media coverage of the events of 1783, the opponents of the lightning rod only figured as anonymous skeptics, easily refuted and often ridiculed, or as uneducated peasants or rioting crowds caught in the web of superstition. Yet there was no clear-cut transition from "superstition" to "reason." The marginalization of popular understandings of nature started on a cultural level. The élites were eager to distance themselves from an allegedly credulous populace, and natural philosophers such as Hemmer were busy establishing themselves as experts.[83]

Despite the big boost in 1783 with hundreds of lightning rods being erected, victory for the Enlightenment was far from complete. People kept blaming lightning rods for attracting and holding thunderstorms to a particular place.[84] As mentioned before, the bans on ringing the bells had to be reissued; nevertheless, bell ringing continued well into the nineteenth century in many places. To trust in technology instead of God was no easy conversion. In some places the lightning rod was blessed instead of the bells.[85] In 1816, the "year without summer" in Europe due to the eruption of the Tambora volcano in South East Asia in April 1815, the lightning rods were blamed again for disturbing the weather.[86] The year 1783 was "a year of marvels," but not everything changed at once.

Notes

1. The public excitement about the first balloons in prerevolutionary France has been well described, e.g., by Michael Lynn, *Popular Science and Public Opinion in Eighteenth-Century France* (Manchester, England: Manchester University Press, 2006), ch. 6.

2. J. A. Kington, "July 1783: The Warmest Month in the Central England Temperature Series," *Climate Monitor* 9, no. 3 (1980): 69–73; and Gaston R. Demarée and Astrid E. J. Ogilvie, "Bons Baisers d'Islande: Climatic, Environmental, and Human Dimensions. Impacts of the Lakagígar Eruption (1783–1784) in Iceland," in *History and Climate: Memories of the Future*, ed. P. D. Jones et al. (New York: Kluwer, 2001), 219–46, at 228. In central England, 1783 was the warmest summer in the period; John Grattan and Mark Brayshay, "An Amazing and Portentous Summer: Environmental and Social Responses in Britain to the 1783 Eruption of an Iceland Volcano," *The Geographical Journal* 161, no. 2 (1995): 125–34.

3. Anton Pilgram, *Untersuchungen über das Wahrscheinliche der Wetterkunde durch vieljährliche Beobachtungen* (Wien, 1788), 211.

4. Anonymous, "Aus Rinteln; Blitzableiter daselbst," *Journal von und für Deutschland* 2, no. 1 (1785). 176–80, at 179.

5. Richard Hamblyn, *The Invention of Clouds. How an Amateur Meteorologist Forged the Language of the Skies* (New York: Farrar, Straus, and Giroux, 2001), 49, quoting from "Accounts from the Country of Damage Done by the Late Storms," *Gentleman's Magazine* 53, no. 2 (1783): 621–22, at 621.

6. *Augsburgische Stats- und Gelehrte Zeitung*, July 3, 10, 24; August 2, 7, 1783. For an overview of quotations from many different parts of Europe see Demarée and Ogilvie, "Bons Baisers d'Islande" (cit. n. 2), 226–29.

7. John Grattan has assembled forty pages of excerpts from English newspapers from the summer of 1783. For reports and observations from Swiss natural philosophers and meteorologists, see Christian Pfister, "Die Lufttrübungserscheinung des Sommers 1783 in der Sicht schweizerischer Beobachter," *Informationen und Beiträge zur Klimaforschung* 7 (1971): 23–29; and Christian Pfister, *Agrarkonjunktur und Witterungsverlauf im westlichen Schweizer Mittelland 1755–1797* (Berne: Geographisches Institut der Universität Bern, 1975), 85–89.

8. Franz Cölestin von Beroldingen, *Gedanken über den so lange angehaltenen ungewöhnlichen Nebel* (Braunschweig, 1783), 3.

9. Gottlieb Christoph Bohnenberger, *Beyträge zur theoretischen und praktischen Elektrizitätslehre*, vol. 1 (Stuttgart, 1793), 129. When the German physician Christoph Wilhelm Hufeland wrote his autobiography decades later, he still remembered the "hot dry summer in the year of 1783 in which after the earthquake in Calabria a dry high smoke filled the entire air"; *Hufeland, Leibarzt und Volkserzieher. Selbstbiographie von Christoph Wilhelm Hufeland*, ed. Walter von Brunn, 3rd ed. (Stuttgart: Lutz, 1937), 58. The first edition was published posthumously in 1863.

10. Ulrich Bräker, *Tagebücher 1779–1788*, ed. Heinz Graber et al. (Munich/Berne: Beck/Haupt, 1998), 408–26.

11. Ibid., 409, 410, 412.

12. Ibid., 410.

13. Ibid., 418, 417.

14. Ibid., 426. For Bräker's view of nature I follow Holger Böning, *Ulrich Bräker: Der arme Mann aus dem Toggenburg; eine Biographie* (Zurich: Orell Füssli, 1998), 108f.

15. Georg Christoph Lichtenberg, *Briefwechsel*, vol. 2, ed. Ulrich Joost and Albrecht Schöne (Munich: Beck, 1985), Lichtenberg to Amelung, July 3, 1783, 640f, at 640.

16. Ibid., Lichtenberg to Schernhagen, July 31, 1783, 671; Lichtenberg to Wolff, July 13, 1783, 653; and Lichtenberg to Wolff, July 21, 1783, 669.

17. Beroldingen, *Gedanken* (cit. n. 8); cf. Lichtenberg, *Briefwechsel* (cit. n. 15), Lichtenberg to Schernhagen, September 4, 18, 1783, 697, 701f. When Lichtenberg had to comment on the dry fog of 1783 in the third edition of the *Erxleben*, a textbook on natural philosophy, he deliberately listed "only a few" (seven) titles dealing with the phenomenon; Johann Christian Polykarp Erxleben, *Anfangsgründe der Naturlehre mit Anmerkungen von G. Ch. Lichtenberg* (Göttingen, 1784), 665.

18. Beroldingen, *Gedanken* (cit. n. 8), 7–15; Johann Ludwig Christ, Von der außerordentlichen Witterung des Jahrs 1783, in *Ansehung des anhaltenden und heftigen Höherauchs . . . wie auch etwas von den Erdbeben* (Frankfurt a.M., 1783), 8–11.

19. Anonymous, *Ueber die Entstehung und Beschaffenheit des ausserordentlichen Nebels in unsern Gegenden* (Frankfurt a.M./Leipzig 1783), 15–23.

20. In 1783 and 1784 the issues 23 and 24 of the *Journal de Physique* are replete with articles on the strange fog, among them Toaldo's "Dissertation sur le Brouillard extraordinaire de 1783; accompagnée de nouvelles vues sur l'origine de ce Brouillard; par M. le Chevalier de Lamanon," *Journal de Physique* 24 (1784): 3–17; Toaldo's ideas were discussed in German in some letters by Michele Torcia, translated from Italian: "Briefe des Herrn Michael Torcia an den Professor Toaldo zu Padua, von dem Höherauch des vergangnen Jahres zu Neapel und in Calabrien," *Deutscher Merkur* (April 1784): 3–16; reprinted in *Allerneueste Mannigfaltigkeiten: Eine gemeinnützige Wochenschrift*, 4 (1785): 349–60; Jérôme de Lalande, "Lettre sur l'état actuel de l'Atmosphère. Aux auteurs du Journal," *Journal de Paris*, no. 182 (July 1, 1783): 762–63; Lalande's article was soon translated into German: *Magazin für das Neueste aus der Physik und Naturgeschichte* 2, no. 2 (1783): 97–99.

21. Bertholon dedicates an entire chapter to the strange fog of 1783, discussing the different explanations before he offers his "electrical conclusion"; Pierre Bertholon, *De l'électricité des météores: ouvrage dans lequel on traite de l'électricité naturelle en général, & des météores en particulier; contenant l'exposition & l'explication des principaux phénomènes qui ont rapport à la météorologie électrique, d'après l'observation & l'expérience*, vol. 2 (Paris, 1787), 128–47; German translation: *Die Elektricität der Lufterscheinungen . . .*, vol. 2 (Liegnitz, 1792), 87–97.

22. *Nürnbergische Oberpostamtszeitung*, July 21, 1783, quoted in *Physikalisches Tagebuch für die Freunde der Natur* 1 (1784): 18.

23. For refutations of the skeptic, see *Augsburgische Stats- und Gelehrte Zeitung*, July 19, 1783; *Nürnbergische Oberpostamtszeitung*, July 21, 1783; *Physikalisches Tagebuch für die Freunde der Natur* 1 (1784): 18f; *Magazin für das Neueste aus der Physik und Naturgeschichte* 2, no. 2 (1783): 96; and Christ, "Von der außerordentlichen Witterung" (cit. n. 18), 30. The lightning rod skeptic whom Alzate had to deal with in New Spain was anonymous too; see Fiona Clark's essay in this volume, 80.

24. The questions were published in, for example, *Augsburgische Stats- und Gelehrte Zeitung*, July 2, 1783; L. P. von Weiskirch, *Gedanken über den Gebrauch der Wetter-*

ableiter (Wien, 1783); and Jakob Langenbucher, *Augsburgisches Intelligenz-Blatt*, no. 28, July 14, 1783; Hemmer, *Augsburger Ordinari Postzeitung*, July 17, 1783. Hemmer was not the only natural philosopher at the time to suggest "earthquake-rods." Similar claims were also made by French researchers such as Bertholon; see Peter Heering's contribution in this volume. For Hemmer as meteorologist, see Albert Cappel, "Das Wetter und seine Aufklärer: Johann Jakob Hemmer in Mannheim," *Photorin, Jahrbuch der Lichtenberg-Gesellschaft* 10 (1986): 14–26.

25. Yet the practice of "antilightning prayers" was as heavily criticized as the ringing of the bells. In the second half of the eighteenth century Protestant theologians thought of God as a loving not as a punishing God; "antilightning prayers" were simply a sign of a bad conscience; Heinz Dieter Kittsteiner, "Das Gewissen im Gewitter," *Jahrbuch für Volkskunde* 10 (1987): 7–26. For the engravings on the bells, see the essay by Christian Fuhrmeister in this volume, 144.

26. See the introduction of this volume and Andrew Dickson White, *A History of the Warfare of Science with Theology in Christendom* (New York/London: D. Appleton and Company, 1928).

27. *Augsburgische Stats- und Gelehrte Zeitung*, July 24, 1783; *Erlanger Realzeitung*, no. 61, 1783; *Berlinische Monatsschrift* 1, no. 2 (1783): 149f, at 150, footnote. The following year Johann Nepomuk Fischer mentioned this statistic in his *Beweiß, daß das Glockenläuten bey Gewittern mehr schädlich als nützlich sey: Nebst einer allgemeinen Untersuchung ächter und unächter Verwahrungsmittel gegen die Gewitter* (München, 1784), 12. For reports about people killed while ringing the bells, see *Augsburgische Stats- und Gelehrte Zeitung*, July 11, 19, 22, 1783.

28. *Augsburger Ordinari Postzeitung*, August 11, 1783; *Berlinische Monatsschrift* 1, no. 2 (1783): 480; *Vossische Zeitung*, no. 153 (1783), quoted in Eberhard Buchner, *Das Neueste von gestern: Kulturgeschichtlich interessante Dokumente aus alten deutschen Zeitungen*, vol. 3 (München: Langen, 1912), 338f.

29. Landesarchiv Speyer, B2, 2746, 255; Stadtarchiv Zweibrücken, H 6, No. 1240.

30. H. Brück, *Die rationalistischen Bestrebungen im katholischen Deutschland, besonders in den drei rheinischen Erzbistümern in der zweiten Hälfte des achtzehnten Jahrhunderts* (Mainz, 1865), 105; and Johann J. Scotti, *Sammlung der Gesetze und Verordnungen, welche in dem vormaligen Churfürstenthum Trier über Gegenstände der Landeshoheit, Verfassung, Verwaltung und Rechtspflege ergangen sind. Vom Jahre 1310 bis zur . . . Auflösung des Churstaates Trier am Ende des Jahres 1802*, vol. 3 (no. 778) (Düsseldorf: Wolf, 1832), 1326.

31. Bistumsarchiv Augsburg BO 7262, No. 43; Anton Gulielminetti, *Klemens Wenzeslaus, der letzte Fürstbischof von Augsburg und die religiös-kirchliche Reformbewegung* (Neuburg a.D.: Selbstverlag, 1911), 535f; and Barbara Goy, *Aufklärung und Volksfrömmigkeit in den Bistümern Würzburg und Bamberg* (Würzburg: Schöningh, 1969), 185.

32. Joseph Mack, *Die Reform- u. Aufklärungsbestrebungen im Erzstift Salzburg unter Erzbischof Hieronymus von Colloredo. Ein Beitrag zur deutschen Kulturgeschichte der Aufklärungszeit* (München: Böck, 1912), 57–59; and Joseph Ernst Ritter von Koch-Sternfeld, *Die letzten dreissig Jahre des Hochstifts und Erzbisthums Salzburg* (Nürnberg, 1816), 134.

33. *Sammlung der Kurpfalz-Baierischen allgemeinen und besonderen Landes-Verord-nungen*, vol. 2. *Von Polizey- und Landesverbesserungs-, Religions-, Kirchen- und Geistlichkeits-, Kriegs- und vermischten Sachen*, ed. Georg Karl Mayr (München, 1784), 1485f.

34. Peter Hersche, "Populärer Widerstand gegen die Aufklärung im katholischen Raum," *Transactions of the Ninth International Congress on the Enlightenment*, vol. 3, *Studies on Voltaire and the Eighteenth Century*, vol. 348 (Oxford: Voltaire Foundation, 1996), 1524–27, at 1526.

35. Goy, *Aufklärung und Volksfrömmigkeit* (cit. n. 31), 188f; Hubert Weitensfelder, *Industrie-Provinz: Vorarlberg in der Frühindustrialisierung 1740–1870* (Frankfurt a.M./New York: Campus, 2001), 63f, 395; I. Bernard Cohen, "Prejudice against the Introduction of Lightning Rods," *Journal of the Franklin Institute* 253 (1952): 393–446, at 400f; and Harald Fähnrich, "Verbot des Wetterläutens (1819–1835)," *Die Oberpfalz* 86 (1998): 219–20. See also the reference in Christian Fuhrmeister's essay in this volume, 144.

36. Goy, *Aufklärung und Volksfrömmigkeit* (cit. n. 31), 187; Johann J. Scotti, *Sammlung der Gesetze und Verordnungen, welche in den ehemaligen Herzogthümern Jülich, Cleve und Berg und in dem vormaligen Großherzogthum Berg über Gegenstände der Landeshoheit, Verfassung, Verwaltung und Rechtspflege ergangen sind*, vol. 2 (no. 2207) (Düsseldorf: Wolf, 1821), 670.

37. *Augsburger Ordinari Postzeitung*, August 23, 1783.

38. Fritz Dross, "Gottes elektrischer Wille? Zum Düsseldorfer 'Blitzableiter-Aufruhr' 1782/83," in *Landes- und Reichsgeschichte. Festschrift für Hansgeorg Molitor zum 65. Geburtstag*, ed. Jörg Engelbrecht and Stephan Laux (Bielefeld: Verlag für Regionalgeschichte, 2004), 281–302.

39. Usually 1769 is given as the date for the first lightning rod in Germany, but the archival evidence shows it was in 1770; Carl Heinz Dingedahl, "Die Bleidecker Mettlerkamp und die ersten Blitzableiter in Hamburg," *Hamburgische Geschichts- und Heimatblätter* 9, no. 11 (1976): 261–66.

40. *Wienerblättchen*, August 28, 1783, 131; J. A. Günther, "Versuch einer Geschichte der Gesellschaft in den ersten 25 Jahren nach ihrer Einrichtung," *Verhandlungen und Schriften der Hamburgischen Gesellschaft zur Beförderung der Künste und nützlichen Gewerbe* 1 (1792): 52–96, at 75 (footnote); for Augsburg and Stuttgart (the castle of Hohenheim) *Augsburgische Stats- und Gelehrte Zeitung*, June 28, 1783.

41. Koch-Sternfeld, *Die letzten dreissig Jahre* (cit. n. 32), 136; Friedrich Gottlieb Busse, *Beruhigung über die neuen Wetterleiter* (Leipzig, 1791).

42. Georg Christoph Lichtenberg, "Gutachten über den Blitzableiter zu Mandelsloh," ed. Ulrich Joost, *Lichtenberg-Jahrbuch* (1994): 72–80. The letters on the temple of Solomon were first published in October 1783 in the *Göttingisches Magazin der Wissenschaften und Literatur* 3, no. 2 (1783): 735–68. Afterward they appeared in *Journal de Physique*, and by that trajectory they eventually even made it to Spanish Mexico. For the latter, see Fiona Clark's essay on Alzate and his Gazeta in this volume.

43. *Augsburgische Stats- und Gelehrte Zeitung*, June 28 and September 27, 1783. For Hemmer's activities, see Adolf Kistner, "Württembergische Blitzableiteranlagen von Joh. Jak. Hemmer," *Mannheimer Geschichtsblätter* 21 (1920): 132–37. The famous mechanic

and minister Philipp Matthäus Hahn erected a lightning rod in late June or early July 1783 on his house in Echterdingen near Stuttgart; Philipp Matthäus Hahn, *Die Echterdinger Tagebücher: 1780–1790*, ed. Martin Brecht (Berlin: de Gruyter, 1983), 59, 60, 63.

44. Marsilio Landriani, *Dell'utilità dei conduttori elettrici* (Milan, 1784). I am quoting after the German translation: Marsilio Landriani, *Abhandlung vom Nutzen der Blitzableiter* (Vienna, 1786). On Italy, see Paola Bertucci's essay in this volume.

45. Landriani, *Abhandlung vom Nutzen* (cit. n. 44), preface, xii–xvi; the list is on 247–68. Obviously, Landriani's compilation reflects his own personal network. Nevertheless the data provided are very valuable. See, for example, the analysis by Michael B. Schiffer, *Draw the Lightning Down: Benjamin Franklin and Electrical Technology in the Age of Enlightenment* (Berkeley: University of California Press, 2003), 201–3.

46. Ibid., Genoa: 249–50; Lucca: 251–53; Zurich: 255–56; and Ansbach: 265–68.

47. *Augsburgisches Intelligenz-Blatt*, no. 32, August 11, 1783.

48. See the introduction of this volume.

49. For the famous lightning rod trial, see Jessica Riskin, "The Lawyer and the Lightning Rod," *Science in Context* 12, no. 1 (1999): 61–99. The story of the committee of the académie is told by Peter Heering in this volume.

50. *Augsburger Ordinari Postzeitung*, July 29, 1784, quoting from a report from Paris dated July 17, 1784. According to Basil F. J. Schonland, *The Flight of Thunderbolts* (Oxford: Clarendon Press, 1950), 8: "in 1786 the Parlament of Paris found it necessary to reissue an edict" banning the ringing of the bells in thunderstorms.

51. Francesco Zambaldi to Giuseppe Toaldo, January 5, 1787, quoted in Giampiero Bozzolato, *Giuseppe Toaldo: Uno scienziato europeo nel Settecento veneto* (Padova: Brugine, 1984), 227. I thank Paola Bertucci for this reference.

52. Demarée and Ogilvie, "Bons Baisers d'Islande" (cit. n. 2), 220, 240.

53. Richard B. Stothers, "The Great Dry Fog of 1783," *Climatic Change* 32 (1996): 79–89, at 79; John P. Grattan, R. Rabartin, S. Self, and T. Thordarson, "Volcanic Air Pollution and Mortality in France 1783–1784," *Comptes Rendus Geoscience* 337, no. 7 (2005): 641–51, at 644; Gaston R. Demarée, Astrid E. J. Ogilvie and De'er Zhang, "Further Documentary Evidence of Northern Hemispheric Coverage of the Great Dry Fog of 1783," *Climatic Change* 39 (1998): 727–30, at 729.

54. John P. Grattan and D. J. Charman, "Non-Climatic Factors and the Environmental Impact of Volcanic Volatiles: Implications of the Laki Fissure Eruption of A.D. 1783," *The Holocene* 4, no. 1 (1994): 101–6.

55. Claire S. Witham and Clive Oppenheimer, "Mortality in England during the 1783–4 Laki Craters Eruption," *Bulletin of Volcanology* 67, no. 1 (2004): 15–26, at 18; and Grattan et al., "Volcanic Air Pollution" (cit. n. 53), p. 648.

56. Bräker, *Tagebücher* (cit. n. 10), 416, 418, 422; the quote is from *Koblenzer Intelligenzblatt*, July 14, 1783, cited after Demarée and Ogilvie, "Bons Baisers d'Islande" (cit. n. 2), p. 235. For the wine harvest in Austria and Hungary, see ibid. and Elisabeth Strömmer, *Klima-Geschichte: Methoden der Rekonstruktion und historische Perspektive. Ostösterreich 1700 bis 1830* (Wien: Deuticke, 2003), 208.

57. All personal communication. I would like to express my sincere gratitude to Gaston Demarée, Richard Stothers, and John Grattan for sharing their insights with me.

58. "Meteorological Imaginations and Conjectures," *Memoirs of the Literary and Philosophical Society of Manchester* 2 (1785): 357–61, at 359–60. The paper was actually written in May 1784 and read on December 22, 1784. It was translated into German as "Meteorologische Gedanken und Vermuthungen," *Magazin für das Neueste aus Physik und Naturgeschichte* 4, no. 1 (1787): 110–15. Crediting Franklin: Hamblyn, *Invention of Clouds* (cit. n. 5), 53.

59. Saemundur M. Holm, *Vom Erdbrande auf Island im Jahre 1783* (Copenhagen, 1784); cf. Otto Mäussnest, "Die isländische Vulkaneruption und das Geheimnis des Hahlrauchs," *Photorin. Mitteilungen der Lichtenberg-Gesellschaft* 9 (1985): 52–59, at 54. Already in May 1784 the Geneva naturalist Jean Senebier referred to an unnamed natural philosopher who claimed that the Icelandic volcano eruption was responsible for the fog: Jean Senebier, "Observation sur la Vapeur qui a regné pendant l'été de 1783, faite à Genève," *Journal de Physique* 24 (1784): 404–11, at 410.

60. Jacques Antoine Mourgue de Mont-Redon, "Recherches sur l'origine et sur la nature des vapeurs qui ont régné dans l'Atmosphère pendant l'été de 1783," *Mémoires de l'Académie Royale des Sciences* (1784): 754–73, at 761; Grattan et al., "Volcanic Air Pollution" (cit. n. 53), 642; and Vincent Courtillot, "New Evidence for Massive Pollution and Mortality in Europe in 1783–1784 May Have Bearing on Global Change and Mass Extinctions," *Comptes Rendus Geoscience* 337, no. 7 (2005): 635–37, at 635.

61. J. R. Salis-Marschlins, "Einige Bemerkungen über den allgemeinen Dampf oder Heerrauch, der im Junius und Julius dieses Jahrs sich auch in unserer Gegend verbreitet hat," *Der Sammler. Eine gemeinnützige Wochenschrift für Bündten* 47 (1783): 393–98, at 393, quoted in Pfister, *Agrarkonjunktur* (cit. n. 7), 88.

62. Hamblyn, *Invention of Clouds* (cit. n. 5), 53, 46.

63. See Lichtenberg's fourth edition of Johann Christian Polykarp Erxleben, *Anfangsgründe der Naturlehre mit Anmerkungen von G. Ch. Lichtenberg* (Göttingen, 1787), 733.

64. Louis-Sébastien Mercier, *Mon bonnet de nuit suivi de Du théâtre*, ed. Jean-Claude Bonnet (Paris: Mercure de France, 1999, orig. publ. 1784–85), at 568, 570. Similar éloges on the achievements of the time were published by several French authors around the same time, in late 1783 and early 1784; Lynn, *Popular Science* (cit. n. 1), 123.

65. *Göttingisches Magazin der Wissenschaften und Literatur* 3, no. 6 (1784): 930–53. Before the year was over, the speech was reprinted at least twice: *Physikalisches Tagebuch für die Freunde der Natur* 1 (1784): 182–87; *Augsburgisches Intelligenz-Blatt*, no. 50, December 13, 1784. The modern edition is Georg Christoph Lichtenberg, *Schriften und Briefe*, ed. Wolfgang Promies, vol. 3 (Munich/Vienna: Hanser, 1972), 63–72, at 63.

66. Staatsarchiv Gotha, Oberkonsistorium Stadt Gotha Nr. 95. The inscription is probably by Ludwig Christian Lichtenberg himself. Robespierre's famous pleading in the trial concerning the reinstallation of a lightning rod in St. Omer dates from May 1783 and was printed in the late summer of that year. For his triumphant tone, see Engelhard Weigl, "Entzauberung der Natur durch Wissenschaft dargestellt am Beispiel der Erfindung des Blitzableiters," *Jahrbuch der Jean-Paul-Gesellschaft* 22 (1987): 7–40, at 31–33.

67. Hans Blumenberg, *Die Genesis der kopernikanischen Welt* (Frankfurt a.M.: Suhrkamp, 1975), 642.

68. See the introduction of this volume, 7.

69. Maximus Imhof, *Grundriß der öffentlichen Vorlesungen über die Experimental-Naturlehre, welche auf Veranstalten der Churfürstlichen Akademie der Wissenschaften in München...*, vol. 2 (Munich, 1795), 408; similar remark in *Wienerblättchen*, August 28, 1783, 131.

70. For comets see Jean-Marie Goulemot, "Démons, merveilles et philosophie à l'Age classique," *Annales. Economies, Sociétés, Civilisations* 6 (1980): 1223–50. The case of the Lisbon earthquake of 1755 is well known.

71. Lalande, "Lettre sur l'état" (cit. n. 20); *Gentleman's Magazine* 53, no. 2 (1783): 613; and *Magazin für das Neueste aus der Physik und Naturgeschichte* 2, no. 2 (1783): 97–99. Cf. Demarée and Ogilvie, "Bons Baisers d'Islande" (cit. n. 2), 239. Reference to a similar climatic phenomenon in 1560 was made by Ehrhardt, "Ist auch schon ehemals ein Höherauch gesehen worden?" *Journal von und für Deutschland* 2, no. 2 (1785): 199f.

72. Grattan and Brayshay, "Amazing and Portentous Summer" (cit. n. 2), 131.

73. *Augsburgische Stats- und Gelehrte Zeitung*, July 5, 1783.

74. *Gentleman's Magazine* 53, no. 2 (1783): 613; Lichtenberg, *Briefwechsel* (cit. n. 15), Lichtenberg to Wolff, July 13, 1783, 658.

75. F. Verdeil, "Mémoire sur les brouillards électriques vus en Juin et Juillet 1783 et sur le tremblement de terre arrivé à Lausanne le 6 juillet de la même année," *Mémoires de la Société des Sciences Physiques de Lausanne* 1 (1784): 110–37, at 110; N. E. Tscharner, "Besondere physische Bemerkungen," *Neue Sammlung physisch-ökonomischer Schriften* 3 (1785): 332–38, at 333f; quoted in Pfister, *Agrarkonjunktur und Witterungsverlauf* (cit. n. 7), 87.

76. For an interesting sermon justifying the use of lightning rods, see Hinrich Heeren, *Ueber die Verehrung Gottes im Gewitter. Eine Predigt am 20. Aug. 1783, als in der Ordnung der biblischen Erklärungen Hiob 37, der Text und zu Vorstellungen dieser Art viel Veranlassung war* (Bremen: Försterische Buchhandlung, 1783), esp. 22f. Interestingly, the "utility" of thunderstorms is "proven" by the "fact" that they drove away the dry fog, 16. For the profit of the instrument-makers, see Dingedahl, "Die Bleidecker Mettlerkamp" (cit. n. 39).

77. The same strategy was pursued in other European countries as well. For France, see the essay by Peter Heering in this volume.

78. Fritz Dross did a keyword search in an online database of German periodicals from the eighteenth century (www.ub.uni-bielefeld.de/diglib/aufklaerung). He came to a similar conclusion comparing different decades: In the 1780s terms such as "lightning," "thunder," and composites of these words are mentioned by far the most with a clear peak in 1783; Dross, "Gottes elektrischer Wille?" (cit. n. 38), 291.

79. *Augsburgisches Intelligenz-Blatt*, no. 31, August 4, 1783.

80. *Augsburger Ordinari Postzeitung*, July 10, 1783; in parts identical with the article in *Augsburgische Stats- und Gelehrte Zeitung*, July 10, 1783; *Wienerblättchen*, August 17, 1783, 3–4; *Stats-Anzeigen*, 1783, 502.

81. This is well argued by Dross, "Gottes elektrischer Wille?" (cit. n. 38), 300–302.

82. Riskin, "The Lawyer and the Lightning Rod" (cit. n. 49), 77.

83. This case has been made with respect to the changing perception of comets from signs of impending evil to mere natural phenomena; Goulemot, "Démons, merveilles" (cit. n. 70).

84. This happened in Augsburg in the summer of 1791. A war of pamphlets ensued that had to be stopped by the magistrate to prevent public uproar; see Oliver Hochadel, "'Hier haben die Wetterableiter unter den Augsburger Gelehrten eine kleine Revolution gemacht.' Die Debatte um die Einführung der Blitzableiter in Augsburg (1783–1791)," *Zeitschrift des Historischen Vereins für Schwaben* 92 (1999): 139–64, at 157–59.

85. Eva Labouvie, lecture at Vienna University, March 2004.

86. *Neue Zürcher Zeitung*, July 9, 1816. Yet also in this case, the critics of the lightning rod only surface in official documents and are denounced as evil-minded people.

Nothing Ventured, Nothing Gained

Lightning and Enlightenment in the *Gazeta de Literatura de México* (1788–1795)

Fiona Clark

THE INTRODUCTION OF ANY NEW IDEA into a society, especially one that challenges common or long-held beliefs, often involves a finely balanced presentation of the potential risk versus the probable gains involved in the process. The individual or group propounding the new idea must seek to place their arguments in a recognizable context, present an acceptable basis for their authoritative stance, and underline the possible ramifications of any form of rejection. Such was the situation in late-colonial Mexico as the secular priest and editor José Antonio Alzate y Ramírez (1737–99) struggled to find acceptance among the reading public of the *Gazeta de Literatura de México* (1788–95) for what was, in the opinion of some, the highly suspect introduction of an instrument known as a lightning rod (*barra eléctrica* or *pararayo*).[1] The arguments and criticism ringing out from the pages of the periodical paint a picture of a society that was torn between the fear of natural elements (linked in part to superstitious beliefs) and the unknown and untested powers of the new technology. This fear, as Alzate is quick to point out, was no greater than the fear evident on occasion among the inhabitants of some of the greatest cities in Europe, including Paris.[2] It was an assessment that placed the Spanish American reaction to more extreme natural phenomena on a par with their European contemporaries. Yet by the mid 1790s it seems possible that the only lightning rod existent in Mexico City was the self-constructed model belonging to Alzate y Ramírez.[3]

The appearance of the lightning rod in Mexico was not the result of a spontaneous discovery or even the passive adoption of an idea into a scientific vacuum. It formed part of an established history of studying natural resources, building upon long-used skills, and transforming ideas to suit the particularities of Mexican reality. As Antonio Lafuente has argued regarding the globalization of science through the process of adaptation, from the point of view of the receiver, "the pre-existing cultural base has been enriched (and deformed) by something different and external."[4] It is the interactive, not the passive, nature of the process that allows for "novelty" to be accepted and used to advantage. In line with this argument, recent scholarship has demonstrated an increasing interest in the use of language, methods, and structures in studies of the history of science in the so-called peripheral regions of the world. Terms such as "western" and "nonwestern" or "colonial" and "metropolitan," long used as opposing forces and frequently portrayed as antagonistic and incompatible by nature, are now being opened up for broader discussion in an attempt to understand the intricacies of the spread of knowledge in the early modern period.[5] Within the wider field of Spanish American history, the work of such figures as Patricia Aceves Pastrana, David Brading, Jorge Cañizares Esguerra, Juan José Saldaña, and Alberto Saladino García, to name but a few, have continued to address the many issues arising from the complexities of understanding scientific movements within their local and international contexts. This essay builds upon the arguments developed within this recent scholarship on the Spanish Americas to show how Alzate, working within an international framework of scientific exchange, used the *Gazeta de Literatura* as an interactive vehicle for renewal on a local level.

The lightning rod is only one aspect of the scientific discourse taking place in Mexico in the eighteenth century falling under the broader banner of "physics" or the "new philosophy." Evidence of interest in the new theories in this field can be seen from the early seventeenth century onward with references to Copernicus, Galileo, and Descartes, among others. By the mid-eighteenth century, a small group of Creole scientists was working to propagate new ways of approaching education in this area, with particular interest in the natural world.[6] Included alongside Alzate we find such individuals as Juan Benito Díaz de Gamarra (1745–83), José Ignacio Bartolache (1739–90) and Antonio de León y Gama (1723–77).[7] Among their interests in physics lay the desire to test the new theories being expounded, largely in Europe, and to question the universality of their application. For Alzate and Bartolache, the periodical press proved to be the most efficient means of circulating their ideas.[8]

The Mexican Pliny

Born in 1737, son of a native Spanish father and Mexican-born mother, Alzate stands as one of the foremost enlightenment figures of eighteenth century Mexico.[9] This fact refers not only to his prolific career but also to his insatiable desire for scientific investigation and the propagation of new ideas. A controversial polemicist, even in terms of his relationship with the various viceroys of New Spain, he was a formidable adversary. While expressive in his praise of those peo-

Figure 4.1.   José Antonio Alzate y Ramírez. From the 1831 Puebla edition of the *Gazeta de Literatura de México*, vol. 1. Courtesy Nettie Lee Benson Latin American Collection, Austin.

ple or ideas he felt deserving, he was also inclined to interpret opposing views as personal affronts and react accordingly with mockery, ridicule, and sarcasm. Trained in the Royal Pontifical University of Mexico in the 1760s, Alzate was to serve under some of the most illustrious state and ecclesiastical authorities of late-colonial Mexico. In this capacity he performed a series of administrative roles communicating directly with various Archbishops and Viceroys, undertaking a wide variety of reports into subjects as diverse as topography, cartography, meteorology, statistics, agriculture, technology, and mining.[10]

During the course of these duties Alzate also played a role in several of the scientific expeditions sent from Europe to the Americas. Already a member of the Basque Society of Friends of the Country, in 1771 he was elected as a corresponding member to the Paris Royal Academy of Sciences as a result of his report on the passage of Venus over the solar disc, a report ensuing from his contact with the ill-fated Chappe D'Auteroche expedition.[11] As the only Mexican to hold such a position during that period it was indeed a mark of esteem. The importance derived from his membership of these widely recognized and acclaimed societies, along with his later incorporation as correspondent of the Royal Botanical Gardens in Madrid, would play a key part in his presentation of ideas in the *Gazeta de Literatura*, including his arguments concerning the lightning rod.

The *Gazeta de Literatura* itself was the culmination of what is arguably his most important contribution to eighteenth century Mexican society, the establishment of the literary-scientific periodical press.[12] It was one of only two periodicals printed in Mexico City at the time. The second, the *Gazeta de México*, acted as a source of official information within the viceroyalty, regularly printing news on Mexican and European events as well as official announcements. As Rafael Moreno was to remark regarding this last of Alzate's publications, "any page of the periodical leaves the impression that New Spain is a material and spiritual community, a geographical and mental unity, despite the differences in race and the separations on the diverse perceptions of politics and religion, imposed by geography and language."[13] Created as a means to instruct and inform the local readership, the periodical continually drew their attention to the country in which they lived, weighing in the balance the utility or appropriateness of foreign innovations to their particular reality.[14]

Within the first two pages of volume I, Alzate uses three phrases that place his periodical and its contents firmly on the map in terms of geography and nationhood: "la Metrópoli del Nuevo Mundo," "la voz México," and "nuestra Nación Hispano Americana." By this means he sets his work in the geographical context of Mexico City, determines for whom he is speaking—the Mexican people, and clarifies that which he is representing, the Hispano-American nation. These efforts toward the popularization of basic scientific ideas were conceived in pursuit of a universal scientific culture. This culture was one that would not be dominated by Europe but would become "nationalized," in this case "Mexicanized," when applied to the local situation. For Alzate, science, or natural philosophy, held the key to making the world a better and more humane place in which to live; as such, theory without application was of little to no importance.

The Lightning Rod: A Literary Response

The present discussion focuses on six specific issues of the *Gazeta de Literatura* published over four years, dating from February 1790 to July 1793.[15] Although Alzate deals on an extended basis with other topics, they either tend to be much wider-reaching categories such as topography or they are particular studies that Alzate had previously written and later sought to publish.[16] The continued reappraisal of the state of knowledge and implementation of such a "useful art" as the lightning rod therefore underlines Alzate's belief in its potential benefits for Mexican society and the need to protect the architectural legacy for future generations.[17] In terms of style and format we find that the first three texts and the last section of the final article are presented in the form of correspondence, appearing either as a response by Alzate to an unknown author, or translations of correspondence taken from the foreign periodical press. The final three articles are largely presented as reports incorporating summaries and references, again, from works originating in foreign publications. Although the following study is topic-specific, limited to those texts dealing with the lightning rod, the three areas highlighted in our discussion are fundamental examples of the overall approach to content material embraced by Alzate as editor. With Alzate's editorial approach in mind, the discussion that follows is divided into three sections structured according to exclamations or exhortations he used to encapsulate his ideas and goals. Each appears in Latin, a tool that Alzate infrequently adopted, and—as with other similar examples in the *Gazeta*—tends toward a somewhat ironic engagement with the educated Mexican elite.

In conclusion to the penultimate lightning rod article, Alzate reminds his readers of the very practical task at hand by proclaiming, "Tractent fabrilia fabri."[18] Alzate thus underlines the fact that whatever the subject in question, whether the lightning rod or the cultivation of the cochineal beetle, his end desire was for each individual to find their particular means to practice the enlightened ideal of serving the good of the society in which they lived. That each workman should use his own particular craft was indicative of the editor's belief in the potential collective gains to be had from fomenting diversity in education and production in all levels of society as well as his clear understanding of his own role within this. Among the many tasks to which Alzate turned his own hand during the second half of the eighteenth century, the craft for which he has gained greatest renown was undoubtedly, as previously mentioned, the establishment of the early literary-scientific periodical press in Mexico. Although many interesting and valuable studies have been published concerning the various themes that appear in Alzate's work, little investigation has so far been undertaken into the origins of the material he used and the methods involved in transforming and presenting these ideas for public consumption.[19] Our first area of exploration, using the few references included within the articles, seeks to shed light on this particular question by tracing, as far as possible, the close links between the periodical press in Mexico and Europe in the matter of the lightning rod. Exploring this question will open up new avenues for understanding how ideas were transmitted and adapted across the eighteenth-century republic of letters, and will deepen our appreciation of Alzate's modus operandi within the context of transatlantic literary networks.

Having formed a clearer understanding of the means by which the material in question was presented and whence it came, we will consider the practical issues linked to the lightning rod and demonstrate how they are clearly a direct practical response to local needs and not a platform for theoretical exposition per se. In so doing we must try to understand the immediate challenges that confronted Alzate and impeded his goal of practically implementing new ideas and encouraging greater entrepreneurial thinking. Despite the exasperation that would lead him to exclaim "Quantum est in rebus inane!,"[20] Alzate's practical mentality drives him to continue in his attempts to engage his readers. The apparent lack of response among the readership was not only a question of difficulties relating to the establishment of public safety on a local level, it was also a foundational challenge to the broadest of Alzate's aims, and one which we will touch upon in our final considerations, the formation of a national common consciousness.

Emblazoned on the first page of the *Gazeta* we find a call to the Mexican readership: "Indocti discant, et ament meminisse periti."[21] Following in the wake of this challenge, the all-encompassing aim of the *Gazeta* was to instruct on a variety of levels. In one sense, as we have seen, Alzate was choosing to follow the enlightened ideal by which education would lead to the greater improvement of society, the betterment of the human condition. Yet he was simultaneously educating, encouraging, and challenging his readers in an attempt to create a greater awareness of, and delight in, the reality of the nation and country in which they lived, whether through depicting natural resources, economic and industrial potential, or the wealth of history that lay within their reach. As such, his treatment of the lightning rod and the natural phenomena and resources linked to it are clear examples of how a "useful art" became a tool in engendering a form of national common consciousness. Let us turn then to consider the first of these areas: the craft of periodical publication.

Each Workman to His Proper Craft

Unlike many other authors in Spanish America at the time, Alzate carried out his editorial responsibilities alone. In contrast to many of its contemporaries, the *Gazeta de Literatura* was not the tool of an economic or scientific society or, after 1789, the combined work of a group of like-minded individuals.[22] Although Alzate intimates that he was originally to be joined by "three lovers of literature" in undertaking the publication of the *Gazeta*, this plan—for reasons he chooses not to disclose—failed to come to fruition.[23] Instead, he was joined for a short period by José Mariano Mociño,[24] who was later to gain fame in the botanical expeditions led by Martín de Sessé, and the lesser-known Mariano Castillejo.[25] After 1789 it appears that Alzate was working alone in the preparation of the periodical.

The lack of a group identity had practical implications that went beyond merely the financial, for it left Alzate open to direct criticism as the only recognized author. Nowhere does Alzate jump to the defense faster than when accused of mishandling the purposes of his own periodical. When attacked for perceived failure to comply with the form of content suggested by the title *Gazeta de Literatura*, he is quick to refer back to European publications, as he obviously expects this to bear greater weight with the reader than any other argument. If his critics would only take time to examine the index of a periodical such as the Spanish *Memorial Literario* or the French *Diario de Física*, he says, they would see immediately the breadth of subjects included in the content.[26] Even in the early stages of the publication, Alzate was forced to defend himself, arguing that because he has no income or title to gain from his work, he has no reason to favor one individual over another but is free to publish his observations freely.[27] These criticisms were not only uncomfortable on an individual level in terms of personal attack; they also had potentially harmful ramifications for the continuation of the periodical. If linked to more delicate subjects of church and government, the periodical stood on insecure ground due to the lack of a protective patron or authoritative group to counteract a backlash by the long arm of the authorities. Given that two of Alzate's first publications had already ceased to exist as a result of government intervention, it is not surprising that Alzate is careful to affirm from the first number of the *Gazeta* that he will not be dealing with matters of state or religion. Instead he claims he will concentrate on the natural sciences, technology, and all manner of discoveries that are to the benefit of the human good.[28]

The outworking of such goals, however, was dependent on establishing a voice of authority, an authority that the self-taught Alzate could not automatically assume. Despite his links with governmental and ecclesiastical authorities in Mexico, Alzate did not hold a position of power within the local established institutions of knowledge. Although he had trained in theology at the Pontifical University, his background in the sciences came as a result of personal study and interest and, as such, lacked the official seal of approval. As a result, and perhaps in response to this situation, his links with European societies and institutions were the key to recognition on both an international and a local level and the foundation from which to launch his arguments. His access to the European publications linked to these same institutions was a further step toward claiming their literary authority. This process is demonstrated in his discussion of the merits of lightning rods where he instructs the reader who is skeptical of his arguments not to "give credit to the words of the author of the Gazeta: but let the doubters examine those authors he cites, and if, after this, they remain obstinately set against what is written and what has been proven, they will held be responsible for the death of many and for the costs that are an inevitable part of restoring buildings damaged by such a powerful meteor."[29] With this instruction in mind it becomes clear that his links with various European centers of learning are further enhanced through his translations of the discoveries and advances promoted within these circles.

Translation was in fact a tool frequently used to this end throughout the pages of the *Gazeta de Literatura*. From the outset, Alzate has unequivocally stated his intention to provide these "extracts, copies, and translations" as a means of informing and educating his readers by providing material that would otherwise lie beyond their sphere of access.[30] As García Gorrosa and Lafarga have demonstrated in their study of eighteenth-century Spain, the role of translator, especially of specialized texts such as the scientific, presented any number of potential obstacles and difficulties.[31] Fundamental among these was the challenge of finding an individual who was competent in the use of languages and had the background knowledge of the field of technology or natural science in question. These concerns find no expression in Alzate's work. His own observations, experimentation, and expansive reading in technology and science, when added to the practical nature of his interests, lead him to focus on the translation of those works that fit within his realm of experience. As a result, very few if any of the articles printed in the *Gazeta* are of a purely theoretical nature. Moreover, given that his principal aim was to instruct his readers, any concerns regarding the finesse of the literary style of a translated work fell far beneath the demand for simple, clear expression of the information at hand. The *Gazeta* was clearly not the place for exalted prose.[32] Given the tendency against the adoption of particular literary styles, it is interesting to note that Alzate employs a new technique in these lightning rod texts, one that does not appear in any other issue of the *Gazeta*, that is, to use the voice of a figure of great authority, in this case Benjamin Franklin, and channel his criticism through the same.[33]

In relation to translation, we find two substantial texts among the six relating to this study: the first comprising correspondence by the Italian mathematician

Pistoi to Abbé Jean-Baptiste François Rozier (1734–93), and the second, correspondence between Johann David Michaelis (1717–91) and Georg Christoph
Lichtenberg (1742–99). Two further short passages appear in the later issues,
one a summary and response to a report by Jacques de Romas (1713–76) and the
other a list of references relating to a work published by Abbé Giuseppe Toaldo
(1719–97).[34] As is frequently the case, these translations and adaptations were
published without any clear indication whence they came, either in footnotes or
in the editorial commentary. It is only with the appearance of the Romas text that
Alzate makes mention of the point of origin, and even then he merely states that
he has read the original report in the *Diario de Física*.[35]

Nevertheless, a more detailed study of these various extracts alongside other
examples that appear throughout the pages of the *Gazeta de Literatura* has
shown that there is strong evidence to indicate that Alzate's main, and often unnamed, source of material was the Parisian periodical published by Rozier, the
aforementioned *Diario de Física*, or *Journal de Physique*. This fact is not surprising because Alzate had indicated the extent of his admiration for Rozier's
work in the periodical publication that preceded the *Gazeta de Literatura*, the
Observaciones sobre la Física, Historia Natural y Artes útiles (1787–88).[36] It is no
secret that this title was taken directly from Rozier's *Observations sur la Physique,
sur l'Histoire Naturelle et sur les Arts*, or *Journal de Physique* as it came to be more
commonly known after the first introductory volumes.[37] Allusions have been
made to the links between Alzate and Rozier's work, but the full extent of this
connection is only now beginning to come to light.[38] It is perhaps ironic then that
the two unreferenced but longer articles can more easily be tracked to the *Journal* than the shorter referenced summaries that appear later.[39]

A closer comparison of the Spanish and French texts in the two longer translations (Pistoi, Michaelis-Lichtenberg) reveals that Alzate is concerned less with
a faithful replication of the articles than with conveying core ideas to his readers.
That said, however, his version of Pistoi's correspondence is almost a word-for-
word translation of the French, providing the text in its entirety. Where slight de-
viations occur, we find that Alzate has chosen to slightly emphasize the positive
aspects of the lightning rod, such as the acclaim with which it was received, or he
omits aspects of the physical description, such as Pistoi's account of the pointed
end of the rod. This omission of, or apparent lack of interest in, whether the rod
should be pointed or blunt is also an element in his treatment of the translation
of the Michaelis-Lichtenberg correspondence. There is no indication of the extent to which Alzate was aware of the ongoing debates regarding the benefits or
apparent dangers of each design, or even the political implications thereof.[40]

The Pistoi article demonstrates how the translations, as well as providing the
reader with information, form a launching point for Alzate's own ideas. On this
occasion his comments are clearly demarcated and separated from the translation

by way of a rather somber title: "Translator's Warning."[41] The few paratextual elements, in this case footnotes, are not remarks relating to the text itself but direct the reader to Alzate's comments. By so doing, the thread of ideas forms a continuum in the reader's mind, thus paralleling Alzate's arguments with his European contemporaries.

By contrast, in his presentation of the Michaelis-Lichtenberg correspondence, the editor has decided against including the letters in their entirety and instead adapts the text, which is already a translation from the German original. On this occasion short paragraphs or individual sentences are selected from a much longer and more involved discussion. In so doing, Alzate omits the majority of the theological content of the debate, including references to Old Testament translation and interpretation. On the one hand, it is possible that this was a decision based on reader interest and an attempt to refine the text to focus on the central concern of his article, the practical use of the lightning rod. Yet on the other hand, it is worth remembering that the publication of any discussion connected to theological issues entailed great delicacy in late eighteenth-century Mexico never to mention work connected to a German Protestant theologian such as Michaelis. The first scenario seems to be the more likely explanation because he later informs the reader that he has chosen to omit the third letter by Michaelis because, despite its erudition, it is not linked to the topic of the current discussion.[42] How far Alzate would have been aware of the theological stance of either Lichtenberg or Michaelis is unknown, but his use of the information relating to the lightning rod indicates how "scientific" ideas could be extracted from a more theological context and presented without difficulty in the Spanish American colonial press. Indeed, Alzate makes no reference in his later reflections to his reasons for including this series of correspondence except to call on the skeptical reader to believe those authors he has published in the *Gazeta*. As such, the translation serves as backing for his comments and an indication to the reader that Alzate's arguments are well founded.

How Much Futility in the World!

How then did these articles form a direct practical response to the local situation? As we have already seen, the key to Alzate's vision of scientific advancement lies in its beneficial potential. In the first article on the lightning rod, the main concern expressed by the anonymous correspondent relates to the disastrous effects experienced despite the use of a conductor on the Cathedral in Puebla. Alzate takes this problem and uses it as a base for an examination of the Mexican/Spanish mentality.[43] This leads to criticism of those he considers to be charlatans in the field of physics before finally promoting his own observations formed over twenty years, the culmination of which was a design for an effective and inexpensive means of constructing a lightning rod.[44]

Following this format we see him first criticize, then encourage study and meditation, and finally draw practical conclusions and application from such study. He tells his correspondent that if indeed it is true that the Cathedral at Puebla had been destroyed despite the implementation of the necessary precautions, the failure is in all likelihood due to the lack of skill in the one responsible for the installation of the conductor, by all accounts a "foreigner," and not the quality of the device itself. Thus Alzate manages at once to defend the new technology and condemn those foreigners arriving in Mexico with baseless claims to knowledge and expertise.[45] Alzate appears to be somewhat exasperated at this apparent weakness in the Spanish character shown by a tendency to believe statements made by such men, simply, he argues, because they speak Spanish and use the title of physicist or Mathematician. In his opinion, they have no merit other than to have traveled widely with what he terms an "Electrical machine."[46] His readers must realize that these men are, in fact, equally as ignorant as their spectators in terms of understanding how or why these machines are capable of performing such curious tricks.[47]

In contrast, he encourages his readers to study the works of learned scientists, such as Le Roy, Beccaria, Magellans and Bertholon, as well as the collected reports from various academies.[48] Beyond all these stands the work of the "new Prometheus who robbed the heavens of their fire," Franklin.[49] However, Alzate is forced in the footnotes to admit that, despite following the precautions advised by the wise "electricistas," his own inquiries into the use of the electric kite have had almost fatal results:

> In my last experiment, which I will never repeat, I fell into the shadows of death despite taking all the necessary precautions advised by the wise electrical experts. I still experience, and will continue to experience for the rest of my life, a certain weakness in my chest, caused by the explosion of electricity. I communicate this news as a precaution to those who would try to repeat similar experiments. Natural electricity is very active in this country, and the means that are normally established to impede the conduction of the force, such as glass, silk cords and resin, are insufficient for the task.[50]

In a later article on atmospheric electricity in his topographical description of Mexico in 1791 he alludes to his experience when he states that "electricity is very active, and . . . I have abandoned many useful experiments that I had planned to undertake as I was afraid of falling victim to a violent death: other individuals with the appropriate instruments for the task, and with sufficient knowledge of the true discoveries of electrical physics will continue to cultivate this field that has scarcely passed before my eyes."[51] Yet even these circumstances serve as material to illustrate observations on the atmosphere of the valley of Mexico in that those elements normally sufficient for protection are ineffective due to the high levels of active natural electricity.

Franklin is again a theme in the second article, Pistoi's letter, which relates a widely known event of lightning striking the rod of the tower of the Cathedral in Siena on April 18, 1777.[52] Having described the advantageous effects experienced by the installation of a lightning rod on the tower of the Cathedral, Pistoi ends his letter by praising Franklin.[53] He expresses hope that if Franklin were ever to read such an article he would feel satisfaction at the homage paid to him by such a distant town and people who view his lightning rod as the most worthy trophy of his immortal genius, a sentiment that no doubt echoed Alzate's own.

In the "Translator's Warning" that follows, the reader is challenged to consider the contrast of the triumphal discoveries of Franklin with the methods of the "sentinels of the ruined bulwarks" of scholastic philosophy.[54] Although the struggle against individuals who held fast to Aristotelian teaching, particularly in the Royal Pontifical University, has been clearly evidenced from the very first articles of the *Gazeta*, the attack now takes a new angle. It is one that seems not to have been used previously, namely to use Franklin as spokesman for his own arguments by providing the statesman with words that he may have desired to speak but could not. In so doing, Alzate is not only equating his opinions with Franklin's but also using the lightning rod as yet another weapon in his ongoing fight against what he considers to be the obsolete ways of the Scholastics. In his eyes, the struggle over the use of the lightning conductor is equivalent to the sixteenth-century heliocentric disputes.[55]

Engaging in an antischolastic polemic typical of the Enlightenment, Alzate (in the guise of Franklin) questions how much benefit has been derived from the voluminous publications and interminable manuscripts provided by these philosophers in all their years of responsibility for public education. Has any living person ever been freed from death by virtue of their disputes? Has any building ever been released from the effects of lightning by their arguments? He concludes that the use of the conducting rod has saved the lives of millions whereas their scholastic philosophy has been the cause of the death of millions. In their lack of method he equates them to medical students who, taught to dispute all matters, stand by a patient arguing over an imagined illness when in fact the real disease has taken the patient's life. At this point his use of Franklin's voice comes to an end.[56]

Based on the previous observations, Alzate later turns his gaze to Europe, telling his readers that even in England, "a country where they are overly given to arguments," the lightning rod has had its defenders and detractors, yet examples can now be seen installed on the naves of all high buildings and gunpowder factories.[57] He remarks, somewhat ironically, that in Spain, where the government does not resolve an issue relating to its people until it has been well deliberated, a royal order has decreed the installation of lightning rods on its powder factories and ships in the Armada. How can it then be that there is no evidence at all of a lightning rod on any of Mexico's magnificent buildings? "What will the future inhabitants of the Metropolis of the New World say when they read the inscriptions on the buildings or the history of this particular period and understand that a discovery made in the Americas, and adopted by many cultured nations, had been despised in Mexico, or not put into practice, which amounts to the same?"[58]

Yet more criticism lies in store for the sorry Mexican reader when Alzate demands in disbelief "How much can a lightning rod cost?" The obvious fact that people spend large quantities of money having their houses painted, and yet refuse to use conductors, only proves that they distance themselves from that which is most beneficial and despise what could be of greatest use. The root cause can be found, in his opinion, in the dregs of scholastic philosophy.[59]

Following the Michaelis-Lichtenberg correspondence we find further comparisons between Mexico and Europe in regard to the salubrious nature of the atmosphere of the valley of Mexico. Alzate argues that those metals that so easily corrode in Europe last longer in Mexico due to the climate and that they were inexpensive, thereby refuting any reasoning related to material expenses.[60] Mexico was also at an advantage in that so many of its buildings were constructed from highly ferruginous materials. He points out that these materials acted as natural conductors and explains why, given the high level of electric activity in the valley of Mexico, there had been so few fatal accidents. The same benefits, however, do not exist for cities such as Guadalajara and Puebla. Alzate thus narrows the scope of his vision from the European scene to the Mexican and then further to the valley of Mexico and specifically Mexico City. His details are such that he is able to observe the difference in the ferruginous quality of the stone taken from the two quarries near Mexico City and to contrast the protective capacity of those houses built from materials taken from Culhuacan compared to those from Los Remedios.[61] The latter, it appears, contained less metal and were therefore less effective.

The two articles published on May 28 and June 11, 1793, follow the same pattern. Two years have passed since he first presented news of the great discovery of the lightning rod and detailed its use by Royal Order in Spain, and we can sense his disbelief that not one person in Mexico City appears to have acted on his advice. In Alzate's statement the reader is led to understand that Alzate's lightning rod, constructed twenty-two years earlier, is the one and only conductor registered in the city. [62] In light of this fact, and to help those still living in fear of the summer storms, he publishes the previously mentioned extract from a report by

Romas.[63] The summary is interspersed with Alzate's own comments to such an extent that it is at times unclear where Romas ends and Alzate begins. The main thrust of the argument in this case, nevertheless, is to open his reader's mind to a different approach to the challenges posed by the amount of natural electricity in Mexico. This he does by suggesting that they should tackle the dangers posed by lightning, a natural substance, in the same way as they would an illness. They should therefore guard themselves against its effects in the same way as they would guard against smallpox, leprosy, or any other contagious diseases that leads to death.[64] He detracts from the supernatural qualities attributed to lightning and instead attempts to present it as a controllable aspect of nature. Lightning is portrayed as a natural phenomenon that can be overcome through the use of human reason and technological achievements.

This extract spills over into the following issue to include a short list of works published in the *Journal de Physique* linked to a report by Toaldo that he believes would conclusively remove any doubts left in the reader's mind.[65] Toaldo's report prompts him to question whether Mexico City must experience a catastrophe, such as that experienced in Brescia in 1769, before a lightning rod will eventually be installed. Just as his 1791 article criticized men for their lack of interest in what is beneficial, so here he also criticizes society at large for showing greater fascination with the latest fashions and trends while remaining ignorant of a useful and life-saving invention.[66] To counter this lack of interest and in response to the questions posed by an unnamed friend's correspondence, Alzate writes a detailed account of very practical means through which the reader could ensure proper protection from lighting. This advice ranges from precautions in materials used for clothing to how and when one should carry a sword, or open and close shutters during a storm. He concludes by assuring his friend and his readers that the use of a conductor reduces the probability of being struck by lightning to that of around one in ten thousand.[67]

A further short supplement in the final article allows him space in which to clarify any possible confusion with regard to the use of materials acting as conductors and to correct critics where he has been unjustly accused of contradicting his own work.[68] In this final section Alzate chooses to present his information as a direct question-and-answer format in response once more to correspondence from unnamed friends.[69] He thereby emphasizes concrete practical application of knowledge presented in the *Gazeta* and creates a direct link with the reader, possibly in the hope of encouraging further correspondence by a wider network. Despite new references to the Leyden jar experiment that have not previously appeared in the *Gazeta*, these answers simply reinforce the advice Alzate has already provided throughout the articles on the conduction of electricity. Although the commentary demonstrates that Alzate chose terms that are of common use to make the ideas accessible to a wide audience, it is also clear that he was well acquainted with the appropriate scientific terminology.[70] In this way he defended his own position as a man of scientific understanding whose direction lay in the popularization and practical application of knowledge within society.

Final Considerations: Let the Unlearned Learn and the Learned Delight in Remembering

Throughout the articles included in this study Alzate has attempted to educate and inform on two fronts. On one level, using reports from Europe, he sought to introduce new technology designed in the Americas (albeit the English Americas), thereby ending the unnecessary suffering and destruction caused by a natural force. On another level, through his investigations into the natural resources available in Mexico, he attempted to widen the readers' appreciation of their own environment. He thereby endeavored to form a commonality of understanding among the readership of the *Gazeta de Literatura* in a very practical way regarding their environment and society, and through this understanding a deeper appreciation of their identity as a "hispano-american" nation. Thus, while the new technology was recognized as being universally applicable, there were certain considerations specific to the Mexican reality: the high level of electrical activity, the ferruginous materials readily available, and the salubrious atmosphere that protects metals seen to corrode easily in the European environment. All these reports are fruit of his own observations within his locality, readily available to any reader willing to follow in his footsteps. For those with eyes to see and the readiness to broaden their knowledge, there was reason to delight and be amazed.

The references to work by Franklin and the various other European authors stood as the foundation from which Alzate could launch his observations and conclusions. Although translations, adaptations, and summaries of foreign publications are plentiful in the *Gazeta de Literatura*, they are seldom presented in a manner that is dispassionate or detached. As editor, Alzate used his privileged position to create paratextual elements through the insertion of comments, footnotes, and commentaries wherever he felt necessary. The lack of criticism aimed at the texts chosen when reporting on the lightning rod is indicative of his approval of their content and the fact that they were serving to support his own observations. The ideas they express in their reports are useful only insomuch as they enabled him to narrow the focus to the Mexican situation and apply them to the practical needs that are evident in everyday life. As with each of the articles included in this study, Alzate's interest lies in the adaptation of ideas first to his "patria," Mexico, then more particularly, to his direct environment, the valley and metropolis. His descriptions of the climate and the natural resources of the valley, when read as a part of the whole corpus of the periodical, appear clearly as another avenue for the defense of the variety and potential available in the Americas, thus refuting arguments based on climatic determinism or Creole degeneracy and inadequacy. On the one hand, to encourage change, he wished to show his reader that Mexico City was lagging behind its European counterparts in the application of technological advances to its own disadvantage. Yet on the other, he recognized that Mexico was already far in advance of Europe in the natural resources available and through certain forms of effective traditional practices still in use.

In this way Alzate embodies the pragmatic approach to science adopted by so many of the Creole enlightened thinkers in the Spanish Americas. While disseminating ideas that were of benefit to the public, he promoted and protected local interests. As Lafuente has also noted, Alzate had convinced himself that he was part of an army of modernity, not as a convert but as a critic of the knowledge brought forth from Europe.[71] In this way he wrote to instruct his own countrymen with regard to their own potential and also highlighted the need for change and the incorporation of new ideas where they were adaptable and applicable to the Mexican context. The *Gazeta de Literatura de México* therefore serves as a critically aware, pragmatic, and versatile tool for the renewal of ideas and society in Mexico. The articles on the lightning rod are but one facet of the broader discourse that was aimed at practical application and a stimulus toward affirming a growing sense of national identity.

Notes

"Quid tentare nocebit (Nothing ventured, nothing gained)," Ovid *Metam.* I, quoted by Alzate in defense of his actions and ideas regarding the lightning rod in *Gazeta de Literatura de México* III, no. 13 (May 28, 1793): 102. My special thanks go to The Bakken Museum and Library for providing the funding that enabled the preliminary research for sections of this study, and especially to Elizabeth Ihrig for her ongoing help and support with the necessary materials.

1. All references to the *Gazeta de Literatura de México* (GLM) follow the format of volume, issue number (date): page. Each volume contains two subscription runs. Due to problems with printing houses and lack of quality the numbering of the various issues and the pagination in volume I is problematic for references. I have, therefore, chosen the following format for the references. In the first subscription run of volume I, each issue was numbered 1–24. The issues in the second subscription run were also numbered from 1 onward, rather than continuing from 25–48 as in the other volumes. To differentiate between the two runs, I have categorized the second by using 'b', that is 1b—24b for the second subscription run. Each reference carries the date to avoid any undue confusion. The Spanish citations when included in the footnotes retain the original orthography.

2. Among his studies into various natural elements, the articles published on the subject of the aurora borealis and the fear caused by such phenomena provide Alzate with an opportunity to show how the citizens of Paris, considered to be one of the most learned courts in Europe, had only a few years earlier been terrified by a misunderstanding that Saturn had disappeared. *GLM* I, no. 6 (November 19, 1789): 42.

3. *GLM* III, no. 13 (May 28, 1793): 100.

4. Antonio Lafuente, "Enlightenment in an Imperial Context: Local Science in the Late-Eighteenth Century Hispanic World," *Osiris* 15 (2000): 155–173, at 156.

5. For a more detailed discussion of the work of Alzate and the *Gazeta de Literatura de México* in terms of peripheries and center, see Fiona Clark, "The *Gazeta de Literatura de México* and the Edge of Reason: When is a Periphery not a Periphery?" in *Peripheries of the Enlightenment*, ed. Richard Butterwick, Simon Davies, and Gabriel Sánchez Espinosa (Oxford: Voltaire Foundation, 2008), 251–64.

6. Two recent publications that directly address the question of the development of the scientific community in Spanish America from the 1500s to the 1700s are Antonio Barrera-Osorio, *Experiencing Nature: The Spanish American Empire and the Early Scientific Revolution* (Austin: University of Texas Press, 2006), and Jorge Cañizares-Esguerra, *Nature, Empire and Nation: Explorations of the History of Science in the Iberian World* (Stanford: Stanford University Press, 2006).

7. For a broader discussion on the field of physics in Mexico in the eighteenth century, see María de la Paz Ramos Lara and Juan José Saldaña, "Newton en México en el siglo XVIII," in *The Spread of the Scientific Revolution in the European Periphery, Latin America and East Asia*, ed. Celina A. Lértora Mendoza, Nicolaïdis Efthymios, and Jan Vandersmissen (Belgium: Brepols, 2000), 91–98. The interest in physics is also explored with relation to the *GLM* in María de la Paz Ramos Lara, "Alzate y la física en sus *Gacetas de Literatura*," in *Periodismo Científico en el siglo XVIII: José Antonio de Alzate y Ramírez*, ed. Patricia Aceves Pastrana (México: Universidad Autónoma de México, 2001), 403–30; and Patricia Aceves Pastrana, "Átomos y Luces en los periódicos de Alzate," 221–49, in the same volume. Within the wider discussion of science in the periodical press, see Alberto Saladino García, *Ciencia y prensa durante la ilustración latinoamericana* (México: Universidad Autónoma del Estado de México, 1996).

8. Although largely focused on medical matters, Bartolache's *Mercurio Volante* (1772–73) is one of the earliest scientific periodicals in Mexico, published at the same time as Alzate's second periodical publication, *Asuntos Varios sobre ciencias y artes* (1772–73).

9. Description taken from an inscription on a portrait of Alzate held at the monastery of Tepotzotlán, México.

10. For more detail, see Roberto Moreno's introduction to and collection of Alzate's early works in José Antonio Alzate y Ramírez, *Memorias y ensayos* (México: Universidad Nacional Autónoma de México, 1985).

11. The Sociedad Bascongada de los Amigos del País (Basque Society of Friends of the Country) founded in 1763 was a precursor to many such societies across Spain and its Spanish America in the eighteenth century. They were established to stimulate progress in agriculture, industry, and commerce, and they published a breadth of information on

scientific, technological, and economic topics. No society was ever established in Mexico City, yet there were a considerable number of members such as Alzate across Mexico. Alzate's father had been a member of the same society. See Robert J. Schafer, *The Economic Societies in the Spanish World (1763–1821)* (Syracuse, N.Y.: Syracuse University Press, 1958). Chappe d'Auteroche died of typhus alongside many of his team during the expedition to map the transit of Venus in 1769. His reports were returned to Paris by Pauly along with a report on the transit written by Alzate. This report, along with a map of the viceroyalty of New Spain; samples of seeds, fruit, plants, minerals, and artifacts; and a letter on the natural history of Mexico, were sent to the Paris Royal Academy of Sciences. As a result Duhamel proposed Alzate's nomination for membership and in 1771 he was elected as a corresponding member to Pingré, with the new title Socio Correspondiente de la Real Academia de las Ciencias de París. After 1786 Alzate's name no longer appeared on the list of correspondents for the Academy, although he does not seem to have become aware of this fact until some years later. In the footnotes added to the published version of Alzate's report, Cassini remarks: "Le zèle de Don Alzate y Ramyrez à nous communiquer tout ce qui peut se trouver d'intéressant dans un pays si nouveau pour nous, ses qualités personnelles, ses connaissances particulières, ont mérité les & excité la reconnaissance de l'Académie, qui s'est empressée de le lui témoigner, en l'admettant au nombre de ses Correspondants." Quoted in Rafael Aguilar y Santillán, "Una carta interesante de Alzate," in *Memorias de la Sociedad Científica 'Antonio Alzate'* XXIII (México: Imprenta del Gobierno Federal, 1905), 87. Regarding his role as corresponding member, see also Patrice Bret, "Alzate y Ramírez et l'Académie Royale des Sciences de Paris: la réception des travaux d'un savant du Nouveau Monde," in *Periodismo Científico* (cit. n. 7), 123–206.

12. His initial periodical publications, the *Diario Literario de México* (March–May 1768) and the *Asuntos Varios sobre ciencias y artes* (November 1772–January 1773) were of very short duration. They were followed some fifteen years later by the *Observaciones sobre la física, historia natural y artes útiles* (May–October, 1787) and finally the *GLM*, running from 1788 to 1795. The *Diario Literario* was discontinued by order of the then viceroy, Marqués de Croix, for including material that was "offensive to the law and the nation" (Archivo General de la Nación, México, *Historia*, t. 399, fol.1., May 15, 1768). There is no clear indication as to the cause of the termination of the *Observaciones*, yet in the Prologue to the *GLM*, he alludes to two possible reasons: a lack of necessary materials, or "involuntary" obstacles. *GLM* I, no. 1 (January 15, 1788): 1.

13. Rafael Moreno, "Creación de la nacionalidad mexicana," *Historia Mexicana* 12 (1963): 531–51.

14. See Rafael Moreno, "Alzate, educador ilustrado," *Historia Mexicana* 2 (1953): 371–89.

15. "Contestación a D.M.," *GLM* I, no. 12b (February 20, 1790): 95–96; "Extracto de una Carta del Señor Pistoi Catedrático de Matemáticas en Siena, dirigida al Abate Rosier con fecha de 25 de Abril de 1777," *GLM* I, no. 15b (April 12, 1790): 115–20; "Correspondencia Literaria entre los Señores De Michaelis, Profesor de Lenguas Orientales, y Lichtemberg Catedrático de Filosofía, acerca de un suceso mencionada en la Antigua Historia relativa al establecimiento y utilidad de los Pararayos del Almacen Lit-

erario de Gottinga," *GLM* II, no. 19 (May 17, 1791): 147–53; "Física experimental," *GLM* III, no. 13 (May 28, 1793): 100–102; "Continuación de la antecedente," *GLM* III, no. 14 (June 11, 1793): 103–4; and "Suplemento a la anterior," *GLM* III, no. 15 (July 3, 1793): 112–14.

16. Articles directly related to topography appear consistently throughout nine issues of volume II and two issues of volume III of the *GLM*, while the extended study of the cochineal beetle provides the entirety of nine consecutive issues of volume III covering the entire period from January to September 1794 and also appears in January 1795.

17. Useful arts or "artes útiles" refers to all level of technological innovation and the practical implementation of new ideas.

18. "Each workman to his proper craft," Horace *Epistles* I, in *GLM* III, no. 14 (June 11, 1793): 103.

19. René Avilés Fabila has dealt a little with the aspect of Alzate as translator in "Alzate, escritor, literato, traductor y periodista: Apuntes," in *Periodismo Científico* (cit. n. 7), 595–602. This essay includes texts outside of the *GLM* and focuses mainly on the areas of poetry, theatre, and biblical references, not all of which were necessarily written by Alzate.

20. "How much futility in the world!" in *GLM* II, no. 19 (May 17, 1791): 152.

21. "Let the unlearned learn, and the learned delight in remembering," Horace, in *GLM* I, no. 1 (January 15, 1788): 1.

22. Among the other contemporary Spanish American periodicals that were either edited by or acted as the voice pieces of such societies we find: the *Gazeta de Guatemala*; the *Mercurio Peruano*; and the *Papel Periódico de la Havana*.

23. *GLM* I, no. 15 (December 16, 1788): 39.

24. José Mariano Mociño Suárez Losada (1757–1820) was born of Spanish parents in Temascaltepec. Having received his bachelor of medicine at the University of Mexico in 1787, he enrolled in Vicente Cervantez's course in botany in May of the same year, demonstrating great aptitude for the science. In 1789 he was enlisted as part of the second wave of the botanical expedition to New Spain under Martín de Sessé, the first having taken place in 1778, covering the area from Mexico City to the Pacific Ocean, including Acapulco. Mociño would later take part in the exploration of Nootka Bay near Vancouver, traveling to Spain in 1803 with Sessé where, amid various difficulties, they attempted to process all the information collected over the course of sixteen years.

25. It appears that Castillejo was also Creole by birth, a lawyer matriculated at the Audiencia of Mexico, and owner of the Hacienda de los Cinco Señores. Between 1802 and 1814 he was awarded the non-stipendary post of district attorney of Oaxaca. In 1808 and 1809 he became the object of attack by Peninsular authorities and was ordered to forego his office. In 1814 he was reappointed Subdelegado of Teotitlán del Valle having also been elected as one of Oaxaca's deputies in 1812. He died soon after.

26. By *Diario de Física* Alzate is referring to Rozier's *Observations sur la Physique*, or *Journal de Physique*. This connection will be considered in more detail later in this chapter.

27. *GLM* I, no. 23 (August 14, 1789): 115.

28. See Prólogo, *GLM* I, no 1 (January 15, 1788): 1–7, and particularly 6.

29. *GLM* II, no. 19 (May 17, 1791): 147–153.

30. *GLM* I, no. 1 (January 15, 1788): 2.

31. María Jesús García Gorrosa and Francisco Lafarga, *El Discurso sobre la traducción en la España del siglo XVIII: Estudio y Antología* (Kassel: Edition Riechenberger, 2004).

32. This fact runs contrary to the argument presented by García Gorrosa and Lafarga. But it must be pointed out that their study centered primarily on translation in literary prose and poetry and not on the short, scientific-technological format of periodical publications such as those published in the *GLM*.

33. In other issues we find ironic fake public decrees, *GLM* I, no. 22 (June 25, 1789): 101–4, or simulated philosophical arguments between imaginary characters, *GLM* I, no. 3 (February 15, 1788): 21–32, but these examples do not carry the same voice of authority as the words supposedly spoken by Franklin.

34. "Extrait d'une lettre de M. Pistoi, Professeur de Mathématiques à Sienne, du 25 Avril dernier," *Journal de Physique* 10 (1777): 379–81; "Correspondance entre M. de Michaelis, Professeur en Langues Orientales à Gottingue, et M. Lichtenberg, Professeur en Physique, sur un trait de l'Histoire ancienne, au sujet des Conducteurs: traduite de l'Allemand du Magasin des Sciences de Gottingue, année 1783, cinquième cahier; par M. Eisen, Ministre Luthérien à Niederbern en basse Alsace," *Journal de Physique* 24 (1784): 321–23, continued in *Journal de Physique* 25 (1784): 297–303 and *Journal de Physique* 26 (1785): 101–6.

35. *GLM* III, no. 13 (May 28, 1793): 100.

36. *Observations on Physics, Natural History and the Useful Arts.*

37. The first issue was printed in 1772 under the title: *Introduction aux Observations sur la Physique, sur l'histoire naturelle et sur les arts, avec des planches en taille-douce . . .* (Paris). After 1780 the periodical was edited by a group of French scientists including J. A. Mongez in 1780 and J. C. de la Métherie in 1785.

38. To date, of the sixty-two translations and extracts sourced as originating in the French publication, fifty can be found in the *Journal de Physique*. For more detailed discussion of the various links between the two periodicals as well as the *Journal des Sçavans*, the *Bibliothèque Physico-Économique*, and the *Journal Encyclopédique*, see Fiona Clark, "Read All about It: Translation, Adaptation and Confrontation in the *Gazeta de Literatura de México* (1788–1795)," in *Science in the Spanish and Portuguese Empires, 1500–1800*, ed. Daniela Bleichmar, Paula de Vos, Kristin Huffine, and Kevin Sheehan (Stanford: Stanford University Press, 2009), 147–77.

39. The exact location of these latter texts within the *Journal* has not yet been fully verified, in large part due to how little of the text appears in the *GLM* and the fact that they are randomly interspersed with Alzate's comments.

40. Such an approach stands in sharp contrast to the so-called Purfleet controversy in England in the 1770s as discussed by R. W. Home in this volume.

41. "Advertencia del Traductor" could also be translated as "Translator's Advice," but given the nature of the criticisms involved in the content, "warning" seems to be the more appropriate term.

42. *GLM* II, no. 19 (May 17, 1791): 149.

43. Although on many occasions Alzate refers to "hispano-american" and "Mexican," here he speaks of the "Spanish" mentality. The mixing of terms for national identity is reflective of the world in which he lived, still a part of the Spanish nation yet increasingly voicing a separate identity.

44. Although he makes no mention of directions on the actual construction of the lightning conductor in this article, in a concluding note he invites the reader to see the lightning rod that he has himself built. By this means he hopes to show how the instrument serves both as an electrometer when the storm clouds are at a distance and a conductor as the storm approaches. *GLM* I, no. 12b (February 20, 1790): 96. The practical details appear in the following article where we find Alzate advocating the use of a conductor with pointed end, emphasizing the need for the instrument to be inserted in damp or wet ground, and advising against the use of chains as any point of separation will lead to disastrous ends. *GLM* I, no. 15b (April 12, 1790): 118.

45. *GLM* I, no. 12b (February 20, 1790): 95. This criticism is directly linked to Alzate's ongoing dispute with Vicente Cervantes and Martín de Sessé, the Spanish scientists responsible for the direction of the newly established botanical gardens in Mexico City, and the teaching of botany as part of the university curriculum. The dispute centered on the introduction in Mexico of Linnean taxonomical principles to the exclusion of indigenous terminology. The former, Alzate argued, was incapable of reflecting the variety of plants available in the New World. It stood, therefore, as a European system imposed on the Americas. See José Luis Peset, "La naturaleza como 'símbolo' en la obra de José Antonio de Alzate," *ASCLEPIO* 79 (1987): 285–95; and Patricia Aceves Pastrana, *Química, Botánica y Farmacia en la Nueva España a finales del siglo XVIII* (México: Universidad Autónoma de México, 1993).

46. For further discussion on examples of public display and the new electrical devices, see Simon Schaffer, "The Consuming Flame: Electrical Showmen and Tory Mystics in the World of Goods," in *Consumption and the World of Goods*, ed. John Brewer and Roy Porter (London and New York: Routledge, 1993), 489–526. The area of science and public performance during this period, such as the case of electrical showmen, is still to be researched to any extent within the Spanish American context.

47. *GLM* I, no. 12b (February 20, 1790): 95. Alzate did not travel widely despite professing plans to travel to Spain in his correspondence with Casimiro Gómez Ortega. His work was mainly carried out within the limits of the valley of Mexico.

48. In the footnotes to this first article the reader is further informed that the particular report by Beccaria can be found in the *Encyclopédie Méthodique* published at Yverdon, which Alzate has translated and inserted into a report he was commissioned to write by the government as a result of the fire at the Royal Powder Factory in 1778. In these terms Alzate is demonstrating that his opinions and advice have been sought after in inner circles of political and social standing. *GLM* I, no. 12b (February 20, 1790): 95.

49. In volume II of the *Gazeta* Alzate chooses to include Franklin among the few individuals for whom he writes a eulogy. His words reflect the high estimation with which he regarded Franklin and again refer to him as the Prometheus who stole fire from the

skies. His discoveries are described as forming true physics, which is useful to all men. Moreover, as experience, observation, and example were the sources from which Franklin made his discoveries his merit will never be successfully questioned. *GLM* II, no. 8 (December 13, 1790): 59.

50. *GLM* I, no. 12b (February 20, 1790): 96. This brush with death left a decided mark on the author in terms of experiments that he was later willing to undertake.

51. *GLM* II, no. 29 (October 18, 1791): 230.

52. See also Jessica Riskin, "The Lawyer and the Lightning Rod," *Science in Context* 12, no. 1 (1999): 61–99.

53. The circumstances surrounding this episode at Siena are also discussed in Paola Bertucci's essay in this volume. The impact of such an event across Europe and as far as the Americas demonstrates both the importance of the public display of successful lightning rods and the successful dissemination of published reports, on this occasion through the Academy of Siena and the *Journal de Physique*.

54. "After I had published an article on the usefulness of the lightning rod in the *Gazeta de Literatura* N. 13 there were many who spoke out of place, ridiculing the matter as of little importance; but who were they? Doubtless they were the sentinels of the already ruined bulwarks of scholasticism." *GLM* I, no. 15b (April 12, 1790): 117. In this citation Alzate makes reference to issue no. 13 of the *Gazeta*; it seems, however, that the appropriate issue was in fact no. 12b.

55. *GLM* I, no. 15b (April 12, 1790): 118.

56. *GLM* I, no. 15b (April 12, 1790): 117–18.

57. *GLM* II, no. 19 (May 17, 1791): 151.

58. Ibid.

59. "Preoccupations, the dregs of scholasticism, that still survive to fill us with embarrassment, distract men from the useful outcomes they should be enjoying as a result of the knowledge liberally poured out upon us by the hand of the Omnipotent." *GLM* II, no. 19 (May 17, 1791): 151.

60. He continues to describe a design whereby one lead tube is inserted into another to form a conductor. In this way he attempts to solve the problem posed by doubts over whether the "electric fluid" passes over the surface or through the interior of the conductor. If the current passes through the material, it provides a large quantity of metal to make this process effective, and if it passes over the surface, the extra tubing provides ample surface space to allow for such an occurrence. The pointed rod would be placed into tubes already built into the walls of the house and inserted at least one and a half to two yards below the ground, thus also gaining sufficient protection by the masonry. *GLM* I, no. 15b (April 12, 1790): 119. In a short supplement that follows the final article included in this study, Alzate provides the actual cost of a lightning rod, measuring the expense against the size of the building on which the rod was to be used. He claims that he is not hypothesizing because he has verified this calculation with all the necessary scrupulousness demanded by the public and has consulted with various architects on the matter. *GLM* III, no. 15 (July 3, 1793): 114.

61. *GLM* II, no. 19 (May 17, 1791): 152.

62. *GLM* III, no. 13 (May 28, 1793): 100.

63. "Memoria acerca de los preservativos para libertarse de los funestos efectos del rayo," in ibid.

64. *GLM* III, no. 13 (May 28, 1793): 103.

65. Although at the end of issue 13 Alzate expresses a desire to print the report in full, it does not appear at any stage in the following text.

66. *GLM* III, no. 14 (June 11, 1793): 103.

67. Ibid.

68. *GLM* III, no. 15 (July 7, 1793): 112–14.

69. The use of the letter format, or references to correspondence, was a common device in newspapers and journals and could possibly have been written by the editor himself without an external source. In this issue one letter is identified as coming from Tepeaca, although he refuses to print the author's name, and the final two are anonymous.

70. *GLM* III, no. 15 (July 3, 1793): 112.

71. Lafuente, "Enlightenment in an Imperial Context" (cit. n. 4), 172f.

Part 2
EXPERTISE, CONTROL, AND IMAGERY— LIGHTNING ROD POLITICS

Points or Knobs

Lightning Rods and the Basis of Decision Making in Late Eighteenth Century British Science

R. W. Home

It is a truism among historians of science that analyzing controversies can often shed a particularly clear light on the science of the past. In this essay, I shall focus on one of the more notorious and certainly one of the more entertaining disputes in the history of science, namely the heated debate that took place in London in the 1760s and 1770s over the best design of lightning rods. Although the story has been told before, it is worth telling again because there are, I believe, significant lessons to be drawn from this history that existing accounts do not sufficiently bring out.[1] These have to do particularly with the emerging claims of science to expert knowledge and, within the scientific community itself, with the establishing and exercising of intellectual authority. To be sure, there were also larger political issues involved, with the dispute between Britain and its American colonies providing the background to the controversy and advocacy of pointed or blunt conductors being linked by many commentators to the revolutionary and loyalist causes respectively.[2] In my view, however, so far as the lightning rod controversy was concerned, this was mere background noise. The more interesting issues, those that help us understand what was going on in the controversy, related more narrowly to the standing and structure of the British scientific community at the time.

The controversy ranged most of the leading British electrical investigators of the day, who agreed with Benjamin Franklin in favoring the use of pointed conductors, against a much smaller group spearheaded by another prominent Fellow of the Royal Society, Benjamin Wilson, who favored conductors with rounded ends. Many of those who have written on the subject have assumed that Franklin and those who agreed with him were unambiguously in the right and that Wilson, in adopting the position he did, was simply being obtuse.[3] I believe, however, that approaching things in this way obscures a number of interesting historical issues. (It is also at odds with current scientific thinking on the subject—though this is not to say that current thinking supports Wilson's position, either!) Rather than assuming that the position adopted by Franklin and his supporters emerged victorious simply because it was natural and right, I propose taking a more symmetrical approach that does not judge ideas retrospectively on the basis of later thinking on the subject. In other words, I shall look seriously at the arguments presented by the losing side in the controversy and focus on the mechanisms by which Wilson's position became discredited. Not, I hasten to say, that I wish to grant Wilson more credit than he deserves. Yet at the very least, we must, I think, acknowledge that there was a genuine argument going on, and that the controversy was not simply a matter of the lone, misguided figure of Wilson tilting futilely at Franklin and the massed forces of eighteenth century scientific orthodoxy that supported him. On the contrary, Wilson garnered a significant measure of contemporary support. Hence, even if we eventually decide (as well we might) that he was misguided, we should also have to conclude that this was not necessarily apparent at the time.

Opening Salvos

At the heart of Franklin's famous "sentry box" experiment to establish the electrical nature of lightning lay the very first discovery he reported to his correspondents in England after he commenced the study of electricity, his observation that pointed conductors drew off electricity from nearby charged objects. In an amazing intellectual leap, Franklin reasoned that a metal rod pointed at the sky in stormy weather should therefore draw off charge from electrified clouds in its vicinity and thereby, if it was insulated, become electrified. The charge it acquired should be detectable by the usual methods, as indeed it proved to be. Not content, however, with showing how the scientific conclusion might be established, in the same paper in which he described his famous experiment, Franklin also suggested that advantage might be taken of this property of points to protect buildings from destruction by lightning. If "upright rods of iron made sharp as a needle" were erected on the highest parts of buildings, he asked, and a conducting line was run from their lower ends down the outside of the building into the ground, "would not these pointed rods probably draw the electrical fire silently out of a cloud before it came nigh enough to strike, and thereby secure us from that most sudden and terrible mischief?"[4]

Franklin later claimed a second and quite different advantage for the lightning rods he wanted to see erected on buildings, namely that they would provide, in the event of a lightning strike, a preferred conducting route that the discharge could follow harmlessly into the ground instead of passing through the material of the building itself and causing destruction wherever it was forced to leap across breaks in the conducting path. So far as I am aware, no one ever denied that conductors fitted to buildings could function in this latter way, or denied the value of fitting buildings with conducting paths for the lightning to follow to the ground if by mischance it struck them. Certainly Benjamin Wilson was as one with other members of the Royal Society's various committees on this and stressed with them the need not only to fit buildings with such conductors but also to pay close attention to the continuity and proper grounding of these.

Having conceived the idea of the lightning rod, Franklin was able to give it wide publicity throughout the American colonies, where electrical storms were much more frequent and destructive than in most parts of Europe, in his enormously popular annual volumes of *Poor Richard's Almanac*. As a result, significant numbers of buildings in America were soon fitted with conductors. In Europe, however, it took longer for the idea to catch on.

In Britain, a turning point in the introduction of lightning rods came in 1764, following a lightning strike on the steeple of St Bride's Church in central London that caused considerable damage. The Royal Society's *Philosophical Transactions* that year included a detailed account by Edward Delaval of the effects of the strike, the damage from which he attributed to discontinuities that he discovered in the chain of electrical conductors from the metal vane and cross at the top of the spire down to the water table in the ground on which the church stood. Delaval took issue with Franklin's notion that a metal wire would suffice as the conducting line because he thought this would be too liable to melt and thus create a gap in the conducting path. Instead, he recommended a thick metal bar. However, he said this was the only point on which he differed from Franklin.

Delaval's paper was accompanied by one by the leading English electrical investigator of the day, William Watson, who likewise completely endorsed both the value of erecting pointed lightning rods on buildings and Franklin's reasoning concerning their action. "The expectation of the utility of this apparatus," he wrote, "is presumed to be the preventing of the accumulation of electricity in its neighbourhood, by affording a constant and easy passage to the electricity of the clouds surcharged therewith." Rather than defending the value of erecting pointed conductors—he simply took this for granted—his principal concern was to avoid any suggestion that such conductors acted by attracting the lightning. In the process, he somewhat obscured his earlier Franklinian argument that a pointed conductor would prevent the accumulation of charge, now stressing instead how it would conduct strikes harmlessly to ground: "I have avoided introducing the term *attraction* here, operating as an active principle; as I consider the apparatus purely passive, and only affording, from the aptness of its parts to that purpose, an easy and uninterrupted passage to the lightning, and thereby preventing its violent effects."[5]

Figure 5.1.   Benjamin Wilson, FRS (n.d.). Courtesy National Portrait Gallery, London.

The 1764 volume of the *Philosophical Transactions* also included a paper by Benjamin Wilson in which he revealed for the first time his disagreement with his colleagues over the use of lightning rods.[6] While fully in accord with Delaval and Watson concerning the need to furnish buildings with apparatus that would lead lightning harmlessly away when it struck, he felt it was inviting trouble to erect conducting rods—especially pointed ones—above the roof line. The flow of electricity in the discharge was so great, he declared, that we certainly should not go out of our way to invite it into our buildings. Of course, once the electrical equilibrium between a cloud and the earth was disturbed, it would inevitably be restored, via the path of least resistance—that is, there would be a lightning strike somewhere. Hence, "as the lightning must visit us, some way or other, from necessity, to restore the *equilibrium*, there can be no reason to invite it at all"; we should concentrate instead on making sure that when it does strike, it is led harmlessly away. Pointed projecting conductors, being the most effective means of inviting a strike, ought to be particularly avoided. Instead, Wilson recommended erecting "a rounded bar of metal" inside the highest parts of the building and continuing this down the wall into the ground.[7]

No more was heard at the Royal Society on the subject of lightning conductors until 1769, when the authorities at St Paul's Cathedral sought advice on how to protect their church. Following its normal practice, the society formed a committee to make recommendations. The members were Watson, Delaval, Wilson, Franklin—once again living in London as agent for the American colonies—and John Canton. Perhaps as a concession to Wilson, the committee in its report made no recommendations about erecting conductors above the existing structure. Instead, noting that the roof was made of lead and that the highest points of the cupola and other projecting parts of the building were metallic, it focused merely on ensuring that there were good conducting links from the metal in the upper, projecting structures to the roof, and from this via the building's lead downpipes to sewers deep in the ground.

The Battle Is Joined

In 1772, the question of lightning rods came before the Royal Society again when the Board of Ordnance, somewhat belatedly responding to reports of lightning having struck a powder magazine at Brescia, Italy, in 1769, sought the society's advice on how to protect its own powder magazines at Purfleet on the lower reaches of the Thames, a principal armament store for the Royal Navy.[8] The society set up a new committee comprising Henry Cavendish as chairman, Watson, Wilson, Franklin, and the society's assistant secretary, John Robertson.[9] When questions were raised about the committee's standing, with doubts being expressed concerning the weight its recommendations would carry, the board responded decisively by reassuring members that "the intention of the Board of Ordnance is, minutely to execute whatever you shall judge proper and necessary for the preservation of His Majesty's Powder Magazines."[10]

After inspecting the buildings at Purfleet, the committee submitted a short report.[11] There were five parallel magazines, long low buildings near the river; in addition, on elevated ground 150 yards away was a lofty hip-roofed building in which the Board of Ordnance held its meetings. The committee recommended sinking wells at each end of each magazine and others at the sides of the complex such that there was at least four feet of water in each one. Lead pipes were to rise from the bottom of each well to ground level where they should be "strongly joined" to upright iron bars fastened to the walls of the magazines by lead straps and extending at least ten feet above the ridges of the buildings. The bars should also be connected to the lead coping of the roof. They should end in sharply tapered copper points. The iron sections should be painted to prevent rusting, which would otherwise soon destroy their conductivity and endanger the buildings they were supposed to be protecting. Similar measures should be taken to protect subsidiary buildings in the magazine complex. The board-house was already well protected, the committee thought, by the lead copings of its high, pointed roof that extended down to lead gutters, from which lead pipes descended at each end of the building into wells in which the water was at least forty feet deep. Hence, they recommended no new structures in this case, except a pointed iron rod on the summit, communicating with the existing metal furnishings of the building. Even this was an afterthought, according to Wilson, added when he pointed out the seeming inconsistency between the committee's reliance elsewhere on pointed conductors and their initial decision in this case, as in the case of St Paul's Cathedral, to rely on the lead roof, when securely connected to ground, to protect the building.[12]

Consistent with the views he had stated previously, Wilson filed a report dissenting from that part of the committee's report that recommended that each conductor should terminate in a point:

> Such conductors are, in my opinion, less safe than those which are not pointed. Every point, as such, I consider as solliciting the lightning, and, by that means, not only contributing to increase the quantity of every actual discharge, but also frequently occasioning a discharge where it might not otherwise have happened. If, therefore, we invite the lightning, while we are ignorant what the quantity, or the effects of it, may be, we may be promoting the very mischief we mean to prevent. Whereas if, instead of pointed, we make use of blunted conductors, those will as effectually answer the purpose of conveying away the lightning safely, without that tendency to increase or invite it.[13]

Wilson backed up this argument with a more extended paper that was published in the *Philosophical Transactions* immediately after the formal committee documents, and also separately.[14] In this he tried to introduce some quantitative considerations, arguing that it appeared from his experiments that a pointed conductor was twelve times as effective as a blunt one in drawing off a charge from an insulated conductor: it followed, he thought, that if a thunder-cloud acted on a point at a distance of 1,200 yards, it would not act on a blunt conductor until the distance was reduced to 100 yards. Erecting a pointed rod rather than a blunt one increased the risk of a lightning strike in the same ratio. It therefore "surely behoves us to proceed with caution" and to avoid attracting unnecessary strikes in this way: "why should we have recourse to a method, which is at best uncertain; and which some time or other may be productive of the most fatal effects?" Wilson acknowledged what "seems to have been once the opinion of Dr Franklin," that pointed rods "draw off, and conduct away, the light-ning imperceptibly and by degrees, without causing any explosion," but did not linger over it since it was confuted, he thought, by Franklin's own reports of pointed conductors being struck by a full lightning discharge. If pointed rods worked as Franklin supposed, he argued, they should have drawn off the elec-tricity before any discharge could occur.

The other members of the committee were clearly not persuaded by Wilson's "either–or" argument here, but they did not spell out the basis of their disagree-ment. Instead, they merely assured the president of the Royal Society, Sir John Pringle, that "having heard and considered the objections to our Report . . . we do hereby acquaint you, that we find no reason to change our opinion, or vary from that Report."[15] The various papers were forwarded to the Board of Ord-nance, and pointed rods were erected in accordance with the majority opinion.

To modern eyes, the whole argument between points and knobs is miscon-ceived and rests on difficulties in scaling up from laboratory-based experiments to real clouds. Wilson's easy extrapolation of his 1:12 ratio does not work. But likewise the Franklinists' insistence on the importance of using sharply pointed conductors, which was also based on an extrapolation from laboratory-scale ex-periments, was misplaced. When an electrified cloud passes over a building fur-nished with a lightning rod, the scale is such that it makes little if any difference whether the rod is pointed or blunt: the lightning will tend to strike the nearest projecting object, whatever its shape.[16] Interestingly, some members of the Royal Society's committee (among whom Henry Cavendish is the most likely) may have recognized this, for Wilson referred to "some gentlemen of the committee expressing their opinion, of its being a matter of mere indifference whether blunted or pointed conductors were made use of."[17] Moreover, Wilson himself, earlier in his paper, had argued that "the *longer* the conductors are above any building, the *more danger* is to be apprehended from them; as they will in that case approximate nearer in their effects to those that are pointed."[18] Far from see-ing the skeptical opinion that he reported as a legitimate and positive conclusion to have reached, however, Wilson—convinced by his own arguments that the issue between points and knobs could be and indeed must be decided—saw it as mere procrastination. In tones that certainly would not have endeared him to his colleagues, he personally attacked those upholding such a view for failing to do their duty: "they have not considered this subject, with all the due attention which so important an object deserves," he proclaimed.

The disagreement over lightning rods reached its climax in 1777 when, despite the protective structures installed five years earlier, a bolt of lightning caused minor damage to the parapet of the board-house at Purfleet. The Board of Ordnance immediately sought advice from the Royal Society, which once again appointed a committee to investigate and report. Wilson was again a member, but the others—William Henley, Timothy Lane, Edward Nairne, and Joseph Planta—were new and represented a generational change among British electrical investigators. Inspection revealed that the lightning had struck one of the iron cramps that held the parapet together and had then passed through a few inches of stone and brickwork, which was damaged as a result, before reaching the lead gutter and being carried off from there to earth without causing further damage. The majority of the committee submitted a short report in which, having described the damage, they declared it "so inconsiderable, that it would scarce deserve notice, was it not an evident proof that the metallic communication with the earth hath, in this case, effectually prevented any farther injury."[19] Wilson once again dissented.[20] The fact that the board-house had been struck within a few inches of the grounded conductor was, he said, contrary to Franklin's assertion "which positively says, that in such circumstances the lightning passes in the metals, and not in the walls." Furthermore, on Franklin's principles, the event should never have occurred "because he says, that pointed conductors will draw all the lightning out of the clouds, and carry it away into the earth silently." The latest incident was, he thought, "a providential warning" of which the society ought immediately to take heed "and reject an apparatus which threatens us every hour with some unhappy consequences." The most unhappy possibility of all he felt it his duty to point out, for it concerned "a house, which is of the first consequence in this kingdom, that hath pointed conductors also fixed upon it: I mean the KING'S."

Wilson sent a copy of his dissenting report to the Board of Ordnance, which in turn referred the matter back to the Royal Society, asking whether "any thing more can be done, in order to the preservation of his Majesty's magazines."[21] Wilson also made sure his awful warning was heard where it mattered most, in the threatened royal palace itself. He was able to do so as a result of social connections that had nothing to do with his standing as an experimental philosopher. Although of humble origins, Wilson had built a highly successful career as a portrait painter in mid eighteenth century London. Among those who sat for him were many fellow members of the Royal Society, including Benjamin Franklin, who in 1758 commissioned a likeness that now hangs in the White House in Washington.

Figure 5.2.   Benjamin Wilson's portrait of Benjamin Franklin, 1759. Courtesy White House Historical Association, Washington, D.C. (White House Collection).

Wilson's most lucrative commissions came, however, from upper-class society and the aristocracy—to such good effect that his income is reputed to have reached the astonishing sum of £1,500 a year.[22] In the early 1760s he gained the patronage of the king's brother, the Duke of York, who in due course secured his appointment to the lucrative sinecure of painter to the Board of Ordnance. By the 1770s, Wilson's artistic skills had brought him to the attention of King George III himself, putting him in a position where he was able to speak directly to the king of his concerns about lightning rods that ended in points. The king took note and ordered the Board of Ordnance to make available to Wilson whatever resources he needed to carry out a definitive series of experiments. Thus encouraged, Wilson designed his experiments as a public spectacle on a grand scale, taking over the great hall of the London Pantheon for the occasion. The king himself came to view the experiments, as did members of the Board of Ordnance and of the Royal Society's committee and, at other times, crowds of interested citizens who could afford the sizeable entry fee of two shillings and sixpence charged by the proprietors of the building for public performances by Wilson and his assistants.[23]

To imitate the effect of an electrified cloud, Wilson commandeered a supply of drums from the Board of Ordnance. These he joined together and coated with tinfoil to form a huge prime conductor 155 feet long and 16 inches in diameter that he suspended by insulating cords from the ceiling of the auditorium. To add to the electrical capacity, he attached to one end of the cylinder a wire several hundred yards long that snaked around the room, likewise suspended by insulating cords from the ceiling. A scale model of the board-house at Purfleet was mounted on rails along which it could be drawn beneath the suspended prime conductor, the relative motion between the two being intended to simulate the passing of a thundercloud over the board-house. Projecting metal rods of various lengths, with pointed or rounded ends as required, could be attached to the roof of the model where they connected to a conducting wire that ran down the side of the model and then away to earth. An elaborate series of experiments confirmed to Wilson's satisfaction that points attracted more charge, and at a greater distance, than rounded conductors did. Nothing, therefore, could be plainer than that they represented a greater danger to a building: conversely, rounded conductors were safer, and those that projected least above the roofline of the building were safest.[24]

The reading of Wilson's report on his experiments occupied several meetings of the Royal Society, which afterward set up yet another greatly enlarged committee to consider the matters that had been raised. On this occasion, the committee was chaired by Sir John Pringle himself and included virtually every Fellow who had ever written on electricity apart from Wilson and Franklin, the latter being no longer in London but in Paris promoting the interests of the American colonists now at war with Britain. On March 12, 1778, the committee reported that "having attentively examined the experiments and observations of Mr. Wilson," they had concluded that it was very unlikely that the magazines, as already protected, would suffer any damage from lightning. They were prepared, however, to commit themselves publicly to points. To increase the security of the magazines still further, they said, additional high, pointed conductors should be erected on the roof and corners of each building and carefully con-

A View of the Apparatus and part of the Great Cylinder in the Pantheon.

Figure 5.3. The experiment at the Pantheon. Wilson, *An Account of Experiments* (London, 1778).

nected to ground. In addition, the whole roof of each magazine should be covered with lead, if possible, that should then be grounded: "We give these directions, being persuaded, that elevated rods are preferable to low conductors terminated in rounded ends, knobs, or balls of metal; and conceiving, that the experiments and reasons, made and alledged to the contrary by Mr. Wilson, are inconclusive."[25] The committee's report inevitably led to further squabbles. Before it could even be read, Wilson's ally, Dr. Samuel Musgrave, sought to delay consideration of it for a fortnight during which time it should be open for inspection by interested Fellows. In response, a motion was tabled (and "carried by a great majority" at the following meeting), reaffirming the society's traditional policy of not, as a body, expressing an opinion on any subject. When a committee was appointed to report on some matter, it was agreed that such report "should be considered the opinion of those members only and not of the Society at large."[26] This prompted Wilson to tell the Board of Ordnance that the society, as a body, did not agree with the report, whereupon the board wrote to the society seeking the opinion of the society at large. This led to a further angry debate, the outcome of which was that the board was informed that the society had "never since its first institution given an opinion as a body at large, but constantly by Committees," and that in this instance, "the Society has no reason to be dissatisfied with the report of its Committee."[27]

Officially, this was the end of the matter. It seems that the Board of Ordnance, perplexed by the conflicting advice they had received, did nothing either to implement the new protective measures recommended by the committee or to remove the pointed conductors as recommended by Wilson but simply left the existing conductors in place. At the Royal Society, however, debate continued. In June 1778 Edward Nairne presented a paper describing experiments intended "to shew the Advantage of elevated pointed Conductors," and this was followed immediately by a paper by Samuel Musgrave setting out his reasons for dissenting from the report of the society's committee and at the same time criticizing Nairne's experiments.[28] A month later Wilson himself returned to the fray, rebutting some experiments reported earlier by William Henley that seemed to contradict much of Musgrave's reasoning.[29]

At the society's anniversary meeting in November that year, Sir John Pringle resigned the presidency and was replaced by the king's friend Joseph Banks. It has sometimes been alleged that Pringle's resignation was forced on him by the king's taking Wilson's side in the lightning rod dispute. The king, it is said, having concluded that Franklin-inspired pointed rods were an American conspiracy and having issued instructions that they be removed from his palace, also told Pringle to have the committee's report changed. Pringle replied that he could not change the laws of nature, but then of course had to resign. The story seems to have had its origin in a remark by Pringle's biographer, Andrew Kippis, who reported he had heard it said that Pringle had resigned because he had been "much hurt by the disputes introduced into the Society" over lightning rods. This was picked up and embroidered by Cuvier in his elegy of Pringle and reported as fact thereafter, notwithstanding Kippis's explicit assertion that he never heard from Pringle "any suggestion of the kind that has been mentioned" and that Pringle had actually resigned because of ill health.[30] It is evidently too good a story not to be true!

Thereafter, the society had a brief respite from arguments about lightning rods. In December 1781, however, the subject came up yet again when a new letter was received from the Board of Ordnance after Wilson drew the board's attention to the fact that a poorhouse at Heckingham, near Norwich, had been struck by lightning despite being fitted with conductors.[31] The society's new president, Joseph Banks, immediately appointed a new two-man committee, Edward Nairne and Charles Blagden, to investigate. They quickly concluded that the strike had had nothing to do with the fact that the conductors were pointed but instead had occurred because the conductors had not been properly earthed.[32] Wilson protested to Banks about Nairne's being a member of the committee of inquiry, attacking his veracity as a witness.[33] He also spread a story that Nairne had known of the strike at Heckingham months before and had tried to suppress reports of it. In addition, Wilson sought information from his own local correspondents as to what Nairne and Blagden had observed, with a view to laying this before the king and the Board of Ordnance at "a proper moment when public affairs will admit of it."[34] By this time, however, Wilson had evidently lost all credibility within the scientific community, for Banks would not for a moment countenance his allegations against Nairne, drawing from Wilson the pained response that he could not "perceive the truth of the President's assertion that Mr. Nairne's veracity in respect to the facts and experiments above mentioned is preferred by the public and the Royal Society in general."[35] Thanks to Banks' firmness, this really was the end of the controversy so far as the Royal Society was concerned.

Resolving the Controversy

Traditionally, those who have written about this unhappy episode have represented it as a case study of the evil that arises when political considerations are allowed to intrude into matters of science, where they have no business. Weld gave the classic statement of this position in his nineteenth century history of the Royal Society:

> It was no longer scientific men who were allowed to decide the question, but political partisans. . . . As usual in such cases, the populace, and even the higher classes of society, took up the quarrel without inquiring into its merits, or, indeed, knowing anything about the matter in a scientific point of view. Wilson thus found protection and support, as he would have done against the theorem of Pythagoras, had geometry been the subject of dispute. But the most extraordinary part of this dispute is that George III is stated to have taken the side of Wilson, not on scientific grounds but for political motives.[36]

More recently, Trent Mitchell, in his account of the controversy, has insisted that "the entire debate was political" in the sense that scientific discourse "aims not only to communicate ideas or facts . . . but also to achieve legitimacy through the assent of intended and actual audiences." According to this account, then, the dispute over lightning rods was no different from any other scientific debate, except perhaps by virtue of the colorful way in which it was conducted. At the same time, Mitchell maintains that Franklin, by not yielding in his preference for points, by "exposing the weaknesses in Wilson's theories of electricity," by being "adept at persuading his readers and colleagues of the veracity of his knowledge claims," and by casting his vote when votes were called for—in other words, by not changing his mind when he thought he was in the right, and by continuing to argue for his position—was as guilty as Wilson of "entrenching the factionalism permeating the entire debate."[37]

Mitchell has also made the implausible suggestion that Wilson's theory—which Mitchell characterizes as based on the idea that a "spiritualized Newtonian aether" that received its activity directly from God governed the actions of otherwise inert matter—was analogous to a "monarchical and absolute" social structure, while "Franklin's electricity was a democratic element . . . that sought to attach itself to all objects equally."[38] One might equally argue that Wilson's theory was inherently democratic since it implied that all matter, being uniformly inert, was equal in the eyes of God until activated by Him through the agency of the ether, while Franklin's theory was hierarchical because Franklin held that different substances had different capacities for electricity.

We can surely agree that both sides of the dispute were trying to persuade people of the correctness of their respective views, and that in doing so they employed their rhetorical, literary, and social skills as well as their experimental and analytical ones. "The purposeful manufacture of consensus was," as Mitchell says, "a major concern that preoccupied eighteenth-century experimentalists"[39]—as, indeed, it has always been for those engaging in scholarly or scientific work, who in putting forward their ideas have inevitably been trying to persuade others of the correctness of these. If this is what the claim that science is embedded in politics reduces to, it is surely neither surprising nor new.

The fuss over lightning rods prompts other questions, however, that I believe are more interesting to pursue, that concern the nature and functioning of scientific authority and the ways in which it may be enhanced or diminished. For there is no doubt that, notwithstanding the socially and politically powerful support that Wilson mustered for his position, he lost scientific credit in the dispute. Moreover, it is clear that in the end it was scientific credit, not his social and political support, that mattered. This, I suggest, represented a relatively new state of affairs in English society and therefore warrants further exploration here.

Gaining Scientific Credit

Let us first recall that neither Franklin nor Wilson had received formal training in science. Neither, indeed, had most of the other Fellows of the Royal Society. Both Franklin and Wilson were self-made men who drew their standing as authorities on experimental philosophy from the work they had done in this area, which had convinced their contemporaries that they knew what they were talking about. Both had been awarded the Royal Society's highest award, the Copley Medal, for their research on electricity—Franklin in 1753 following the success of his proposed lightning experiment a year earlier and Wilson in 1760 for "the very many curious Experiments, relating to Electricity, that he has made upon the Tourmaline and other Bodies; and which he has laid before this Society since our last Anniversary Meeting."[40]

When differences emerged over the design of lightning rods, Franklin had the initial advantage, partly because the widespread acceptance of his doctrine of two opposite electricities and his explanation of the Leyden experiment gave him enormous prestige as an electrical theorist, partly because the use of pointed conductors in the manner he proposed was seen to be integral to the success of the famous experiments on lightning at Marly and elsewhere. But Wilson also had a substantial record of achievement behind him, so, when he came out in opposition to pointed conductors, his views had to be taken seriously. His reputation was, however, chiefly as an adept experimenter in a field in which it is notoriously difficult to obtain reproducible results. His early writings had displayed signs of ill-digested thought and over-hasty publication, his attempts at theory-construction attracting criticism even from his friends;[41] and even his work on the tourmaline, well done though it was, had raised as many questions as it resolved. He had also showed himself willing, even at the very outset of his career, to engage in controversy. Even before his first electrical papers were published in 1746, he had lost the support of William Watson, who until then had been encouraging him in his study of electricity, by accusing him at the Royal Society of stealing his ideas.[42]

To some extent, we can surely sympathize with some of Wilson's arguments in relation to lightning rods. In particular, if the reasoning was correct that initially led Franklin to propose the erection of pointed rods on buildings, namely that the points would silently draw the charge off approaching clouds and so prevent the stroke, it was difficult to understand how pointed rods could sometimes nevertheless be struck. And if points did not function in the way Franklin had said, maybe erecting them high above buildings was indeed increasing the risk of a stroke rather than reducing it.

In December 1773, William Henley reported some experiments to the Royal Society in which he tried to clarify the issues involved by using his recently invented electrometer, one of the first reliable instruments of this kind, to measure changes in the degree of electrification of a large insulated conductor as a pointed or a blunt rod was brought toward it. He found that, at distances still too great for a discharge to occur, when a pointed rod was used, the electrometer consistently showed a reduction in the charge on the conductor, whereas no such reduction occurred when the blunt rod was used. Recognizing, however, that the conductor remained significantly charged, he was much more cautious than Franklin had been in the conclusions he drew. In the case of an approaching electrified cloud, he thought a point would probably silently draw off part of the charge before it got within striking distance, "and the stroke itself might thereby, perhaps, be a little lessened."[43]

The Role of the Royal Society

Perhaps the most remarkable feature of the whole story is the authority granted to the Royal Society by all involved to act as arbitrator in this matter. At each stage of the dispute, those who carried the burden of decision making regarding the protection of public buildings turned to the society for advice and, when the question was raised, made it clear that they would follow whatever advice was given. Even Wilson was forced to work within this framework. Finding himself in 1777 outnumbered for a second time on the society's working committee, he sought not to have the matter taken out of the society's hands altogether—he must have known that such a strategy would not succeed—but to persuade members of the committee to change their minds or, if they would not, to have the committee's recommendations debated (and he no doubt hoped overturned) by the society at large. Later, in the case of the lightning strike at Heckingham, he sought to undermine the credibility of the particular committee that Banks appointed, not the principle of referring the question to the Royal Society in the first place. Lightning rods were a new form of technology that had emerged from research in experimental philosophy, and that had clear implications for public as well as private affairs. Members of the Board of Ordnance—including Wilson's close supporter Daines Barrington—clearly felt that (a) they needed expert advice about them; (b) this needed to be provided in a formal way; (c) the relevant expertise lay within the Royal Society, making it the appropriate source of such advice; and (d) they were bound to follow whatever advice they were given, rather than any opinions of their own on the matter.

The increasing tendency of governments at this period to rely on technical expertise has been thoroughly documented in the case of France.[44] By contrast, Gascoigne has noted that in eighteenth century Britain, "virtually the only agency which could provide government with scientific advice was the Royal Society," but that the government "had traditionally not found much need to trouble the . . . Society." He has also, however, noted that, even in Britain's less regulated governmental structures, by the early 1770s this situation was gradually changing, with the Royal Society's advice being crucial in persuading the government to mount expeditions to observe the transits of Venus across the face of the Sun in 1761 and 1768—including James Cook's first famous expedition to the South Pacific, 1768–71—and with the society's being entrusted with dispersing the funds allocated for these expeditions. In the 1770s, in addition to its providing advice about lightning rods, the society's advocacy lay behind the government's funding of an expedition to the Arctic under Constantine Phipps in 1773 and Cook's third Pacific voyage, on which he embarked in 1776. Thereafter, the society and especially Joseph Banks, its long-serving president, were increasingly drawn into the machinery of government as a source of expert advice on technical matters.[45]

Denouement

It is in the light of these considerations that we need to view Wilson's spectacular experiments at the Pantheon. Trent Mitchell has emphasized the manipulation of spectacle and audience involved here, with the king, members of the Board of Ordnance and of the Royal Society's committee, and the general public all being called in as witnesses. As subsequent events demonstrated, however, the outcome depended on one part only of this audience, namely the acknowledged experts on the subject, the members of the Royal Society's committee. The Board of Ordnance, which ultimately had to decide what was to be done, had already indicated that it would defer to the committee's judgment. Even the story concerning Sir John Pringle's resignation confirms the committee's role: while the king was evidently convinced by Wilson's experiments and was free to change the conductors on his own house from pointed to blunt if he wished, he could have the conductors at Purfleet changed only if the committee recommended this.

If the role of the Royal Society's committee is crucial, then we need to look more closely at factors that might have influenced its decision. Several scholars in recent years have noted the problems created for natural philosophers in various parts of Europe during the second half of the eighteenth century by the very popularity of their calling, which led to a dramatic increase in the number of practitioners making a living by offering courses of lectures in "experimental philosophy," illustrated by entertaining demonstrations and increasingly associated with radical politics, to an ever-expanding public. Natural philosophers whose self-image included a commitment to advancing understanding sought to distinguish themselves from mere "scientific salesmen"—indeed, according to Simon Schaffer, the problem became so acute that "the most important desideratum for the survival of natural philosophy was the institution of some system of policing of its own practitioners."[46] If so, Wilson's decision to open his experiments to the general public and to acquiesce in the charging of an admission fee was, to say the least, ill considered, for he thereby opened himself to the accusation of being a mere showman—though surely not one tarred with the brush of radical politics—and put his status as a natural philosopher at risk.

Considerations of this kind may well have played a part in Wilson's downfall. This was not, however, the only or, in my view, the most damaging factor at work.

A more important consideration, I believe, was the fact that Wilson lost his scientific credibility, for this had a crucial effect on the weight that members of the committee were likely to have placed on the different experimental evidence before them. Such evidence is never clear-cut; it is the interpretation placed upon it that counts. Wilson, as we have seen, had built a high reputation as an experimenter. The fact that his views about pointed conductors were rejected in 1772 shows that, even then, the other experts in the field were not always convinced by the interpretations he put upon experiments. In 1777, these same people—the people who counted in the decision-making process—also lost faith in the experiments he performed.

This disillusionment with Wilson became general knowledge when the *London Evening Post* on September 16, 1777, reported a "great dispute" at the Pantheon between him and one of the leading electrical investigators of the younger generation, Lord Charles Mahon. We are told that Mahon "repeated several experiments of his own to prove his assertions, and by invariably succeeding in them, at the same time that those of Mr. Wilson failed repeatedly, his lordship proved this to demonstration. . . . Mr. Wilson went to the other end of the room, as if to avoid seeing Lord Mahon's experiments. He afterwards said that he had *not changed his opinions*, and would publish his own hypothesis."[47]

Worse was to follow. On October 2, 1777, a group of the Royal Society's experts on electricity attended the Pantheon to observe Wilson's experiments and concluded that they were based on a deception. The news spread like wildfire, as may be seen, for example, from a letter dated October 7, 1777, from that indefatigable Portuguese scientific intelligencer Jean Hyacinthe de Magellan to the secretary of the St. Petersburg Academy of Sciences, Johann Albrecht Euler.[48] In an earlier letter, Magellan, who had been living in London for many years and was a very well-regarded figure on the international scientific scene, had told the St. Petersburg scientists about the dispute over lightning rods.[49] Now he described observing Wilson's experiments at the Pantheon, together with a large crowd of *"gens instruits."* The experiments, he said, were fraudulent, and Wilson, upon whom the St. Petersburg Academy had recently bestowed a medal for his experiments with phosphori, was a charlatan—Magellan repeated the word several times in the course of the letter—whose apparently convincing demonstration that pointed rods drew a discharge from the suspended conductor at a greater distance than did rounded ones depended on an undisclosed breaking of the connection between the model of the board-house and the earth at a crucial moment in the experiment.

Given Magellan's way of working, we may be sure that he was reporting the collective views of the group with whom he observed Wilson's experiments—he mentions Henley and Nairne explicitly as among those present—and that he would have reported the news to his scientific correspondents all over Europe, not just to St. Petersburg. Wilson's imposture having been discovered, he wrote, it was necessary to spread the news everywhere so that the philosophical world should no longer be troubled by such deceits. Others were writing at the same time, notably the Viennese experimental philosopher Jan Ingenhousz, currently visiting London, who wrote to Paris about *"la charlatanerie et la mauvaise foi de ce coquin,"* in terms sufficiently vituperous that Franklin declared them "too angry to be made use of by one philosopher when speaking of another, and on a philosophical question."[50]

Wilson, as noted earlier, had not shirked controversy even at the outset of his career in science. Perhaps pride — understandable in someone who from such humble beginnings had advanced himself so considerably in society — caused him to be unduly short with people on occasion. Perhaps what Franklin's grandson William Temple Franklin imputed to him in relation to the dispute over lightning rods was a larger problem with him: "The makers of experiments are very liable to be deceived by them, and to flatter themselves with a belief that they have made great discoveries, when there is no solid foundation for such belief."[51] Wilson's autobiography certainly suggests that he was rather a prig, prickly of temper, puffed up with self-importance, and inordinately proud of hobnobbing with members of the aristocracy — a man very different in character from Franklin and his easy-going friends who made up the so-called Club of Honest Whigs, and unlikely to have been regarded sympathetically by them.

Whatever the cause, Wilson's behavior undoubtedly cost him friends. In his autobiography, he tells how he earned the enmity of one influential member of the Royal Society, Lord Charles Cavendish, by tactlessly pointing out to him in front of a large audience that an experiment he was demonstrating was not new but had been described by Newton in his *Opticks* sixty years earlier.[52] Even as the argument over lightning conductors built up during the 1770s, Wilson was engaged simultaneously in a bitter dispute with Joseph Priestley and others over various experiments on phosphori. The preface to the second (1776) edition of Wilson's *A Series of Experiments on the Subject of Phosphori*, for example, includes an extraordinary diatribe against Priestley, much of which is clearly based on a misreading of something Priestley had written.[53] Soon afterward the Italian physicist Giambattista Beccaria, who also wrote extensively on phosphori, found reason to accuse Wilson of lack of candor in reporting some of his experiments.[54]

Thus by the mid-1770s, there would have been a number of experimental philosophers, both in London and abroad, who would have been pleased to bring Wilson down a peg or two. Perhaps Magellan, whose outlook on life was much more akin to Franklin's than to Wilson's, and Ingenhousz were among them; this would certainly account for the assiduity with which they spread the news of Wilson's supposed perfidy among their correspondents abroad. Enmity alone does not normally suffice, however, to bring someone down in science. It was the nature of the crime of which Wilson was accused that destroyed his reputation. As the careers of both Franklin and Wilson attest, the republic of letters was open to all, but the progress of science relied heavily on trust. Above all, people depended on their colleagues reporting their experiments carefully, providing all relevant details. This was especially so in cases like Wilson's experiments at the Pantheon, which were on a scale such that others were unlikely to be able to repeat them for themselves. In such a case, even the suspicion of deceit was enough to destroy all credibility. Wilson's standing in science rested heavily, as we have seen, on his reputation as a highly skilled experimenter. Once doubts were cast on his experiments, his standing was gone.

Opinionated and argumentative Wilson may have been, yet I find it hard to believe he would have stooped to deliberately cheating. I think he had too much confidence in the rightness of his cause for that. I suspect that William Temple Franklin's suggestion is closer to the truth. Maintaining an adequate earth connection to the model board-house as it moved under the suspended prime conductor in the experiments at the Pantheon would have been a tricky business, and perhaps Wilson was not careful enough in ensuring that the connection remained good. Confident in the correctness of his reasoning, he would not have been surprised by the results of his experiments and so would not have been prompted to recheck the earth connection. If so, the mistake cost him dearly. His opponents, confident of their own contrary understanding of the science involved, would have felt that there had to be something wrong that would account for his results. And when they discovered what it was, their dislike of Wilson would have made it all too easy for them to think the worst and assume that this was no careless mistake, but that Wilson was deliberately trying to deceive.

Certainly Wilson does not appear to have understood the damage that had been done to his reputation by the events at the Pantheon. Although well aware of Magellan's letter-writing campaign, for example, he merely expressed puzzlement in his own letters as to why Magellan would have attacked him in this way.[55] And he was clearly hurt and astonished by Banks' later insult about his veracity. Deliberately or not, however, Wilson was perceived as having broken the rules, and that was that. While the famous controversy about lightning rods was settled through experiments, it was not the actual content of the experiments but the way in which they were conducted—or rather, were perceived to have been conducted—that decided the matter.

Notes

1. C. R. Weld, *A History of the Royal Society, with Memoirs of the Presidents*, vol. 2 (London, 1848), 92–102; H. G. Lyons, *The Royal Society, 1660–1940: A History of Its*

Administration under Its Charters (Cambridge: Cambridge University Press, 1944), 192–94; John L. Heilbron, *Electricity in the 17th & 18th Centuries: A Study of Early Modern Physics* (Berkeley and Los Angeles: University of California Press, 1979), 380–83; and Trent A. Mitchell, "The Politics of Experiment in the Eighteenth Century: The Pursuit of Audience and the Manipulation of Consensus in the Debate over Lightning Rods," *Eighteenth-Century Studies* 31 (1998): 307–31.

2. See, for example, Richard Anderson, *Lightning Conductors: Their History, Nature, and Mode of Application* (London: Spon, 1880), 41; I. Bernard Cohen, ed., *Benjamin Franklin's Experiments: A New Edition of Franklin's Experiments and Observations on Electricity* (Cambridge, Mass.: Harvard University Press, 1941), 137f; and Basil F. J. Schonland, *The Flight of Thunderbolts*, 2nd ed. (Oxford: Clarendon Press, 1964), 29f.

3. Cohen, *Benjamin Franklin's Experiments* (cit. n. 2), 134ff.

4. Ibid., 220f.

5. Edward Delaval, "An Account of the Effects of Lightning in St. Bride's Church," *Philosophical Transactions* 54 (1764): 227–34; and William Watson, "Observations upon the Effects of Lightning," *Philosophical Transactions* 54 (1764): 201–27.

6. Benjamin Wilson, "A Letter . . . with Some Observations on the Effects of Lightning," *Philosophical Transactions* 54 (1764): 246–53.

7. Wilson's view seems not to have attracted much attention at first, although other electrical investigators would of course have been aware of it. In the first edition of the *Encyclopaedia Britannica*, published in 1771, the article "Electricity" includes a section, "Method of securing buildings and persons from the effects of lightning," that draws heavily on Franklin's recommendations and on the work of his Italian follower Giambattista Beccaria in advocating "long sharp points communicating with the earth." No mention is made of Wilson's opposing ideas on the subject. The article sets out a curious (and, so far as I am aware, novel) theory of the way in which pointed conductors act in preserving buildings. Such conductors, it is said, when presented to the ragged, hanging-down parts of electrified clouds closest to the ground, silently draw the charge from these. Once uncharged, the hanging-down parts are attracted upwards by the main electrified body of the cloud, and this "may leave the distance so great as to be beyond the reach of striking." See *Encyclopaedia Britannica*, vol. 2 (Edinburgh, 1771), 483f.

8. Wilson, in his unpublished autobiography, a typescript copy of which is at the National Portrait Gallery, London, reports (pp. 46f) that he and Franklin had separately been consulted by friends at the Board of Ordnance; that Franklin, without telling him of his own involvement, had urged him unsuccessfully to change his report and recommend pointed rather than blunt conductors; and that it was only after this that the Board referred the matter to the Royal Society.

9. According to Wilson, Delaval was also appointed to the committee but was absent from London at the time it met; Benjamin Wilson, "Observations upon Lightning, and the Method of Securing Buildings from Its Effects," *Philosophical Transactions* 63 (1773): 49–65, at 65. There is no record in the Royal Society's *Journal Book* or council minutes of Delaval's being appointed, but it is unlikely that Wilson would have made such a public claim if it were not true.

10. Charles Frederick on behalf of the Board of Ordnance to members of the committee, n.d., Royal Society Archives, MM.3.72.

11. "A Report of the Committee appointed by the Royal Society," *Philosophical Transactions* 63 (1773): 42–47.

12. Wilson, "Observations upon Lightning" (cit. n. 9), 62.

13. "Mr. Wilson's Dissent to Part of the preceding Report," *Philosophical Transactions* 63 (1773): 48.

14. Wilson, "Observations upon Lightning" (cit. n. 9); and Benjamin Wilson, *Observations upon Lightning, and the Method of Securing Buildings from Its Effects* (London, 1773).

15. Henry Cavendish et al., "A Letter to Sir John Pringle . . . on pointed Conductors," *Philosophical Transactions* 63 (1773): 66.

16. In their essay in this volume, Moore, Aulich, and Rison suggest that, according to the latest research, slightly blunt rods are, if anything, more effective than pointed ones in promoting the onset of a discharge; in other words, contrary to what everyone involved in the eighteenth century dispute believed, an elevated blunt rod will draw a lightning strike from a passing cloud more effectively than will a pointed one. Thus, Wilson was right in holding that a blunt conductor protects a building better but wrong because it does so by doing precisely what he thought should at all costs be avoided!

17. Wilson, "Observations upon Lightning" (cit. n. 9), 63.

18. Ibid., 51.

19. "Report of the Committee," *Philosophical Transactions* 68 (1778): 236–38.

20. Ibid., 239–42.

21. Ibid., 242f.

22. H. Randolph, ed., *Life of General Sir Robert Wilson* (London, 1862), 15. Sir Robert Wilson was Benjamin Wilson's son; his biography includes an extended account of Benjamin Wilson's life that draws heavily on the latter's unpublished autobiography.

23. Mitchell, "The Politics of Experiment" (cit. n.1), 322. Wilson's report of his experiments was published in the *Philosophical Transactions* 68 (1778): 245–312, and separately as *An Account of Experiments made at the Pantheon, on the Nature and Use of Conductors* (London, 1778).

24. Ibid., 56.

25. "Report of the Committee," *Philosophical Transactions* 68 (1778): 313–17, at 317.

26. Royal Society, *Journal Book Copy*, vol. XXIX, minutes of meetings of March 12 and 19, 1778.

27. Public Record Office, London, W.O. 47 92, minutes of the Board of Ordnance, July 1, 1778. See also Weld, *History of the Royal Society*, vol. 2 (cit. n. 1), 98.

28. Edward Nairne, "Experiments on Electricity, being an Attempt to Shew the Advantage of Elevated Pointed Conductors," *Philosophical Transactions* 68 (1778): 823–60; Samuel Musgrave, "Reasons for Dissenting from the Report of the Committee Appointed to Consider Mr. Wilson's Experiments," *Philosophical Transactions* 68 (1778): 801–22.

29. Benjamin Wilson, "New Experiments upon the Leyden Phial, Respecting the Termination of Conductors," *Philosophical Transactions* 68 (1778): 999–1012. Wilson's

paper was also published separately as an appendix to his account of his experiments at the Pantheon (see n. 23).

30. Andrew Kippis, "Life of the Author," in John Pringle, *Six Discourses* (London, 1783), lvi–lvii; Georges Cuvier, "Éloge de M. Banks," *Mémoires de l'Académie des Sciences de l'Institut de France*, 5 (1821–22), *Histoire* 204–30, at 220f; Weld, *History of the Royal Society*, vol. 2 (cit. n. 1), 64, 101f; Heilbron, *Electricity in the 17th & 18th Centuries* (cit. n. 1), 382.

31. Wilson's role behind the scenes is revealed in a letter written to him by Daines Barrington on December 26, 1781; British Library, Add. MS 30094, f. 206.

32. "Proceedings Relative to the Accident by Lightning at Heckingham," *Philosophical Transactions* 72 (1782): 355–78.

33. British Library, Wilson to Banks, Add. MS 30094, f. 208.

34. Ibid., Wilson to Gamble, f. 212.

35. Ibid., Wilson to Banks, f. 210.

36. Weld, *History of the Royal Society*, vol. 2 (cit. n. 1), 100.

37. Mitchell, "The Politics of Experiment" (cit. n. 1), 316, 318, 319, 325.

38. Ibid., 316f.

39. Ibid., 320.

40. Royal Society, Journal Book, entry for December 1, 1760. See also Benjamin Wilson, "Experiments on the Tourmalin," *Philosophical Transactions* 51 (1759): 308–39; and R. W. Home, "Aepinus, the Tourmaline Crystal, and the Theory of Electricity and Magnetism," *Isis* 67 (1976): 21–30.

41. British Library, J. Smeaton to Wilson, September 24, 1746, and February 4, 1746–47, Add. MS 30094, ff. 22, 29–30.

42. Wilson, unpublished autobiography (see note 8).

43. William Henley, "Experiments Concerning the Different Efficacy of Pointed and Blunt Rods, in Securing Buildings against the Stroke of Lightning," *Philosophical Transactions* 64 (1774): 133–52, at 138.

44. C. C. Gillespie, *Science and Polity in France at the End of the Old Regime* (Princeton, N.J.: Princeton University Press, 1980).

45. John Gascoigne, *Science in the Service of Empire: Joseph Banks, the British State and the Uses of Science in the Age of Revolution* (Cambridge: Cambridge University Press, 1998), 23–25.

46. Simon Schaffer, "Natural Philosophy and Public Spectacle in the Eighteenth Century," *History of Science* 21 (1983): 1–41, at 11. See also Simon Schaffer, "The Consuming Flame: Electrical Showmen and Tory Mystics in the World of Goods," in *Consumption and the World of Goods*, ed. John Brewer and Roy Porter (London and New York: Routledge, 1993), 489–526; John Money, "Teaching in the Market-place, or 'Caesar adsum jam forte: Pompey aderat': The Retailing of Knowledge in Provincial England during the Eighteenth Century," in *Consumption and the World of Goods*, ed. John Brewer and Roy Porter (London and New York: Routledge, 1993), 335–77; J. C. C. Rupp, "The New Science and the Public Sphere in the Premodern Era," *Science in Context* 8 (1995): 487–507; Oliver Hochadel, *Öffentliche Wissenschaft: Elektrizität in der deutschen Aufklärung* (Göttingen: Wallstein, 2003); and Oliver Hochadel, "The

Business of Experimental Physics: Instrument Makers and Itinerant Lecturers in the German Enlightenment," *Science and Education* 16 (2007): 525–37.

47. Quoted by William Temple Franklin, ed., *Memoirs of the Life and Writings of Benjamin Franklin, LL.D*, vol. 1 (London: H. Colburn, 1818), 322.

48. Archives Department, St. Petersburg Branch, Russian Academy of Sciences, Magellan to Euler, October 7, 1777, f.1, op. 3, Nr. 63, ll. 80–81.

49. Ibid., Magellan to Euler, June 27, 1777, f.1, op. 3, Nr. 63, ll. 67–69.

50. Both Ingenhousz's letter and Franklin's comment are quoted in Franklin, *Memoirs*, vol. 1 (cit. n. 47), 323f. The editor also gave a detailed account of these stirring events in which he referred to Wilson as a "soi-disant" philosopher whose "pretended improvement, founded on deceptive experiments, was completely destroyed by the discovery and exposure of the tricks he had employed to obtain a partial success." In later editions of this work, following representations from Wilson's son, Gen. Sir Robert Wilson, the accusations were softened and a footnote added, as follows: "Since the publication of the first edition of these memoirs, the editor has been satisfied that there was no evidence to justify a charge of *intentional fraud* against Mr. Wilson. The makers of experiments are very liable to be deceived by them, and to flatter themselves with a belief that they have made great discoveries, when there is no solid foundation for such belief." William Temple Franklin, ed., *Memoirs of the Life and Writings of Benjamin Franklin, LL.D*, vol. 2 (London: H. Colburn, 1833), 79. See also the correspondence of Sir Robert Wilson with W. T. Franklin, British Library, Add. MS. 30094.

51. Franklin, *Memoirs*, 79.

52. Wilson, unpublished autobiography (see note 8), 49.

53. Benjamin Wilson, *A Series of Experiments on the Subject of Phosphori*, 2nd ed. (London, 1776), iv–viii.

54. Giambattista Beccaria, *A Letter to Mr. Wilson, Concerning the Light Exhibited in the Dark by the Bologna Phosphorus* (London, 1776).

55. British Library, Wilson to J. Allamand, April 9, 1779 (draft), Add. MS 30094, f. 194.

Styles of Experimentation and the Attempts to Establish the Lightning Rod in Pre-Revolutionary Paris

Peter Heering

Introduction

THE AIM OF THIS ESSAY is to use the example of lightning rods to discuss different experimental approaches used to establish a consensus about the utility and appropriate design of this device. In doing so my major concern will be to discuss what was considered adequate to establish "facts" with respect to this topic. The discussion is limited to Paris and to the period before the outbreak of the French Revolution. In this essay, I shall first describe an initial period that was in particular dominated by the authority of Jean-Antoine Nollet. His rejection of the lightning rod made it necessary for subsequent researchers to establish the utility as well as the safety of this device.

In the second part of this essay, I shall discuss these attempts with particular focus on researchers who were not members of the Paris *Académie des Sciences* and thus could not use the reputation of an academician. In doing so, I will, in particular, analyze their approaches to demonstrate that lightning rods were safe and reliable. In the third part, I will discuss the approaches within the Paris Academy, namely of Jean Baptiste Le Roy, the successor of Nollet as the leading French researcher in the field of electricity. I will argue that the strategies employed by Le Roy were very similar to the ones used by other researchers. In the final section, I will discuss the electrical research of Charles Augustin Coulomb, who was to become the new French authority in the field of electricity. I shall argue that Coulomb's electrical research was another attempt to demonstrate the utility of the lightning rod. At the same time, the analysis of his

I am indebted to Christine Blondel and John Heilbron, who made valuable comments on earlier versions of this essay. I would also like to thank Falk Rieß for many critical questions and thoughtful comments.

experimental practice reveals that the experiments were carried out in an entirely different style of experimentation.[1] This aspect can be taken as a development in the scientific practice that meant a break with standard procedures and that can, at least in part, be explained by ascribing a political meaning to this new style of experimentation.

Setting the Stage

The Marly experiment of 1752, which can be taken as being the first relevant demonstration of the principle of lightning rods, was carried out in France along with other experiments such as Romas' "electrical kite" performed there at about the same time (see fig. 6.1). One might expect that numerous lightning rods would be erected in a place where these experiments were carried out. However, this was not the case.

In the *Histoire de l'Académie Royale des Sciences* for the year 1773, complaints can be found that in this respect France had significantly fallen behind America, Great Britain, Italy, and Austria.[2] This is of course not only a description of the actual situation but it is used as an argument for the erection of lightning rods in France.[3] The most important person to object to the erection of lightning rods was Nollet, who was characterized in 1781 by Aimé-Henry Paulian as the person "the electrical researchers should look upon as their Chef."[4] In a report Nollet published in 1764, he "declared that the rods placed outside of a building are useless and dangerous."[5] Nollet accepted that lightning was considered an electrical phenomenon, but he questioned whether lightning rods could serve as protection because their dimensions were too small to conduct the enormous amount of electrical fluid from such a stroke.[6] Nevertheless, it was not only Nollet who had strong objections to the introduction of lightning rods. In 1766 the Abbé Poncelet suggested that the erection of lightning rods should be prohibited by the police.[7] In this context, it has to be remarked that the term "lightning rod" could have a dual meaning: either a device that was grounded and would serve as protection of the edifice, or a means to get atmospheric electricity into a cabinet to perform experiments (see fig. 6.2).[8]

Toward the end of Nollet's life, his influence at the *Académie des Sciences* evaporated, and in 1772, two years after Nollet's death, Benjamin Franklin was elected "*associé étranger*."[9] Another five years later, Guyton de Morveau argued in his article "Tonnerre" in a supplement to the *Dictionnaire* that lightning was an electrical phenomenon and that "one can no longer ignore the fact that the electrical fluid searches metals in preference to all other substances, and when it reaches these metals, it follows the direction offered by them. If the metals lead

courege fecit

Figure 6.1. Demonstration of the electrical nature of lightning. Romas claimed that he had performed the electrical kite experiment earlier than Franklin; however, he also acknowledged that Franklin was first with respect to publication. Plate from Jacques de Romas, *Mémoire, sur les moyens de se garantir de la foudre dans les maisons; suivi d'une lettre sur l'invention du cerf-volant electrique, avec les pièces justificatives de cette même lettre* (Bordeaux, 1776).

Figure 6.2.   Experiments with an ungrounded lightning rod. Frontispiece from Poncelet, *Nature*.

this fluid, which is terrible when concentrated, into water or moist ground, it disperses peacefully and regains its equilibrium."[10]

One might expect that at least from then on lightning rods were erected in large numbers, but this was not the case. In 1781, eleven lightning rods were found in France, none of them in Paris; the first public building protected with a lightning rod was the Paris Louvre in 1782.[11] In this period lightning rods were erected more or less on an individual basis.[12]

Establishing the Utility of Lightning Rods

One of the case studies that has been analyzed in detail is the lightning rod erected by Charles Dominique de Vissery de Blois-Valé at St. Omer and the juridical controversies that followed. Another indication of the difficulties establishing lightning rods is that the first trial in 1780 ended with the ruling that the lightning rod had to be removed. Although I shall not recapitulate Riskin's study, which is very much focused on the relation between jurisprudence and natural philosophy, she introduced one of the people to be looked upon—at least from the sheer number of publications on the subject—as "a leading advocate and designer of lightning rods":[13] the Abbé Pierre Bertholon. Having a medical background, he became one of the central figures at the Montpellier Academy and published several monographs and articles on medical and atmospheric electricity, promoting lightning rods in the latter.

Bertholon employed two strategies to demonstrate the utility of this device. First he demonstrated that atmospheric electricity could not be distinguished from frictional electricity. For this purpose Bertholon charged a conductor with electricity from an electrical kite and performed experiments such as the ignition of spirit, Franklin's bells, the killing of small animals, and the melting of small wires.[14] Having established this identity, Bertholon used this knowledge in the field of natural philosophy to make the utility of lightning rods plausible.[15]

But this is only one aspect of his strategy; more important for him seemed to be another way of demonstrating the utility of lightning rods: He cited numerous cases where lightning struck buildings that were either damaged or protected by lightning rods.[16] These very detailed accounts are obviously intended to place the reader in the position of a virtual witness, a strategy that was rather common in the second half of the eighteenth century.[17] Consequently, one can find numerous attempts to promote lightning rods in this manner. However, the number of descriptions does not mean that there were as many cases; several incidents were described by various authors.

The attempt to demonstrate the utility, efficiency, and security of lightning rods is only one aspect in Bertholon's writings. As this became more and more accepted, he focused on two other aspects that he discussed in several publications with similar strategies: the protection from lightning strokes going from the earth to the clouds and the protection against other phenomena such as hailstorms and earthquakes. Both issues were discussed controversially by his contemporaries and make clear in retrospect that the situation was not as simple as one might suppose.

From his writings it is obvious that Bertholon was (like Franklin) arguing based on a one fluid theory, from which concept resulted the question of the direction in which lightning strikes. Initially the theory proposed that a stroke of lightning is a discharge from a cloud to the earth; however, Bertholon cited many cases where the observation was different.[18] The direction of the lightning is determined by the ramification that is considered to be at its end. For these strokes of lightning, a different design of lightning rod seemed to be necessary: "It would indicate the highest degree of nonsense if the edifices are to be protected against the ascending lightning with common lightning rods."[19] Bertholon proposed a different construction: "Since the direction of the stroke of lightning is upwards, it is necessary to place the points in this direction; they will then tap the electric fluid or the matter of the lightning; and the opposite end will discharge it in silence in the atmospheric air in the form of sparks."[20] This was, however, not the final design that can be seen as a combination of both types of lightning rod.

For Bertholon, the protection of the edifices against lightning was not the only positive effect of lightning rods. Because they were supposed to dissipate a stroke of lightning and to discharge electrified clouds (or the earth) silently, they would also have other consequences. According to Bertholon, hail was an electrical phenomenon, as were earthquakes.[21] His position is similar to that formulated by Charles Richardon: "consequently, since it is proven that the lightning conductors, either while neutralizing, or by tapping the electric matter of storm clouds, prevent the explosion of the lightning, it is also proven that they prevent the formation of hail."[22]

Although Bertholon was not the only researcher to link atmospheric electricity and the formation of grains of hail, other researchers such as Monge expressed their doubts: "The increase electricity produces in the evaporation of water, and the cooling which results from this are not that considerable, thus, if they contribute sometimes, which is possible, to the production of the phenomenon [of hail], they can neither be looked upon as being the principal cause nor as the necessary one."[23]

Besides the strategies already mentioned, other demonstrations were employed. The best known was of course the thunder-house,[24] which was also promoted by Bertholon; however, a much earlier version had been published in the *Journal de Physique* in 1773. Various versions of this device were commonly used; a similar although slightly different device was propagated by Boyer-Brun in 1786.[25] He combined a Leyden jar with a lightning rod and a spark gap; in the latter, a powder cartouche was inserted which was to detonate when the electricity in the air was sufficiently high.[26]

These devices were used to demonstrate the ability of lightning rods to protect edifices. Demonstrations were very popular in the late eighteenth century. This can even be seen from Paris guidebooks of this period; Thiéry mentions in his two-volume edition of 1787 some nine institutions where physical or electrical experiments and courses were offered.[27] One of these courses was offered by Rouland, nephew and successor of Sigaud de la Fond, who continued his famous courses. From Sigaud's publications it is clear that he promoted lightning rods. His strategies seem to be similar to those of Bertholon: apart from mentioning several incidents of strokes of lightning, Sigaud discussed a thunder-house as well as experiments that were supposed to demonstrate the identity of atmospheric and frictional electricity: "In general, there is no electric phenomenon which one cannot imitate perfectly when one makes a communication between the cord of an electrical kite, or an iron bar elevated high above a house, and an insulated conductor and one can use this in the same way one uses the ordinary conductor of an electric machine."[28]

Although practitioners as well as public lecturers played an important role in shaping public opinion, it was the Paris *Académie des Sciences* that was the ultimate scientific authority. After Nollet's death, Jean Baptiste Le Roy, one of his pupils, was in the position to become his successor. However, Le Roy, who was known to be a Franklinist, did not perpetuate Nollet's positions. Thus, it is not surprising that Le Roy was also promoting lightning rods. His first publication on the utility of lightning rods was read at the Paris Academy in 1770, and started with the following statement: "The history of physics offers us so many examples of significant discoveries which have been disputed over a long period of time. This disease of the intellect, which leads us to reject all that is new or contrary to our ideas, shows that we cannot be too cautious about this hurdle."[29]

The construction of the arguments in Le Roy's paper shows several similarities with the approaches discussed above. First, he discussed several electrical phenomena he considered relevant. Second, he described some observations on strokes of lightning. Third, he showed that "the consequences of the electric phenomena and of the effects of thunder tend to establish in the most solid manner the utility of the metal bars in protecting the edifices from them."[30] Finally, Le Roy described the best way of designing lightning rods.

Three years later, Le Roy read his next memoir on lightning rods, this time at a public meeting of the Academy.[31] Whereas Le Roy, in his first paper, was mainly concerned with demonstrating the utility of lightning rods, he now discusses their appropriate design that is to be seen in the context of the discussions that took place at the Royal Society in connection with the Purfleet controversy.[32] Consequently, Le Roy analyzes whether a point or a knob at the end of a conductor is more adequate. In doing so, he presents a needle with a point and another terminating in a knob to a charged conductor. In the case of the point, he observes a luminous point at a distance of some three feet and a spark at a very small distance, while in the case of the knob, he observes only a spark (see fig. 6.3). From these and some other experiments Le Roy "proves in a crucial way the advantage of rods terminating in points."[33]

To summarize the discussion so far, the provincial researchers and the public demonstrators in Paris as well as the Paris academicians discussed lightning rods

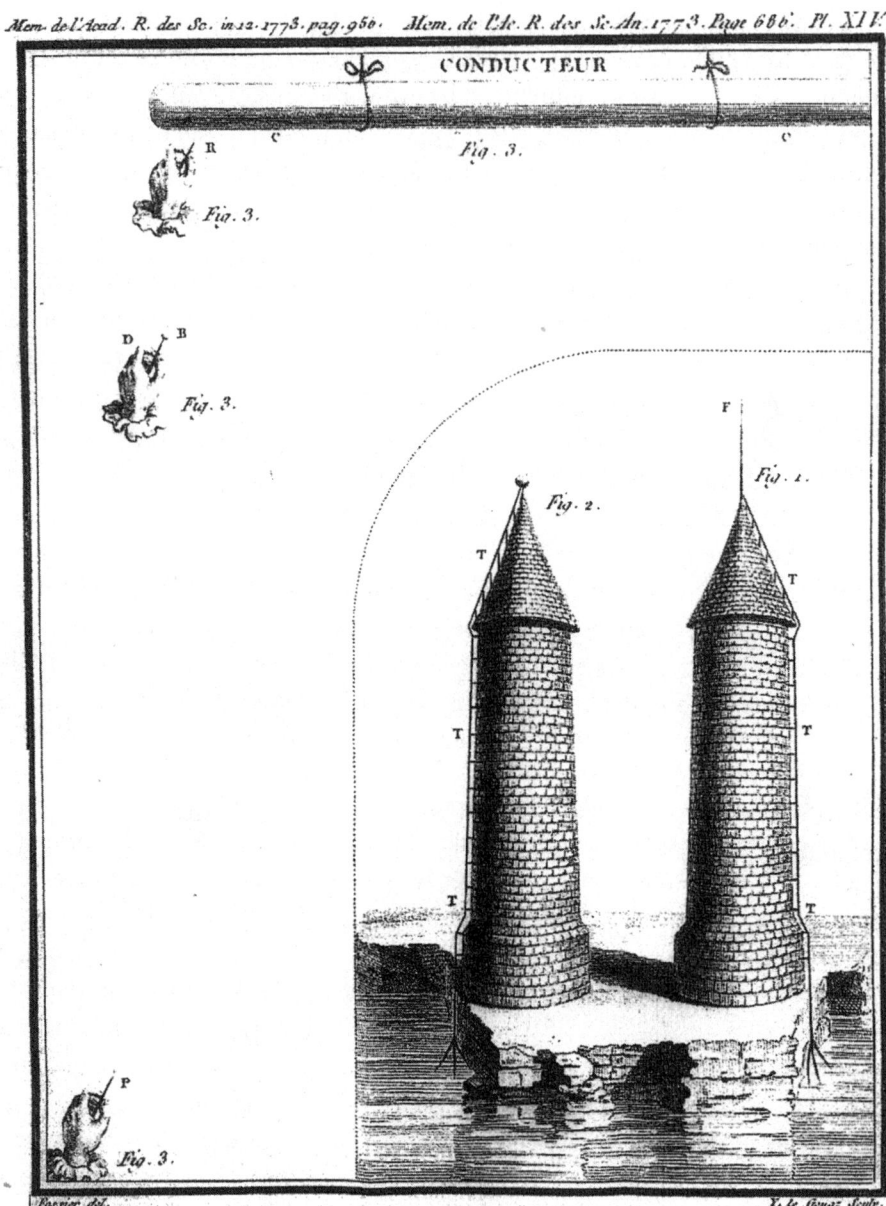

Figure 6.3.   Le Roy's experiments to compare points and knobs. Plate from Le Roy, "Forme," with permission of the Landesbibliothek Oldenburg.

and their design on a similar basis using similar strategies. At first, the identity between atmospheric and frictional electricity was demonstrated. This made it possible to use some of the established experiments in the latter field to make obvious the necessary properties of lightning rods. Moreover, some newly developed experiments were also carried out to demonstrate the utility as well as the adequate design of lightning rods; these can be labeled as imitative experiments.[34] Both kinds of experiments were designed to be performed in front of an audience, which should be convinced by eye-witnessing the outcome of the experiment. Incidents of lightning strokes were analyzed and—like those where a lightning rod was struck—used as the basis for arguing in favor of the erection of lightning rods. In these cases, the description of the incident was made in a way that placed the reader in the position of witness.

Apart from these similarities among the experiments, another aspect is worth mentioning. The design of a lightning rod was determined in a manner that was common in engineering, a characterization introduced much later by Arago: "'If,' says he, 'we take the dimensions to be given to conductors from experience, and if those which we adopt have been found to resist the strongest lightning recorded for over a century, what more can reasonably be asked for?"[35]

Analyzing the Lightning Rod with a New Style of Experimentation

A completely different approach to analyzing lightning rods began with a more or less standard procedure. In 1784 the French minister of war, Philippe Henri de Ségur, asked the *Académie des Sciences* to prepare a report on the possibilities of protecting the planned powder magazine in Marseille against the dangers of lightning. Ségur's demand was taken by the academy as a "subject of such importance, not only because of the nature of the object, but also for the fortunate effect which will result."[36] As customary, the academy formed a committee, and it is useful to take a closer look at its constitution. One of the members was—not surprisingly—the "father of the lightning rod," Benjamin Franklin, who had lived in Paris since 1777. Likewise, it seems to be self-evident that Jean Baptiste Le Roy became one of the members of this committee, not only because he was considered the leading French electrician but also because he had acted on comparable committees.[37] Alexis-Marie de Rochon had not published anything in the field of electricity; however, he was, together with Le Roy, responsible for the Cabinet du Roi.[38] In a similarly indirect way, the choice of the fourth member of this committee can be explained. Even though Pierre-Simon de Laplace had carried out some research on electricity together with Lavoisier, he may have become a member of this committee due to his status at the academy. The nomination of the fifth and final member can be explained in a more direct manner. Charles Augustin Coulomb was a member of the academy who was through his training very familiar with the design of military buildings and fortifications. Moreover, Coulomb was transferred to Paris by the French military to enable him to make the latest scientific achievements available to the Corps du Génie. Thus, although Coulomb himself had not made any contributions to the field of electricity up to then, he was a very plausible choice as a member of the committee due to his military background.

On April 24, 1784, the committee presented its report. First, the theoretical basis of lightning rods was described: "It is necessary to know or to determine its [the lightning rod's] extension, in order to decide if one or several should be erected, which is a point important to establish before attempting anything else. Unfortunately, electrical experiences have not yet taught us anything that can guide us to know the extension of the sphere of action of the point of a conductor."[39]

Lacking "proper scientific knowledge" the commission had to rely on observations: "but since we have been arming the edifices [with lightning rods], several observations have taught us that the parts of the edifices which are found at a distance of more than 45 feet from the point of the conductor are thunderstruck, from which it follows that they should be placed in a manner in which their sphere of action does not have to defend other than the parts situated at a smaller distance."[40]

The uncertainty of the committee was not limited to the lightning rod's sphere of action; they were also lacking general rules for the height and the diameter of the conductor. Thus, because of these incertitudes, the only criterion they could propose for the dimensions of the rods was that "this thickness should be more than sufficient so that it shall transmit the fulminant matter of the most violent thunder clap."[41]

Having demonstrated that the general scientific knowledge on the proper design of lightning rods was meager and insufficient, the commission turned to give their advice for the actual project.

Some three weeks later Ségur addressed the academy once more in communicating some changes to the original plans that were proposed by Ravel and Pierron.[42] Three of these were considered by the committee as being of little importance because they were in accordance with the advice given in the first report.[43] The only change they considered problematic was the idea of raising the bars of transmission. One of the reasons for their objection was the fear that "it could happen, by some unknown cause, that an electrical spark could spring from the conductor."[44] This statement shows again that, although the members of the committee agreed on the utility of lightning rods, some uncertainties remained.

From my point of view, these difficulties in giving proper advice on a problem in the military context seem to be the starting point of Coulomb's electrostatic research.[45] Moreover, this interpretation could also explain why Coulomb turned so quickly to the field of electricity.[46] When Coulomb's six memoirs on electricity are read in the context of the discussions at the academy, a research program to determine the best way of protecting powder magazines from lightning can be seen: In the first and second memoir, Coulomb introduced his prin-

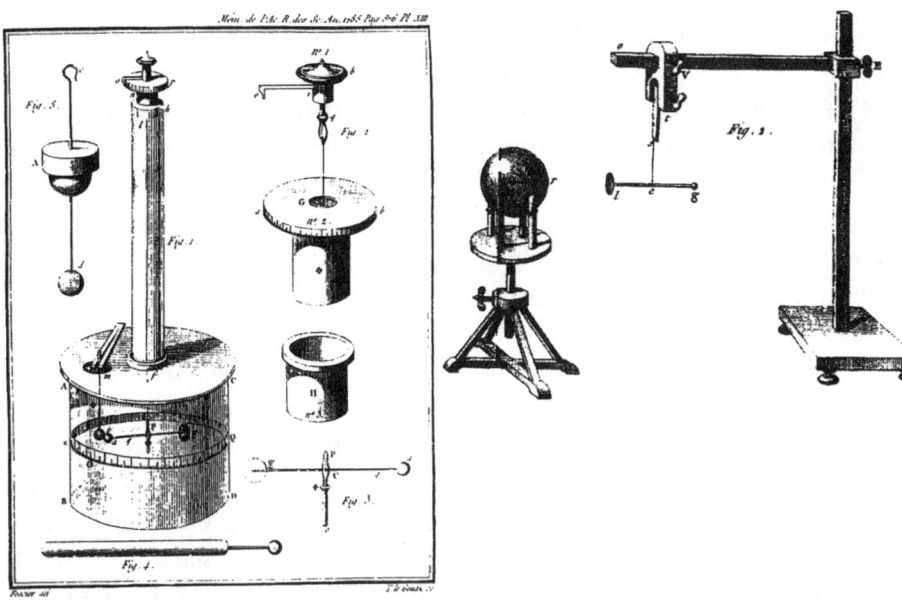

Figure 6.4.   Coulomb's torsion balance and his electrical pendulum. Plates from Coulomb, "1st EM" and "2nd EM," with permission of the Landesbibliothek Oldenburg.

cipal instruments: the torsion balance and the electrical pendulum, a device analogous to the gravitational pendulum where the time of oscillations is taken as a measure of the force (see fig. 6.4).[47] Both devices and the related measuring procedures were introduced by demonstrating their ability to be used as tools to prove the inverse square law in the case of electrostatic repulsion and attraction. In doing so Coulomb showed their potential as well as their reliability, stressing in the presentation also their sensitivity and "precision."[48]

In his third memoir, Coulomb analyzed charge leakage and its time dependence. This memoir forms a basis of the following three where the effects of charge leakage now only had to be calculated. Coulomb formulated his motivation for these investigations explicitly: "in the following, the knowledge of this law [of charge leakage] is absolutely necessary in order to be able to submit the other phenomena of electricity to calculus. As the examinations planned to investigate these phenomena cannot be executed in an instant, they are not comparable with each other without knowing the alteration which they experience during the time which is elapsing from one to the other."[49]

In the first part of the fourth memoir, Coulomb determined "that this [the electrical] fluid does not spread itself in any body by a chemical affinity or by a selective attraction but that it distributes between different bodies placed in contact merely by its repulsive action."[50] In the second part of this memoir, Coulomb established that the charge of any body is only on its surface. Again this result was important for the following experiments because Coulomb was now able to determine the "electrical density in every point of a conducting body."[51]

This was done in the next memoir when Coulomb first determined the charge distribution between two spheres with different diameters. He then described experiments for the determination of the "density of the electrical fluid on different points of the two spheres being in contact."[52] He measured the charge by taking it off at different points of the surface. Subsequently, Coulomb examined systems of three spheres in a row, where the sphere in the middle was first smaller than the two outer ones and afterward of the same size. In the second part, Coulomb described experiments where "we have determined exactly by examination the way in which the electrical fluid distributes over surfaces, whether they are plane or curved, and over bodies of different shapes."[53] Again, for the evaluation of these experiments, the preceding proof of the charge being entirely on the surface was necessary. Coulomb could now assume that "one can suppose each [charged] sphere covered by an infinity of small conducting, electrically charged spheres."[54] Obviously the determination of the charge distribution over the surfaces of bodies was Coulomb's crucial concern. This was also the focus of the sixth memoir, which Coulomb named as a continuation of the fifth. In this final memoir Coulomb intended to investigate the following aspects:

1. How electricity is distributed between any given number of similar spheres placed in contact so that all their centres are in a straight line;
2. How the electrical fluid is distributed over the different parts of an electrified cylinder;
3. How it is distributed between a large sphere and a row of small spheres in contact with the large sphere;
4. In which ratio the electrical fluid spreads between a large sphere and some cylinders of different diameters and different lengths which are successively placed in contact with the sphere.[55]

From the formulation of these aspects in particular, as well as from the plate published together with Coulomb's memoir (see fig. 6.5), it becomes obvious that the spheres can be taken as being the first step toward determining the charge distribution on cylinders. Coulomb described experiments where he investigated the charge distribution of up to twenty-four spheres in a row.[56]

In concluding his sixth memoir, Coulomb explicitly stated that this was also the end of his electrical research: "The calculations of this article, although very imperfect, are almost sufficient for every practical purpose in relation to electricity, where it might be necessary to employ calculus."[57] This is evidence that Coulomb's research was motivated by aspects of applicability.

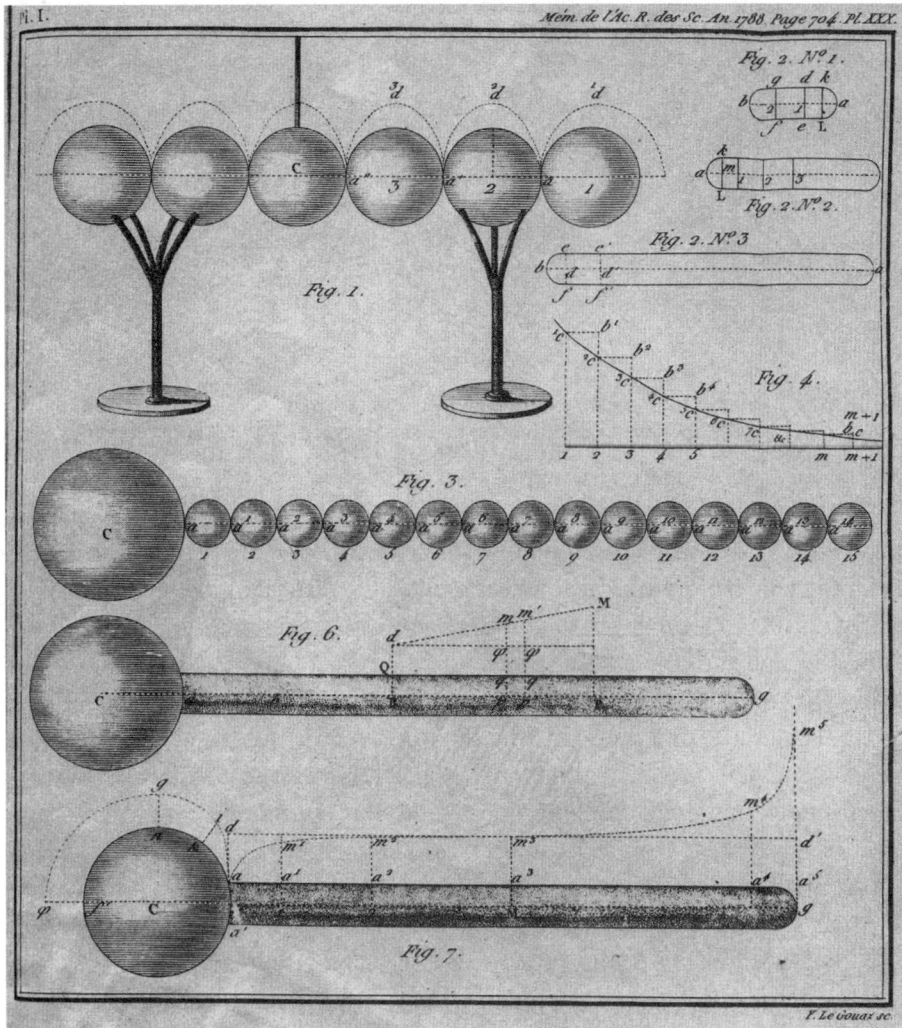

Figure 6.5.   Plate showing experiments with spheres and a cylinder. C. A. Coulomb, "6th EM," with permission of the Landesbibliothek Oldenburg.

The background of Coulomb's research program can also be seen from the application of his results, which he discussed at the end of his paper. First he described the "application of the preceding formula to an example analogous to lightning rods."[58] In the following he discussed the "application of the preceding result to the effect of lightning rods."[59] In this context Coulomb also discussed an experiment in which a charged sphere was moving toward a grounded cylinder (see fig. 6.6).

In this experiment, the sphere corresponded to the thundercloud, the cylinder to the lightning rod. These are—apart from the "electrical kite"[60]—the only

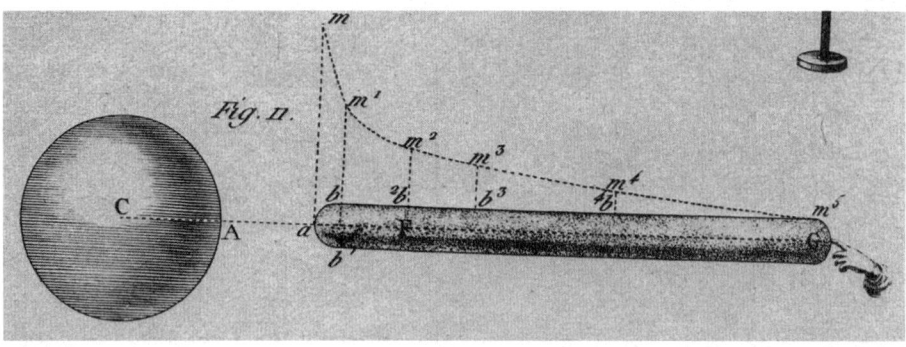

Figure 6.6.   Coulomb's experiment to imitate the thundercloud approaching a lightning rod. C. A. Coulomb, "6th EM," with permission of the Landesbibliothek Oldenburg.

applications Coulomb presented in the whole series of his electrical research. From the context formulated above, it is plausible that Coulomb applied, at the end of his research program, the results to the original problem.

In the context of the investigations into the protection of powder magazines against lightning, Coulomb's research program can be summarized as follows: In the first and second memoirs, the measuring devices were introduced. The next two memoirs contain measurements that served as a basis for the following experiments. In the two final memoirs—which is essentially one memoir published in two parts—Coulomb finally was able to come to a solution for the initial problem in giving a mathematical description of a lightning rod.

At this point of the discussion, it could be questioned whether this analysis of Coulomb's research warrants closer examination, particularly because neither Coulomb nor any of his contemporaries or historians of science placed his electrical research within this context. There are at least two remarks that might be useful additions in this respect. During his time at the Académie Royale des Sciences, Coulomb was still an officer of the Corps du Génie; moreover, when he applied for a transfer to Paris to become a member of the academy, he explicitly stated that this would enable the military to benefit from the latest scientific achievements.[61] If this is taken to be not only a rhetorical statement but also a serious one, it is plausible that Coulomb focused his research on questions relevant for the Corps du Génie.

Moreover, another work on lightning rods prepared in England had a comparable background. In 1779 Charles Stanhope published his *Principles of Electricity* in which he also introduced the inverse square law and—earning him more fame—the returning stroke. Like Coulomb, Stanhope was an officer in the artillery and worked in a military context. In the German translation of his monograph, the relation between Stanhope's military position and his electrostatic research was pointed out explicitly: "The protection of powder magazines against a lightning stroke is a problem of the highest importance and the problem of laying out good powder magazines is one of the most excellent in the sphere of the knowledge in artillery."[62] However, it is only the translator—J. F. Seeger—who made the military relevance explicit; neither Stanhope nor his contemporaries considered this to be necessary, and even Seeger considered it to be almost self-evident.[63]

It is obvious that Coulomb's style of experimentation, as well as his way of analyzing the problem of protecting a powder magazine against lightning, constitutes a new quality. Contrary to the approaches described in the first part of this essay, Coulomb neither argued with observations nor used demonstrations or imitative experiments. The differences become even more obvious when the experiences encountered during the replication of the experiments are taken into consideration.[64] It became obvious that the apparatus were designed to be very sensitive.[65] The experiments are therefore accident sensitive and require the experimenter to move slowly in the room and—particularly when working with the electrical pendulum—even to breathe very slowly and controlled. Consequently, these experiments cannot be performed in front of an audience. Moreover, the experiments are quantitative and are labeled as being precise. The results of the experiments are to be communicated in the form of mathematical relations (see fig. 6.7).

These characteristics have at least one aspect in common: it is no longer possible to be an eyewitness of the experimental results. On the contrary, the public is not only excluded from the experimental space because of the sensitivity of the apparatus; it is also excluded from becoming a virtual witness by the form of presentation of the results. The mathematical formulation of the experimental result limits the audience once more since specific training is necessary to understand the outcome of the experiments.

Because Coulomb's experiments show several aspects of a newly developed style of experimentation—this is found not only in Coulomb's work; other studies by members of the Paris Academy correspond to this style of experimentation—a reason for this development may be required. In conclusion, I would like to offer two explanations, both of which are not to be taken as being monocausal: As a graduate of Mézières—a military school that was well known for its elaborate mathematical instruction—Coulomb had specific training that enabled him to design experiments requiring the mathematical knowledge necessary to analyze the data produced. At the same time, this made it possible for Coulomb to overcome the traditional engineering-like approach and to attempt the development of an adequate mathematical description.

However, as I have argued elsewhere,[66] the development of this new style of experimentation cannot be explained simply by the different training that

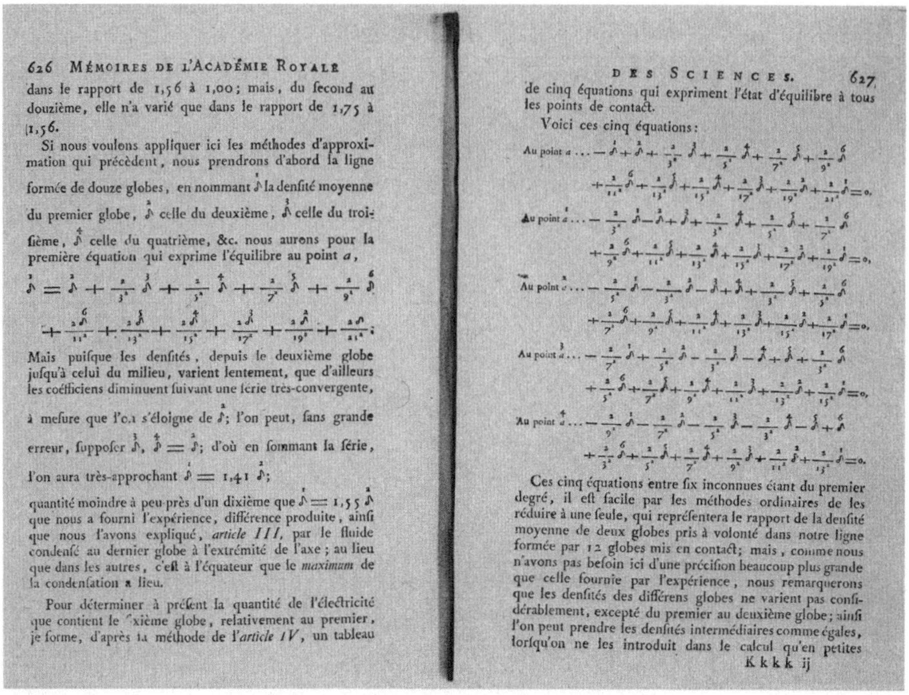

626 MÉMOIRES DE L'ACADÉMIE ROYALE

dans le rapport de 1,56 à 1,00; mais, du fecond au douzième, elle n'a varié que dans le rapport de 1,75 à 1,56.

Si nous voulons appliquer ici les méthodes d'approximation qui précèdent, nous prendrons d'abord la ligne formée de douze globes, en nommant $\overset{1}{\delta}$ la denfité moyenne du premier globe, $\overset{2}{\delta}$ celle du deuxième, $\overset{3}{\delta}$ celle du troifième, $\overset{4}{\delta}$ celle du quatrième, &c. nous aurons pour la première équation qui exprime l'équilibre au point *a*,

$$\overset{1}{\delta} = \overset{2}{\delta} + \frac{2}{3^2}\overset{3}{\delta} + \frac{2}{5^2}\overset{4}{\delta} + \frac{2}{7^2}\overset{5}{\delta} + \frac{2}{9^2}\overset{6}{\delta} + \frac{2\overset{6}{\delta}}{11^2} + \frac{2\overset{5}{\delta}}{13^2} + \frac{2\overset{4}{\delta}}{15^2} + \frac{2\overset{3}{\delta}}{17^2} + \frac{2\overset{2}{\delta}}{19^2} + \frac{2\delta}{21^2};$$

Mais puifque les denfités, depuis le deuxième globe jufqu'à celui du milieu, varient lentement, que d'ailleurs les coëfficiens diminuent fuivant une férie très-convergente, à mefure que l'on s'éloigne de $\overset{2}{\delta}$; l'on peut, fans grande erreur, fuppofer $\overset{3}{\delta}$, $\overset{4}{\delta} = \overset{2}{\delta}$; d'où en fommant la férie, l'on aura très-approchant $\overset{1}{\delta} = 1{,}41\,\overset{2}{\delta}$; quantité moindre à peu-près d'un dixième que $\overset{1}{\delta} = 1{,}55\,\overset{2}{\delta}$ que nous a fourni l'expérience, différence produite, ainfi que nous l'avons expliqué, *article III*, par le fluide condenfé au dernier globe à l'extrémité de l'axe; au lieu que dans les autres, c'eft à l'équateur que le *maximum* de la condenfation a lieu.

Pour déterminer à préfent la quantité de l'électricité que contient le fixième globe, relativement au premier, je forme, d'après la méthode de l'*article IV*, un tableau

DES SCIENCES. 627

de cinq équations qui expriment l'état d'équilibre à tous les points de contact.

Voici ces cinq équations:

Au point a ... $-\overset{1}{\delta} + \overset{2}{\delta} + \frac{2}{3^2}\overset{3}{\delta} + \frac{2}{5^2}\overset{4}{\delta} + \frac{2}{7^2}\overset{5}{\delta} + \frac{2}{9^2}\overset{6}{\delta} + \frac{2}{11^2}\overset{6}{\delta} + \frac{2}{13^2}\overset{5}{\delta} + \frac{2}{15^2}\overset{4}{\delta} + \frac{2}{17^2}\overset{3}{\delta} + \frac{2}{19^2}\overset{2}{\delta} + \frac{2}{21^2}\overset{1}{\delta} = 0,$

Au point $\overset{1}{a}$ ... $-\frac{2}{3^2}\overset{1}{\delta} - \overset{2}{\delta} + \overset{3}{\delta} + \frac{2}{3^2}\overset{4}{\delta} + \frac{2}{5^2}\overset{5}{\delta} + \frac{2}{7^2}\overset{6}{\delta} + \frac{2}{9^2}\overset{6}{\delta} + \frac{2}{11^2}\overset{5}{\delta} + \frac{2}{13^2}\overset{4}{\delta} + \frac{2}{15^2}\overset{3}{\delta} + \frac{2}{17^2}\overset{2}{\delta} + \frac{2}{19^2}\overset{1}{\delta} = 0,$

Au point $\overset{2}{a}$ ... $-\frac{2}{5^2}\overset{1}{\delta} - \frac{2}{3^2}\overset{2}{\delta} - \overset{3}{\delta} + \overset{4}{\delta} + \frac{2}{3^2}\overset{5}{\delta} + \frac{2}{5^2}\overset{6}{\delta} + \frac{2}{7^2}\overset{6}{\delta} + \frac{2}{9^2}\overset{5}{\delta} + \frac{2}{11^2}\overset{4}{\delta} + \frac{2}{13^2}\overset{3}{\delta} + \frac{2}{15^2}\overset{2}{\delta} + \frac{2}{17^2}\overset{1}{\delta} = 0,$

Au point $\overset{3}{a}$ ... $-\frac{2}{7^2}\overset{1}{\delta} + \frac{2}{5^2}\overset{2}{\delta} - \frac{2}{3^2}\overset{3}{\delta} - \overset{4}{\delta} + \overset{5}{\delta} + \frac{2}{3^2}\overset{6}{\delta} + \frac{2}{5^2}\overset{6}{\delta} + \frac{2}{7^2}\overset{5}{\delta} + \frac{2}{9^2}\overset{4}{\delta} + \frac{2}{11^2}\overset{3}{\delta} + \frac{2}{13^2}\overset{2}{\delta} + \frac{2}{15^2}\overset{1}{\delta} = 0,$

Au point $\overset{4}{a}$ ... $-\frac{2}{9^2}\overset{1}{\delta} - \frac{2}{7^2}\overset{2}{\delta} - \frac{2}{5^2}\overset{3}{\delta} - \frac{2}{3^2}\overset{4}{\delta} - \overset{5}{\delta} + \overset{6}{\delta} + \frac{2}{3^2}\overset{6}{\delta} + \frac{2}{5^2}\overset{5}{\delta} + \frac{2}{7^2}\overset{4}{\delta} + \frac{2}{9^2}\overset{3}{\delta} + \frac{2}{11^2}\overset{2}{\delta} + \frac{2}{13^2}\overset{1}{\delta} = 0.$

Ces cinq équations entre fix inconnues étant du premier degré, il eft facile par les méthodes ordinaires de les réduire à une feule, qui repréfentera le rapport de la denfité moyenne de deux globes pris à volonté dans notre ligne formée par 12 globes mis en contact; mais, comme nous n'avons pas befoin ici d'une précifion beaucoup plus grande que celle fournie par l'expérience, nous remarquerons que les denfités des différens globes ne varient pas confidérablement, excepté du premier au deuxième globe; ainfi l'on peut prendre les denfités intermédiaires comme égales, lorfqu'on ne les introduit dans le calcul qu'en petites

Kkkk ij

Figure 6.7. Some of Coulomb's results in his sixth memoir. With permission of the Landesbibliothek Oldenburg.

Coulomb and others received at Mézières.[67] Nevertheless, another aspect that can be taken as relevant is politics. This new style of experimentation was developed by scientists who were Royalists on the eve of the French Revolution. Moreover, the characteristics of this style of experimentation (particularly if compared with those of the style of experimentation that was established during the Enlightenment) show a remarkable correspondence with the political ideas held by Royalists. When these characteristics are summarized, one of the key aspects is that the outcome of the result was not to be understood by everyone; instead the audience was limited to a few experts who had the ability to understand the experimental findings. Everyone else had to accept the conclusion of these experts; subsequently, the concept of public demonstration was given up or even rejected in favor of a more "professional" but also exclusionary approach.[68] Retrospectively, this style of experimentation can be seen as an advancement compared with the experimental practice established during the Enlightenment, but it also meant a significant break with some of the ideas connected with the Enlightenment.

From this point of view, Coulomb's style of experimentation can be interpreted as a reaction to the obvious social tensions, which tried to show that an authoritarian system of knowledge production could contribute to the development of scientific knowledge. In this respect, Coulomb's work can be seen as a counterdraft to the attempts to establish the utility and safety of lightning rods as described in the first part of this essay. However, as can be inferred from the discussions that took place in nineteenth century France, none of these attempts resulted in a general and complete acceptance or understanding of the operation of the lightning rod.[69]

Notes

1. Style of experimentation is used as an expansion of the conception developed by Ludwig Fleck; see Ludwik Fleck, *Genesis and Development of a Scientific Fact*, ed. Thaddeus J. Trenn and Robert K. Merton (Chicago: University of Chicago Press, 1979).

2. Anonymous, "Sur l'usage des Barres métalliques, pour préserver de la foudre," *Histoire de l'Académie Royale des Sciences pour l'Année 1773* (Paris, 1777): 3–6, at 5.

3. A similar argument can be found in other works that were intended to promote technical lightning protection; see, for example, Pierre Bertholon, *Mémoire sur un nouveau moyen de se préserver de la Foudre* (Montpellier, 1777); Pouget, "Reflexions sur les conducteurs électriques, destinés à préserver les Vaisseaux de la foudre," *Assemblée publique de la Société Royale des Sciences, Tenue dans la Grande Salle de l'Hôtel de cette Société, en présence des Etats de la Province de Languedoc, le 27 Décembre 1780* (Montpellier, 1781): 53–67, at 54; Jean Baptiste Le Roy, "Mémoire sur la Forme des Barres ou des Conducteurs Métalliques destinés à préserver les Édifices des effets de la Foudre, en transmettant son feu à la Terre," *Mémoires de l'Académie Royale des Sciences pour l'Année 1773* (Paris, 1777): 671–86, at 673; see also Felix Sestier, *De la Foudre*, vol. 1 (Paris, 1866), 448f.

4. Aimé-Henri Paulian, *Dictionnaire de Physique*, vol. 2, 8th ed. (Nimes, 1781) 365. "Nollet's attitude was part of his general antipathy to Franklin, whose work had achieved so great a fame and popularity in France that his own position as the leader of electrical thought had been greatly damaged. Furthermore, Franklin's unitary theory . . . constituted a flat contradiction of everything Nollet had been teaching on electricity," I. Bernard Cohen, "Prejudice against the Introduction of Lightning Rods," *Journal of the Franklin Institute* 253 (1952): 393–446, at 414. Heilbron discussed the entire dispute between Nollet and Franklin in great detail and made clear that it had its basis in a dispute between several members of the *Académie des Sciences*, namely Nollet and Réaumur on one side and Buffon and Dalibard on the other; see John L. Heilbron, *Electricity in the 17th and 18th Centuries* (Berkeley: University of California Press, 1979), 346–62.

5. Cohen, "Prejudice" (cit. n. 4), 414. Nollet published his doubts against Franklin's lightning rods in his "Mémoire sur les effets du tonnerre comparés à ceux de l'électricité; avec quelques considérations sur les moyens de se garantir des premiers," *Mémoires de l'Académie Royale des Sciences pour l'Année 1764*, 408–51; cf. Cohen, "Prejudice," 394.

6. Cf. Jessica Riskin, "The Lawyer and the Lightning Rod," *Science in Context* 12, no. 1 (1999): 61–99, at 77.

7. Poncelet, *La Nature dans la Formation du Tonnerre et la Reproduction des Êtres Vi-vans* (Paris, 1766), 117f.; see also Sestier, *De la Foudre* (cit. n. 3), vol. 2, 447. An earlier criticism of Franklin's work had been published already in 1753; see Charles Rabiqueau, *Le Spectacle du Feu élémentaire ou Cours d'Electricité expérimentale* (Paris, 1753).

8. In 1753 the latter version caused the death of the St. Petersburg professor Rich-mann, who was killed when his observational device was struck by lightning. The inci-dent was not only widely discussed, it was also utilized as an indication of the danger of lightning rods; see J. A. Nollet, *Vergleichung der Würkungen des Donners mit den Würkungen der Electricität nebst einigen Betrachtungen über die Mittel sich vor dem er-stern zu bewahren* (*Aus den Mémoires de l'Acad. Royal de Paris 1764*), 2nd ed. (Prag: 1773), 93.

9. Institut de France, *Index Biographique de l'Académie des Sciences du 22 Décembre 1666 au 1er Octobre 1978* (Paris, 1979), 254.

10. On the acceptance of lightning as an electrical phenomenon see Guyton de Morveau, "Tonnerre," *Nouveau Dictionnaire pour servir de Supplément aux Diction-naires des Sciences, des Arts et des Métiers*, vol. 4 (Paris, 1777), 948–52, at 949; Saussure in Jean Lantiers, *Essai sur le Tonnerre considéré dans ses effets moraux sur les hommes; et sur un coup de Foudre remarquable* (Lausanne, 1789), 50; Joseph Aignan Sigaud de La Fond, *Précis Historique et Expérimental des Phénomènes Électriques* (Paris, 1781), 370ff. Although in general a consensus existed that lightning was an electrical phenomenon, a few scientists still held a different view. For example, the public demonstrator and fa-mous balloonist Pilâtre de Rozier considered lightning to be caused by inflammable gases that were ignited by an electrical spark; see Jean-François Pilâtre de Rozier, "Réflex-ions sur les causes de la foudre," *La Vie et les Mémoires de Pilâtre de Rozier: Écrits par lui-même; & publiés par M. T.* (Paris, 1786), 103–9. Guyton de Morveau distinguished between a "*véritable paratonnerre*" and a "*conducteur isolé*"; in the following, I will refer only to the proper lightning rod as most authors who discussed this device explicitly em-phasized the importance of its careful grounding.

11. See Riskin, "Lawyer and the Lightning Rod" (cit. n. 6), 73; Sestier, *De la Foudre* (cit. n. 3), vol. 2, 447.

12. Examples are given in Jean Baptiste Le Roy, "Mémoire sur la nécessité et les moyens d'armer les édifices de Paratonnerres ou de Conducteurs, pour les préserver de la foudre," *Mémoires de l'Académie Royale des Sciences pour l'Année 1790* (Paris, 1797), 583–600, at 589.

13. Riskin, "Lawyer and the Lightning Rod" (cit. n. 6), 73.

14. Pierre Bertholon, *De l'Électricité des Météores: Ouvrage dans lequel on traite de l'Électricité Naturelle en général, & des Météores en particulier; contenant l'exposition &l'explication des principaux phénomènes qui ont rapport à la Météorologie Electrique, d'après l'observation & l'expérience* (Paris, 1787).

15. For this purpose Bertholon also employed experiments with a thunder-house; see ibid.

16. Pierre Bertholon, *Nouvelles Preuves de l'Efficacité des Para-Tonnerres* (Montpel-lier, 1783); Bertholon, *Ouvrage* (cit. n. 14).

17. Placing the reader in the position of a virtual witness was a common strategy that goes back to the seventeenth century—the best-known example in this respect is the controversy between Boyle and Hobbes; see Steven Shapin and Simon Schaffer, *Leviathan and the Air-Pump* (Princeton, N.J.: University Press, 1985).

18. In 1777 the *Académie des Sciences* charged Le Roy and Fougeroux with preparing a report on Bertholon's findings. The report was not entirely favorable but more empirical evidence seemed to be necessary (see Procès-Verbaux 1777, 169r–170r). A comparable position was held by Monge in 1790, who stated explicitly that the question whether lightning always strikes from the sky to the earth or also from the earth to the sky "is still a topic of large uncertainty"; Monge, "Mémoire sur la Cause des principaux Phénomènes de la Météorologie," *Annales de Chimie* 5 (1790): 1–71, at 63.

19. Pierre Bertholon, *Die Elektricität der Lufterscheinungen, worinne von der natürlichen Elektricität überhaupt, und von den Lufterscheinungen besonders gehandelt wird, auch die vornehmsten Phänomene der elektrischen Meteorologie nach Beobachtungen und Erfahrungen aufgestellt und erklärt werden* (Liegniz, 1792), 136.

20. Pierre Bertholon, *Mémoire sur un nouveau moyen* (cit. n. 3), 24. This new design of lightning rod was criticized by some of Bertholon's contemporaries; see Carnus, "Lettre . . . sur les Eudiomètres & les Paratonnerres," *Observations sur la physique, sur l'histoire naturelle et sur les arts* 22 (1783): 223–25.

21. Pierre Bertholon, "Mémoire sur un Para-Tremblement de terre, & un Para-Volcan," *Observations sur la physique, sur l'histoire naturelle et sur les arts* 14 (1779): 111–21; Bertholon, "Mémoire sur la cause électrique des Tremblemens de terre," in *Assemblée publique de la Société Royale des Sciences, Tenue dans la Grande Salle de l'Hôtel de cette Société, en présence des Etats de la Province de Languedoc, le 28 Décembre 1779*, 1780: 58–92. A similar position was held by Lapostolle; see Alexandre Ferdinand Léonce Lapostolle, *Traité des Parafoudres et des Paragrêles en Cordes de Paille* (Amiens, 1820).

22. Charles Richardon, *Nouveau Système d'appareils contre les dangers de la foudre et le fléau de la grêle* (Paris, 1825), 39.

23. Monge, "Cause" (cit. n. 18), 53.

24. For a more detailed discussion of thunder-houses as an example of model experiments, see Willem Hackmann's contribution in this volume.

25. Boyer-Brun, "Lettre . . . contenant la Description d'un Electroscope adaptable aux Para-tonnerres," *Observations sur la physique, sur l'histoire naturelle et sur les arts* 28 (1786): 133–35.

26. The idea of developing a device that could indicate an emerging thunderstorm is not unusual; see, for example, François Para du Phanjas, *Théorie des Etres Sensibles ou Cours Complet de Physique, Spéculative, Expérimentale, Systématique et Géométrique*, vol. 3 (Paris, 1772), 470. Actually, this was an idea that can be traced back to Franklin, who used a combination of an insulated rod and bells as an indicator for an approaching thundercloud; see E. Philip Krider, "Benjamin Franklin and Lightning Rods," *Physics Today* 59, no. 1 (2006): 42–48. Indicating that a thundercloud was approaching was a utility of a rod that even Nollet acknowledged; see J. A. Nollet, *Vergleichung* (cit. n. 8), 93. Therefore, it is not surprising that other researchers advocated this use as well.

27. Luc-Vincent Thiéry, *Guide des Amateurs et des Étrangers voyageurs à Paris*, 2 vols. (Paris, 1786/1787). For a detailed account of public lectures in experimental physics in Paris, see Michael Lynn, *Popular Science and Public Opinion in Eighteenth-Century France: Studies in Early Modern European History* (Manchester: Manchester University Press, 2006), chs. 2 and 3.

28. Sigaud de la Fond, *Précis* (cit. n. 10), 441.

29. Jean Baptiste Le Roy, "Mémoire sur les Verges ou Barres métalliques, destinés à garantir les Édifices des effets de la foudre, avec la manière dont ces Barres doivent être disposées, pour que leur effet soit aussi certain qu'il est possible," *Mémoires de l'Académie Royale des Sciences pour l'Année 1770* (Paris, 1773): 339–69, at 339.

30. Ibid., 342f. A similar way of establishing the utility of lightning rods was used, for example, by Aliés some ten years later; see Pierre-François Aliés, *Mémoire sur la Méchanisme secret et sensible de l'Électricité naturelle* (Montpellier, 1780).

31. Le Roy, "Mémoire sur la Forme" (cit. n. 3). A somewhat different version of this paper was published in the December issue of the *Journal de Physique*; see Le Roy, "Précis du mémoire . . . sur la forme des Barres ou des Conducteurs métalliques, destinés à préserver les Edifices de la foudre, en transmettant son feu à la Terre," *Observations sur la physique, sur l'histoire naturelle et sur les arts* 2 (1773): 437–45. In the introduction to this paper, its importance was highlighted: "We desire to place the excellent dissertation of M. Le Roy under the eyes of the audiences, but the Academy reserves the right to publish these precious works in their proceedings, and until then they are deposited in their archives"; ibid., 437.

32. Le Roy, "Précis du mémoire" (cit. n. 31), 438. This controversy is discussed by R. W. Home in this volume.

33. Ibid., 441.

34. I am using this term with reference to Willem D. Hackmann, "Scientific Instruments: Models of Brass and Aids to Discovery," in *The Uses of Experiment*, ed. David Gooding, Trevor Pinch and Simon Schaffer (Cambridge: Cambridge University Press, 1989), 31–65.

35. Arago in Richard Anderson, *Lightning Conductors: Their History, Nature, and Mode of Application* (London, 1879), 73. Arago compares this with the practice of an engineer who is to build a bridge based on experiences that had been gathered. The engineer "in making his plan, keeps somewhat beyond the dimensions dictated by the greatest floods and the heaviest rains which have ever been observed. . . . Greater precaution or foresight than this cannot be demanded from the constructor of lightning conductors, nor is any needed"; ibid., 74.

36. Procès-Verbaux, April 24, 1784, 90v-95r, at 90v. The explosion of a powder magazine as a consequence of a lightning stroke was far from impossible; there were at least six cases for the period between 1769 and 1782; see Anderson, *Lightning Conductors* (cit. n. 35), 197.

37. In 1781, Franklin and Le Roy had collaborated on another committee that was to prepare the report on the lightning rod that Barbier de Tinan designed for the Strasbourg cathedral; the report was read on May 12, 1781 (see Procès-Verbaux for 1781, 128r–134v). Although the report was favorable, the project was not realized because the

expenses were considered to be too high; see Ant. Fargeau, *"Établissement d'un Para-tonnerre sur la flèche de la Cathédrale de Strasbourg"* (Strasbourg, 1833), 4f. Besides, Le Roy was, together with d'Arcy, responsible for a report on a design of powder magazines that was read on March 2, 1774; see Procès-Verbaux 1774, 61v–66r). Moreover, in 1784 Le Roy was also charged by the First Lord of the admiralty to erect lightning rods on the principal marine buildings in the Atlantic harbors of Brest, Lorient, and Rochefort; see Sestier, *De la Foudre* (cit. n. 3), vol. 2, 447.

38. Thiéry, *Guide des Amateurs* (cit. n. 27), vol. 1, 13. On Rochon, see Danielle Fauque, "Alexis-Marie Rochon (1741–1817), savant astronome et opticien," *Revue d'histoire des sciences* 38, no. 1 (1985): 3–36.

39. Procès-Verbaux, April 24, 1784, 90v–95r, at 91r.

40. Ibid.

41. Ibid., 92r. The adequate proportions of lightning rods were also disputed in the trial on Vissery's lightning rod; see Riskin, "Lawyer and the Lightning Rod" (cit. n. 6), 67. From my understanding this may have resulted from Nollet's reasoning against the lightning rods. Moreover, it is remarkable that this committee used the "engineering strategy" discussed later by Arago; see note 35.

42. Ravel and Pierron had initially proposed the design for the lightning protection of the powder magazines; both were officers, one in the Corps Royal d'Artillerie, the other in the Corps du Génie.

43. The members were the same as those in the first committee except for Le Roy. The report can be found in the Procès-Verbaux for 1784, 171r–174r, and the manuscript of this report is kept in the pochette of the Séance, July 3, 1784.

44. Procès-Verbaux, July 3, 1784, 173r.

45. According to the Procès-Verbaux, Coulomb read the first part of the first memoir on electricity on May 4, 1785, and finished the reading of the sixth memoir on March 17, 1790. For the dates of all presentations of Coulomb's memoirs, see C. Stewart Gillmor, *Coulomb and the Evolution of Physics and Engineering in Eighteenth-Century France* (Princeton, N.J.: University Press, 1971), 232f.

46. In speaking of Coulomb turning quickly to electricity, I refer to Coulomb's memoir on torsion where he made clear that the electrical balance was constructed after the reading of this paper; see C. A. Coulomb, "Recherches théoriques et expérimentales sur la force de torsion, et sur l'élasticité des fils de metal," *Mémoires de l'Académie Royale des Sciences pour l'Année 1784* (Paris, 1787), 229–69, at 255. According to the Procès-Verbaux, Coulomb read his torsion memoir on September 4, 1784, eight months before he presented his first memoir on electricity but only two months after the second report of the lightning rod committee had been read.

47. C. A. Coulomb, "Premier mémoir sur l'électricité et le magnétisme," *Mémoires de l'Académie Royale des Sciences pour l'Année 1785* (Paris, 1788), 569–77; and Coulomb, "Sur l'électricité et le magnétisme, deuxième memoir," *Mémoires de l'Académie Royale des Sciences pour l'Année 1785* (Paris, 1788), 578–611.

48. The term "precision" is used frequently by Coulomb in his memoirs; however, its meaning differs from ours and seems to be more related to sensitivity than to precision. For a detailed discussion of this aspect, see Peter Heering, "Die Coulombschen Experi-

mente mit der Torsionswaage—eine der ersten Präzisionsmessungen?" in *Genauigkeit und Präzision in der Geschichte der Wissenschaften und des Alltags,* ed. Dieter Hoffmann and Harald Witthöft (Braunschweig: PTB, 1996), 303–19. Lucio Fregonese correctly observed "Coulomb's overriding interest in pointing out the quality and the sensitivity of the measurements made possible by his recent torsional studies"; see L. Fregonese, "Two Different Scientific Programmes: Volta's Electrology and Coulomb's Electrostatics," in *Restaging Coulomb: Usages, Controverses et Réplications autour de la Balance de Torsion,* ed. Christine Blondel and Matthias Dörries (Florence: Leo S. Olschki, 1994), 85–98, at 89. This interest is to be understood in the context of Coulomb's research program. The acceptance of the measurement's quality and the setup's sensitivity made it impossible to challenge any of the results Coulomb was going to determine in the following memoirs. Therefore, it seems doubtful to me whether Fregonese's interpretation of "errors of strategy in the presentation" (ibid.) is correct. In the context of Coulomb's research program, mainly the macroscopic effects are interesting. Consequently, I cannot agree with Fregonese's appraisal that "Coulomb certainly intended to show that an inverse square law between the centers of macroscopic spheres implied an inverse square law between all the microscopic elements of electricity distributed on their surfaces" (ibid., 88). This might have been a minor aspect for Coulomb but not the central one.

49. C. A. Coulomb, "Sur l'électricité et le magnétisme, troisième memoir," *Mémoires de l'Académie Royale des Sciences pour l'Année 1785* (Paris, 1788): 616–38, at 612.

50. C. A. Coulomb, "Quatrième Mémoire sur l'électricité," *Mémoires de l'Académie Royale des Sciences pour l'Année 1786* (Paris, 1789): 67–77, at 67.

51. C. A. Coulomb, "Sur l'électricité et le magnétisme, cinquième memoir," *Mémoires de l'Académie Royale des Sciences pour l'Année 1787* (Paris, 1789): 421–67, at 425.

52. Ibid., 437.

53. Ibid., 453.

54. Ibid., 454.

55. C. A. Coulomb, "Sur l'électricité et le magnétisme, sixième memoir," *Mémoires de l'Académie Royale des Sciences pour l'Année 1788* (Paris, 1791), 617–705, at 617.

56. Ibid., 625ff.; see also fig. 6.5.

57. Ibid., 704f. Comparing Coulomb's claim with what he actually achieved clarifies that his aim was limited to aspects that were relevant to his (military) context. This does not only mean that Coulomb did not examine aspects such as the generation of electricity or the storage of electricity in a Leyden jar. Moreover, as König already remarked in his translation of Coulomb's memoirs for the famous series of *Ostwald's Klassiker der exakten Wissenschaften:* "In fact it could be said that Coulomb's investigations were in one aspect incomplete. He investigated the law of attraction and repulsion of electricity [Elektricitätsmengen] under various circumstances but he did not change the surrounding medium; instead he performed all the experiments in air"; see *Vier Abhandlungen über die Elektricität und den Magnetismus von Herrn Coulomb,* ed. Walter König (Leipzig, 1921), 80. In the context of the research program described above such a limitation is highly plausible.

58. C. A. Coulomb, "6th EM" (cit. n. 55), p. 690.

59. Ibid., 692.

60. Ibid., 654.

61. Gillmor, *Coulomb* (cit. n. 45), 39f.

62. Seeger in Lord Mahon, *Grundzüge der Elektrizität* (Leipzig, 1789), iv. Like Stanhope, J. F. Seeger was an officer in the field of artillery; thus, his motivation to translate Stanhope's treatise can also be seen in this military context.

63. This can be seen from the preliminary phrase to the above quotation where Seeger wrote, "To what extent this part of natural philosophy is connected with activities of an officer in the artillery would be another question which I hardly should answer"; ibid.

64. For a detailed discussion of the experiences made in analyzing Coulomb's experiments with the replication method, see Peter Heering, "On Coulomb's Inverse Square Law," *American Journal of Physics* 60 (1992): 988–96; and Heering, "The Replication of the Torsion Balance Experiment: The Inverse Square Law and Its Refutation by Early 19th-Century German Physicists," in Blondel/Dörries, (cit. n. 48), *Restaging Coulomb*, 47–66.

65. Actually, Coulomb made clear at the end of his first memoir that he reduced the sensitivity of his torsion balance to be able to carry out measurements.

66. Peter Heering, "Regular Twists: Replicating Coulomb's Wire-Torsion Experiments," *Physics in Perspective* 8 (2006): 52–63.

67. Among those researchers who were either trained or who taught at the École du Génie at Mézières were "Monge, Coulomb, Borda, Lazare Carnot, Meusnier, du Buat, Malus, etc."; Maurice Caullery, *French Science and Its Principal Discoveries since the Seventeenth Century* (New York, 1934; repr. New York: Arno Press, 1975), 40.

68. With respect to this political interpretation of the change of the style of experimentation, the public séance of September 1784 was crucial in which on the one hand Mesmer's work was rejected, and on the other hand demonstrations of adequate ways of producing scientific knowledge were presented by Lavoisier, Tenon, Rochon, and Coulomb. For a detailed discussion of this séance, see Heering, "Regular Twists" (cit. n. 66).

69. This can be seen in the reports presented at the *Académie des Sciences* in 1808, 1823, 1854, and 1867.

"Eripuit caelo fulmen sceptrumque tyrannis"

The Political Iconography of Lightning in Europe and North America, 1750–1800

Christian Fuhrmeister

Introduction

THE LIGHTNING ROD HAS FREQUENTLY BEEN LABELED one of the most ingenious human answers to a threat posed by nature. Though now ubiquitous and unremarkable, this device is in fact a complex and multifaceted phenomenon, situated at the crossroads of cultural–intellectual history and history of the natural sciences. Hence, the lightning rod has been the subject of studies that stressed, to name just a few aspects, the circumstances of its invention, the sociocultural dimensions of its victory over superstition or its subsequent global distribution. Few authors, however, have attempted to link the manifold implications of this technical device to the larger horizon of the cultural meaning of lightning itself.

Even within the circles of Enlightenment philosophy and science, the rational explanation of thunder and lightning could not completely soothe the multiple fears instigated by and traditionally associated with them. Contrasting sharply with common narratives of unflagging scientific progress, many documents reveal the endurance of older beliefs and practices. As late as 1825, for instance, new church bells were inscribed *Vivos voco. Mortuos plango. Fulgura frango* ("I call the living. I mourn the dead. I break the lightning bolts."), demonstrating that bell ringing to distract lightning from hitting a church or a village was not at all as outmoded as late-eighteenth-century partisans of a consistent progress of mankind would have liked to have it.[1]

The sober scientific explanation that lightning is a discharge of electricity that can be safely channeled and controlled with technology—just as dike engineering can help to control storm tides—did not necessarily lead to the disappearance of older ideas, metaphors, customs, and imageries. These persist and survive;

they are able to govern individual narratives as well as collective and cultural memories for long periods of time (*"longue durée"*). Although the old representations—which illustrated certain forms of thought—get fewer and more subtle, they partly maintain their former significations and partly acquire new meanings. Following concepts of cultural anthropologists or of art historians such as Aby M. Warburg, one has to acknowledge that lightning had actually become a very powerful symbol well before the Enlightenment. Condensing diverse theological, psychological, and cultural tensions, the motif of lightning in fact retained its former momentum in the second half of the eighteenth century. According to Warburg, it is a characteristic trait of such transformations that the energy that was once attached to a specific pictorial form can not only be transferred to different contexts but also employed for different purposes.

Based on the premise that the full scope of the lightning rod can only be understood when we take into account the ideas associated with lightning itself, this essay seeks to demonstrate that the meanings that were attributed to lightning are intricately linked to longstanding traditions. Provided that both the former conventional and the new images of lightning are salient features accompanying the dawn of the lightning rod, this essay investigates in particular how the traditional iconography of lightning was quite suddenly overthrown in the third quarter of the eighteenth century. Because this text studies but a very small episode of the vast cultural history of lightning, it neither addresses the diverse, worldwide dimensions of the topic nor the time span from antiquity to the present.[2] Although limited in scope, this essay isolates one particularly interesting aspect of the cultural history of lightning—namely its political iconography, that is, the different political meanings that were ascribed to lightning in different contexts during the late eighteenth century in western Europe as well as in North America (i.e., in the thirteen East Coast states that founded the United States). However, in accordance with the factual distribution, the majority of my examples are directly related to the French Revolution.

Focusing on those uses of lightning that signify either refusal or acceptance of social upheaval, I am most interested in those visual conceptions that skillfully employ the lightning motif to demonstrate the uncontrollable force of political movements. As far as I can see, this specific usage—predominantly, but not exclusively in the field of the graphic arts—has not only not received any comprehensive scholarly attention in art history and the adjoining disciplines, it has virtually not been dealt with at all.[3] The motif has recently been addressed from the perspective of German language and literature studies, highlighting precisely the relationship between natural phenomena such as earthquakes, eruptions, floods, lightning, and comets, on the one hand, and the multifaceted human reactions to them during and since the Enlightenment era on the other, but we still lack a similar account of the visual paradigm.[4]

Good Lightning?

Lightning was not always perceived as an impending fearful danger or a brutal, destructive negative force. At times, lightning was also seen as a power to which certain positive meanings—in a wide sense—could be attributed. Under certain circumstances, lightning was even welcome and helpful because some social and political projects of the time corresponded in a number of ways to the intellectual efforts and the scientific endeavors that wanted to overcome superstition and establish laws of nature in their place.

The more "electricians" explained the peculiar character of the new, hitherto unknown energy called "electricity"—simultaneously elucidating the phenomenon of lightning—the more its principles were dispersed outside the confinements of purely scientific arguments, and the better its modes of operation became known to a wider public, the easier was it for commentators to draw parallels. In fact, "taming the electrical fire" with the lightning rod was soon regarded as a symbol for the success of the Enlightenment in general. Consequently, we observe a significant increase of the metaphor "lightning" in contemporary discourses in Europe as well as in North America. As metaphorical trends indicate conceptual changes, they imply dynamics and intersections between scientific, social, political, religious, and cultural contexts.[5]

Within a relatively short time, the feelings connected with lightning (and electricity) changed radically. Related to both the reception of Edmund Burke's *A Philosophical Enquiry into the Origin of Our Ideas of the Sublime and Beautiful* (1757) and to early romanticism, people started to perceive, for instance, the aesthetic qualities of a thunderstorm. Natural electricity, on the other hand, was good because it could be employed for healing.[6] The upper classes were interested to attend spectacular presentations, and it was funny and entertaining when traveling "electricians" staged performances throughout the country.[7]

As soon as the phenomenon of natural electricity was no longer understood as the epitome of the "Inexplicable and Supernatural," it simply turned into a powerful agent, barely reminiscent of its former horrors.[8] As such, it became a force that could easily be fused with various aspirations, projects, and developments. As a matter of fact, lightning acquired these other "positive" meanings exactly at that moment in time when its former mysteries were unveiled and explained.

The Cultural Framework of Lightning

To grasp the nature of these shifts, we need to consider briefly the nonscientific concepts associated with lightning when Benjamin Franklin invented the lightning rod in 1752. Since the earliest times, mankind had tried to identify and as-

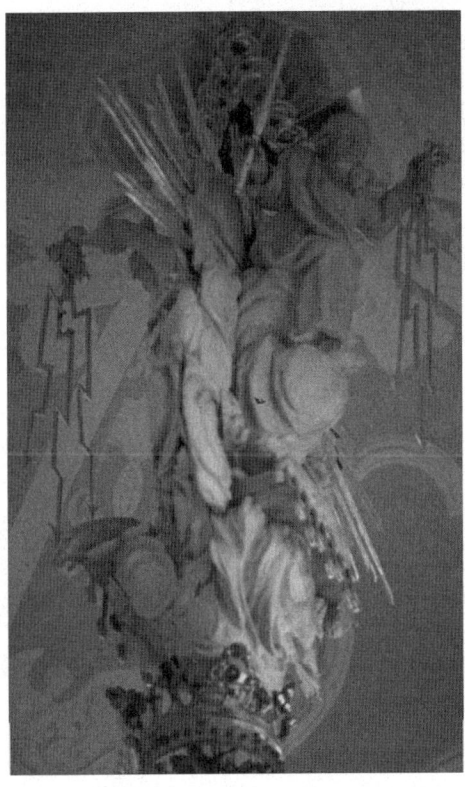

Figure 7.1.   God Almighty hurling down flashes of lightning to destroy Evil (snake embracing the globe, below) and to open the tomb of Christ. Detail of the Chapel of the Holy Sepulchre in the southern transept of St. Michael in Bamberg, 1787–89. Photograph Fuhrmeister 2003.

sociate lightning with various figures; Zeus, Jupiter, and Donar/Thor are typical personifications in the Western hemisphere. John Michael Ruysbrack's statue of "Thuner" (i.e., Thunder), 1728–30, is a case in point—the artist's depiction of the Saxon God or Deity of Thunder neatly illustrates the quintessentially pagan attitude toward lightning.[9]

The dominant perspective of the early eighteenth century, however, took lightning to be God's will in action—or, more precisely, lightning was understood as the just punishment for those who did not comply with the Ten Commandments. From high to low, from elaborated artistry to primitive and popular imagery, from grandiose, spectacular scenes such as the Chapel of the Holy Sepulchre (figure 7.1) to vernacular glass paintings from Tyrol, the motif of lightning was very widely employed.[10]

Following the iconography of "Jupiter tonans,"[11] the almost life-sized figure of God Almighty in Bamberg, hurling down flashes of lightning to destroy Evil and to awaken the Messiah, is—to my present knowledge—one of the last manifestations to illustrate the orthodox tradition. As if the contested issue of lightning bolts had to be settled once and for all, this impressive display authoritatively claims the necessity of a supreme being. Regardless of both the discussions in learned societies and of the lightning rods that had been installed on church towers in the last decades and that had continually withered the persuasive power of purely religious arguments, the installation obviously attempts to reground the origin of lightning in the hands of God alone.[12] It comes as no surprise, then, that in the late eighteenth century, rural German communities often rejected lightning rods as "Teufelszeug" (hellish stuff).[13]

Throughout the Western hemisphere, Christianity had been quite successful in substituting its monotheistic notion of the divine for the ancient pagan deities. Since Luther, whenever lightning flashed, people were inclined to introspection because a direct link between celestial events and human behavior had been established: Heinz Dieter Kittsteiner has lucidly demonstrated how the concept of a bad conscience is tied to Christian morale, and how natural events were interpreted as comments on human dispositions.[14] Consequently, in addition to causing physical harm, lightning evidently triggered enormous fears: It demanded the individual's complete surrender because it was God's own voice and hand that drove flashes to the earth. Seen from this perspective, it is no surprise that the lightning motif was frequently used by those in power to keep them in power. Aegidius Sadeler's portrait of the Holy Roman Emperor Ferdinand II is a case in point: Accompanied by lightning bolts, the victorious emperor triumphs over his enemies.[15] Precisely because both religious and secular leaders unanimously used the lightning motif to demonstrate their uncontested authority (or even attempted to legitimize their position in alluding to God's will), Franklin's analyses of the processes that lead to a discharge of lightning meant a real revolution—and also had enormous implications for the political iconography of lightning.

Political Lightning

Jean-Antoine Houdon's bust of Benjamin Franklin (1778, marble) epitomizes the intersection of physics and politics that results from Franklin's diverse roles as scientist, politician, diplomat, and statesman.[16] The Latin epigram "Eripuit caelo fulmen sceptrumque tyrannis" ("He seized lightning from the heavens and the scepter from the tyrants"[17]) inscribed on one of the copies of the bust conjures the most salient qualities of the man who was dubbed "electrical ambassador" and who was widely acclaimed as a stubborn fighter for liberty.[18] This linkage between sociopolitical standing and scientific discoveries is all the more important since Franklin apparently influenced public opinion in many respects, including the reception of the American Revolution in Europe, more than even George Washington did.[19]

The well-known homage by Jean-Honoré Fragonard, "Au Génie de Franklin" (1778), further illustrates the conjunction between natural science and the political sphere.[20] Assisted by Minerva and Mars, parrying the menacing bolts with a large shield that alludes to the lightning rod, the protagonist skillfully channels the power of lightning to drive away the figures of Tyranny and Avarice. In doing so, he simultaneously instigates and spreads genuinely American ideas in Europe.

Taking Houdon and Fragonard as starting points, we can observe a fairly coherent practice that identifies lightning as a symbol that was associated with liberation and progress, a symbol that was frequently used to signal opposition to the status quo. In fact, we witness the birth of a new visual paradigm: Attacking those powers that had hitherto pretended that nature safeguards and even legitimizes their position, the symbol of lightning was now employed for exactly opposing goals. This custom was already firmly established in the early 1780s, well before the Revolution. Within the Enlightenment movement, the concurrence of Rationalism and politics was nowhere more articulated than in France. The imagery of the French Revolution, particularly in the realm of the graphic arts—engravings, caricatures, leaflets, pamphlets, broadsheets, and so on—thus provides us with many objects of study. However, only a few exemplary images can be mentioned here.

Jacques-Louis Perée's engraving "Les droits de l'homme" ("The Human Rights," 1794/1795[21]) depicts a heroically naked man—his nakedness refers to the pure state of nature. Standing on the earth's bowl, he raises the certified Human Rights to the sky. While the "New Adam" had already been able to smash the incense burner and some coats of arms with his pickaxe, the sky now answers with lightning that destroys the Symbol of the Ancien Régime, the King's Crown.

The engraving by Niquet le Jeune titled "Déclaration des droits de l'homme et du citoyen" (Declaration of the Rights of Man and Citizen) (figure 7.2[22]) offers a telling juxtaposition. On the right, we see a bright scene with free citizens dancing around a pole with the phrygian cap, celebrating the Declaration of the Rights of Man and Citizen. The dark left side shows how flashes of lightning help to establish a new order of society in attacking specifically the "droits féodaux et privilèges," the feudal rights and privileges of the aristocracy.

Similarly, the anonymous colored engraving "La Fayette présente la Liberté à la France" (figure 7.3), depicting a scene of "La Fête de la Fédération" on July 14, 1790, reiterates this visual association. The caption reads: "La Nation Française assistée de Mr. De La Fayette terrasse le Despotisme et les abus du Regne Feodal qui terrassaient le Peuple" ("The French Nation assisted by Mister Lafayette overthrows both despotism and the abuses of the Feudal Reign which had held down the people"). The allegorical figure—the French Nation rather than Liberty herself—has a bundle of lightning bolts in her right hand, ready to take action to free the chained, muscular man signifying the ordinary Frenchman.

Again and again, lightning supports and even legitimizes the attacks on the monarchy and absolutism. An anonymous engraving from 1790, "La Liberté tri-

Figure 7.2.   "Déclaration des droits de l'homme et du citoyen," engraving by Niquet le Jeune. Musée Carnavalet, Paris; photo Édimédia. Reproduction after Michel Vovelle, *La Révolution Française: Images et Recit 1789–1799*, vol. 2, *Octobre 1789—Septembre 1791* (Paris: Messidor, 1986), 299.

omphe et détruit les abus" ("Liberty triumphs and destroys the abuses," figure 7.4), states unmistakably that lightning no longer figures as God's punishment. Instead, the composition denotes that the revolutionaries are in alliance with the strongest possible partner: nature itself. This is most obvious in those many representations that depict the destruction of the Ancien Régime's prime symbol in Paris, the Bastille.[23]

As early as 1783, the destruction of the Bastille was envisioned in a prophetic image (figure 7.5).[24] The destruction of the symbol of despotism that actually took place six years later is depicted in an utterly ironic manner, since Louis XVI is—miraculously—supposed to have ordered to tear down the prison in the early 1780s. This imaginary order should then have resulted in a monument dedicated to him, glorifying this act of liberation. We witness a paradoxical scene that shows Louis XVI uttering words by Voltaire—"Soyez libres: vivez"—while those who have left the Bastille fall down to their knees to thank him. What is of interest to us is the background of the etching that shows lightning bolts attacking the Bastille, thus demonstrating that either nature or God—or both, in this case—fervently support this act of liberation.

Figure 7.3.   Lightning setting the people free. Anonymous colored engraving, Bibliothèque Nationale de France, Paris. Reproduction after Michel Vovelle, *La Révolution Française: Images et Recit 1789–1799*, vol. 2, *Octobre 1789—Septembre 1791* (Paris: Messidor, 1986), 147.

Many artists who dealt with the "storming" of the Bastille followed this formula in later years. In addition to them, numerous other examples attest that "political lightning" formed an indispensable ingredient of the widespread pro-Revolutionary iconography, from the French[25] and Swiss[26] graphic arts to large oil canvases such as "Triomphe de la Civilisation" by Jacques Réattu, 1795.[27]

Within the scope of this volume, the anonymous colored etching "La Chûte en Masse" (figure 7.6), distributed in 1793 in one thousand copies, is a particularly interesting and informative example.[28] This emblematic image neatly demonstrates the intersection of scientific research and technical invention with the cultural and political sphere. It prophesies the fall of the decadent rulers, from William V (Netherlands) and George III (England) to Pope Pius VI and the "Prussian Tyrant" Frederick William II, from Catherine II (Russia) and the

Figure 7.4.   Lightning destroying the feudal rights. Anonymous engraving, Bibliothèque Nationale de France, Paris. Reproduction after Michel Vovelle, *La Révolution Française. Images et Recit 1789–1799*, vol. 1, *De la prérévolution à Octobre 1789* (Paris: Messidor, 1986), 198.

Figure 7.5.   Lightning destroying the Bastille. Reproduction after Rolf Reichardt, "Historisch-politische Bildpublizistik als Anzeiger von Ereignis-Strukturen: Ein Essay anhand französischer Medienereignisse (1660–1804)," in *kritische Berichte*, no. 1 (2000): 38–61, at 49.

Spanish despot Carl IV to Victor Amadeus III (Sardinia and Piedmont). Holding the "Constitution Républicaine" (Republican Constitution) in his left hand, a Sansculotte rotates the disc—identified as the Declaration of Human Rights—with his right hand. Using a common electrical frictional machine, the electrostatic generator, the young revolutionary generates "republican electricity" that causes a commotion sufficient to overthrow the rulers' thrones.[29] The new form of power, as universal as the human rights, obviously participates actively in establishing a new order of society. The "electrical spark of liberty" emanating from below the Phrygian cap atop the machine and growing into a lightning bolt is entirely consistent with "Liberté Egalité Fraternité Unité Indivisibilité de la République," suggesting a concurrence of revolutionary acts in science and politics. What makes this print so interesting is the fact that electricity is no longer represented in terms of allegory but rather as a natural phenomenon under the control of the revolutionary citizen.

It is tempting to assume that images such as "La Chute en Masse" either directly adopted the iconography of North American precursors such as the 1770 print "Political Electricity" or resulted from a widespread metaphorical use that has recently been analyzed in great detail by James Delbourgo.[30] If so, we would encounter one of the rare instances where eighteenth century discourses in Europe were in fact decidedly influenced or even shaped by developments in North America: "As the active powers of nature were conjured as the active powers of politics, the electricity of republican virtue brought a new American body politic to life. . . . Sudden political awakenings now had the force of giant electric shocks." These transatlantic correspondences with regard to enlightened politics certainly deserve more exploration. At any rate, Delbourgo's discussion of the narratives of "electrical politics" and "political electricity" indicates a number of striking parallels between the imageries of electricity (including lightning) and politics in the public sphere.[31]

Lightning did not cease to figure prominently in the later stages of the French Revolution's development. But now it was more often used to denote the victory of those factions within the Revolutionary movement that had overcome the former allies who had now turned into enemies. Hence, different parties or factions within the Revolutionary movement now used the symbol of lightning for their particular ends. An anonymous aquatint shows Jacobins struck by lightning,[32] and a print from summer 1793 titled "Sans Union point de Force—Sans Force point de Liberté" ("Without Union, No Power—Without Power, No Freedom") depicts a "Montagnard," that is, a Sansculotte, on a mountain, who throws lightning against the treacherous vermin in the swamp that symbolize the Girondists in the Convent.[33] Once again, the image of Zeus/God is appropriated but not as radically as the thoroughly secularized image of republican electricity above. Finally, Robespierre's "horrible conspiracy" in summer 1794 instigated an en-

Figure 7.6.   Republican electricity causes "La Chute en Masse," 1793. Anonymous colored etching (engraved by François Marie Isidore Quéverdo, probably based on a design by Dupuis). With permission of the Bibliothèque Nationale de France, Paris.

graving by Peysson that shows lightning emanating from the Phrygian bonnet to attack the snakes that symbolize the traitors.[34]

The best proof for the huge success of the pro-Revolutionary symbol of lightning, however, is the application of the same imagery by the various counter-revolutionary parties, particularly in Germany,[35] Austria,[36] and Britain.

In "Destruction of the French Collossus" (figure 7.7), James Gillray's topic is how the French Army under Napoleon lost the Battle of Abukir against the British.[37] He depicts the French Republic as a giant, like the Colossus of Rhodes. One foot still on the Egyptian pyramids, the giant tries to reach the French homeland, but British lightning effectively dismembers the monster. I see this less as a reversion to the old absolutist imagery but rather as a reaction to the efforts that intended to transform the evocative motif.

Despite these collective efforts to form counterimagery and to reinstall lightning as a means to preserve the old order, most images that used the lightning motif are evidence of—as far as I can see—a decidedly pro-Revolutionary position. After 1800, this disposition in favor of suppressed people did not wane.[38] Interestingly, comments on the freedom of press now often employed the motif, particularly in both France and Germany.

Lelarge's caricature "Les journeaux en Mai 1815" neatly illustrates a situation in May 1815: When Louis XVIII lost power and Napoleon returned, the

Figure 7.7.   James Gillray, "Destruction of the French Collossus." Colored etching, aquatint, 1798. Museum für Kunst und Gewerbe, Hamburg. Reproduction after Werner Hofmann, ed., *Europa 1789. Aufklärung—Verklärung—Verfall* (exhibition catalogue Hamburger Kunsthalle, (Cologne: DuMont 1989), 319.

emperor saw the need to establish better relations with the public. For that purpose, he was advised to reinstall the freedom of press. He did so immediately, but the liberal attitude was in fact superficial. As a consequence, an otherwise unknown engraver named Lelarge produced a satiric caricature of the different journals which were in fact firmly directed by Napoleon's régime.[39] Ironically, only "Le Censeur," masked as Zeus/Jupiter, riding the Pegasus and holding the incorruptible mirror of truth in front of him, is truly independent. His fight against the tight reins of those editors that obey the orders of the "police secrète" is visualized by flashes of lightning.

An anonymous colored lithograph published in 1822 in Berlin with the title "Without Title (Europe 1822)" presents the fight between progressive and reactionary elements: The Symbol of Backwardness, the Cancer, in the lower right, shoulders a number of men who carry flags that name different European antirevolutionary, antibourgeois, and antidemocratic journals.[40] They are the press of the Restoration Era. The "Drapeau Blanc," for instance, founded in 1815 by the Ultraroyalist Party, is surrounded by the "Droit de chasse" (hunting privilege), the "Lettre de Cachet" (the power to send someone to prison without trial) and the "Droit de Seigneur" (feudal masters rights). Their march follows a snuffer, the incarnation of the forces that want to abolish everything associated with Enlightenment, which is topped by the bat, a prime heraldic symbol of darkness. But this group will no longer be able to dominate the European press scene: The Column of Human Rights with the "Law for Everybody" is firmly rooted on a Rousseau-style island, the approaching ship signals progress and "Constitutional Liberty," the wind is blowing and the lightning flashes above help to disperse the bloc of Reactionaries—the "good old times," as it says on the snuffer, are already struck by lightning. Hence, this combined attack against the proponents of the Ancien Régime will doubtlessly be victorious.

Further research on related topics such as abolitionism[41] or, more generally, the American Revolution will certainly yield results that strengthen the arguments put forward here.

Summarizing this case study, it seems safe to say that only multidisciplinary approaches can adequately assess the true impact of the lightning rod. The fields of science, culture, and politics are no isolated paradigms. When we realize how, for instance, caricatures react to technical inventions, we obtain a glimpse of the interwoven—and constantly interacting—patterns of human endeavors. Although common beliefs are eventually challenged and disputed, they still persist, regardless of contradictory findings, experiences, and research results that claim to overthrow the old order of things. Notably, the early history of the lightning rod and the accompanying use of the motif of political lightning bolts is a case in point that testifies to this clash of worldviews.

While the cultural history of both lightning and the lightning rod remains to be written, one thing is clear: Images not only reproduce reality but also act as agents, generating fear, power, and pride, and they may even govern public beliefs. As Rolf Reichardt put it, the value of images "lies not in their depictions of individuals or events but in their symbolic, metaphorical, and allegorical interpretation of collective ideas and the questions of the day. They can show us the ways the contemporaries processed and interpreted their experiences—ways that remain hidden in the written sources."[42] Slightly modifying an important observation by Robert Darnton, images not only record what happened, they are also "an ingredient in the happening, . . . an active force in history, especially during the decade of 1789–1799, when the struggle for power was a struggle for mastery of public opinion."[43] Both genuine works of art and anonymous engravings, broadsheets, leaflets, and caricatures testify that the image of lightning had become a catch icon and a pictorial slogan, to translate Warburg's term "Schlagbild," to which Michael Diers devoted a book in 1997.[44]

Exploring the conjunction notably between natural science and the political sphere, images of lightning assumed a central role in the French Revolution. Precisely because the lightning motif suggested an alliance of the revolutionary forces with nature itself, it witnessed such a huge success among its contemporary audiences. However, its relevance withered in the following decades, to be resumed in the later nineteenth century, when electrical sparks unchained slaves and workers. In this sense, the political iconography of lightning has truly transgressed boundaries of time and space to become an integral part of our collective memories.

Notes

For the most part, this text (dedicated to Sabine and Nora Helene) was written in 2003. While some sentences and references have been edited in 2004–6, I have not tried to truly update my contribution.

1. Karl-Heinz Hentschel, "Man erhitze sich daher zur Zeit der Ungewitter nicht. Dunst und Dämpfe, Kult und Kämpfe: Menschen, Blitz und Donner," in *Hierzuland. Badisches und Anderes von Rhein, Neckar und Main: Organ des Arbeitskreises Heimatpflege Nordbaden/Regierungsbezirk Karlsruhe* 8, no. 15 (1993): 6–23, at 10. The bell that was cast in 1825 for the church of Höchenschwand was the last one bearing these words.

2. See Christopher Blinkenberg, *The Thunderweapon in Religion and Folklore: A Study in Comparative Archaeology* (Cambridge: Cambridge University Press, 1911; repr. New Rochelle, N.Y.: Caratzas Brothers, 1987); J. F. Shipley, "Lightning and its Symbols," *Quarterly Journal of the Royal Meteorological Society* 67, no. 290 (1941): 135–51 (E. Philip Krider drew my attention to this article, and David Rhees provided me with a copy—my thanks to both of them); Hans Prinz, "Gewitterblitze in Mythologie und Wissenschaft," in *Feuer, Blitz und Funke*, ed. Hans Prinz (Munich: Bruckmann, 1965), 8–36; and Claude Gary, *La foudre: Des mythologies antiques à la recherche moderne* (Paris/Milan/Barcelona: Masson, 1995). See also the entries in *Reallexikon zur Deutschen Kunstgeschichte*, vol. 2, 1948; *Lexikon für Theologie und Kirche*, vol. 2, 1958; and *Lexikon der Christlichen Ikonographie*, vol. 1, 1968.

3. See Adolf Rieth, *Der Blitz in der bildenden Kunst* (München: Ernst Heimeran, 1953); and Astrid Winter-Fritzsche, *"Naturgetreue" Blitzdarstellungen in der venezianischen Malerei des 16. Jahrhunderts mit einer Einführung in Deutung und Darstellung des Blitzes* (Frankfurt/Main, Bern, New York: Peter Lang, 1990).

4. Olaf Briese, *Die Macht der Metaphern: Blitz, Erdbeben und Kometen im Gefüge der Aufklärung* (Stuttgart & Weimar: J. B. Metzler, 1998); and Olaf Briese, "Der abgeleitete Blitz: Metapherngeschichte als Mentalitätengeschichte," *Euphorion. Zeitschrift für Literaturgeschichte* 92 (1998): 413–35.

5. See James J. Bono, "Science, Discourse, and Literature: The Role/Rule of Metaphor in Science," in *Literature and Science. Theory and Practice*, ed. Stuart Peterfreund (Boston: Northeastern University Press, 1990), 59–90.

6. See the various studies in Paola Bertucci and Giuliano Pancaldi, eds., *Electric Bodies: Episodes in the History of Medical Electricity* (Bologna: CIS, University of Bologna, 2001).

7. See Simon Schaffer, "The Consuming Flame: Electrical Showmen and Tory Mystics in the World of Goods," in *Consumption and the World of Goods*, ed. John Brewer and Roy Porter (London and New York: Routledge, 1993), 489–526; and Oliver Hochadel, *Öffentliche Wissenschaft: Elektrizität in der deutschen Aufklärung* (Göttingen: Wallstein, 2003).

8. Alain Beltran and Patrice Carré, *La fée et la servante: La société française face à l'électricité, XIXe—XXe siècle* (Paris: Éditions Belin, 1991), 19 ("Inexpliqué et Surnaturel").

9. Commissioned by Lord Cobham and made of Portland Stone, the sculpture is on display in the Victoria and Albert Museum, London, Inv.-Nr. A 10–1985.

10. The front view of the two-sided figure was sculpted by Georg Hofmann, and the similar rear view by Georg Josef Mutschele; see Ewald M. Vetter and Helmut Ricke, "Die Heilig-Grab-Kapelle des Klosters St. Michael in Bamberg," *Anzeiger des Germanischen Nationalmuseums* 1969: 121–49, at 128. An illustration of the anonymous glass painting "Mary as the Good Shepherd with the Lamb of God, Hitting the Evil in the shape of a Wolf with a flash of lightning" (second half of the eighteenth century, today in the Schloss Museum Murnau) is found in Kay Reinhardt and Helmut Bichler, *Blitz und Funke. Zur Kulturgeschichte der Elektrizität* (Marktoberdorf, 2001), (Exhibition Catalogue Stadtmuseum Marktoberdorf), 11. Compare the image of Mary hurling down flashes toward the Turks in a pilgrimage prayer book of 1726, *Süddeutsche Zeitung* 76 (March 31, 2004): 5.

11. See Manfred Beller, *Jupiter Tonans: Studien zur Darstellung der Macht in der Poesie* (Beihefte zum Euphorion, 13) (Heidelberg: Winter, 1979).

12. See the examples given by Engelhard Weigl, "Entzauberung der Natur durch Wissenschaft—dargestellt am Beispiel der Erfindung des Blitzableiters," *Jahrbuch der Jean-Paul-Gesellschaft* 22 (1987): 7–39; and Engelhard Weigl, *Instrumente der Neuzeit: Die Entdeckung der modernen Wirklichkeit* (Stuttgart: Metzler, 1990), 174–200.

13. Ulfried Müller, "Der Bau des Wetter-Ableiters auf der St. Osdag-Kirche in Neustadt-Mandelsloh 1782–1784," *Lichtenberg-Jahrbuch* 1994: 81–92, at 90. I owe this reference to Volker Schümmer.

14. Heinz Dieter Kittsteiner, *Die Entstehung des modernen Gewissens* (Frankfurt/Main and Leipzig: Insel, 1991).

15. Copperplate engraving, 1629, Deutsches Historisches Museum, Berlin, Inv.-Nr. Gr 93/39 (available online: *http://www.dhm.de/sammlungen/gifs/sammlungen/grafik/gr93_39.jpg*; December 14, 2006. I owe this reference to Stephan Klingen, Munich).

16. Tom Armstrong et al., eds., *200 years of American sculpture*, exhibition catalogue, The Whitney Museum of American Art, New York (Boston: David R. Godine in association with the Whitney Museum of American Art), 1976, Ill. 37–41. See H. H. Arnason, "Jean-Antoine Houdon: Benjamin Franklin," in *Metamorphoses in Nineteenth-Century Sculpture*, ed. Jeanne L. Wassermann, Exhibition Catalogue, Fogg Art Museum (Cambridge, Mass.: Harvard University Press, 1975), 63–70; and Willibald Sauerländer, *Ein Versuch über die Gesichter Houdons* (Passerelles, 1), (München and Berlin: Deutscher Kunstverlag, 2002), 33–35.

17. Ernst Benz, "Theologie der Elektrizität: Zur Begegnung und Auseinandersetzung von Theologie und Naturwissenschaften im 17. und 18. Jahrhundert," *Akademie der Wissenschaften und Literatur Mainz, Abhandlungen der geistes- und sozialwissenschaftlichen Klasse*, Jg. 1970, Nr. 12, 1–98, at 20. See Rudolf Drux, "'Über Gewitterfurcht und Blitzableitung': Lichtenbergs Abhandlung im Diskursverbund der Spätaufklärung," *Lichtenberg-Jahrbuch* 1997: 163–78, at 163 and 171. The most thorough and convincing analysis of the epigram attributed to Turgot is given by Alfred Owen Aldridge, *Franklin and His French Contemporaries* (New York: New York University Press, 1957), 124–38.

18. Quoted in Simon Schama, *Citizens: A Chronicle of the French Revolution* (New York: Alfred A. Knopf, 1989), 44.

19. Horst Dippel, "Franklin: An Idol of the Times," in *Critical Essays on Benjamin Franklin*, ed. Melvin H. Buxbaum (Boston: G. K. Hall & Co., 1987), 202–10, particularly 206. A good general overview of Franklin's portraits is provided by Brandon Brame Fortune with Deborah J. Warner, *Franklin & His Friends. Portraying the Man of Science in Eighteenth-Century America* (Philadelphia: University of Pennsylvania Press, 1999) (exhibition catalogue National Portrait Gallery, Smithsonian Institution, Washington, D.C., 1999).

20. Fragonard's sepia drawing has been popularized by an etching by Marguerite Gérard; see Charles Coleman Sellers, *Benjamin Franklin in Portraiture* (New Haven, Conn., and London: Yale University Press, 1962), 284–89; compare Schama, *Citizens* (cit. n. 18), 45; and Maria M. Tatar, *Spellbound: Studies on Mesmerism and Literature* (Princeton, N.J.: Princeton University Press, 1979), illustration after p. 80. The seminal interpretation is Mary D. Sheriff, "'Au Génie de Franklin': An Allegory by J.-H. Fragonard," *Proceedings of the American Philosophical Society* 127, no. 3 (1983): 180–93.

21. See Rolf Reichardt, "Die Bild gewordene Revolution in Frankreich: Eine bildhistorische Dokumentation," in *Freiheit—Gleichheit—Brüderlichkeit. 200 Jahre Französische Revolution in Deutschland*, ed. Rainer Schoch (Nürnberg: Germanisches Nationalmuseum Nürnberg, 1989) (exhibition catalogue), 157–200, at 188; Klaus Herding and Rolf Reichardt, *Die Bildpublizistik der Französischen Revolution* (Frankfurt/Main: Suhrkamp, 1989), 50; Klaus Herding, *Im Zeichen der Aufklärung. Studien zur Moderne* (Frankfurt/Main: Fischer, 1989), 123; and Christian Fuhrmeister, *Beton, Klinker, Granit—Material Macht Politik. Eine Materialikonographie* (Berlin: Verlag Bauwesen, 2001), 40–41.

22. The engraving is today in the Musée Carnavalet, Paris.

23. See, for instance, the examples in Rolf Reichardt, "Die Bildpublizistik zur Bastille 1715 bis 1880," in *Die Bastille—Symbolik und Mythos in der Revolutionsgraphik* (Mainz: Mittelrheinisches Landesmuseum, 1989) (Exhibition Catalogue), 23–70.

24. The etching by Spilbury (based on a drawing by Wortman and Mutlow) was first published as a frontispiece to Henri-Simon Linguet, *Mémoires sur la Bastille* (London: Spilbury, 1783); see Rolf Reichardt, "Historisch-politische Bildpublizistik als Anzeiger von Ereignis-Strukturen. Ein Essay anhand französischer Medienereignisse (1660–1804)," *kritische berichte*, no. 1 (2000): 38–61, at 49. See his thorough analysis of this image in his article "Prints: Images of the Bastille," in *Revolution in Print. The Press in France 1775–1800*, ed. Robert Darnton and Daniel Roche (Berkeley and Los Angeles: University of California Press, in collaboration with the New York Public Library, 1989), 223–51, at 226 and 229.

25. To name four of them: (1) "Souveraineté du Peuple, Destruction du Clergé et de la Royauté" ("The Sovereign French People, Destruction of the Clergy and the Royalty"), 1793, anonymous engraving in the Bibliothèque Nationale, Paris (Herding and Reichardt, *Bildpublizistik* (cit. n. 21), Ill. 194); (2) "Aus den Trümmern des Ancien Régime ersteht die neue Ordnung" ("From the ruins of the Ancien Régime a new order arises"), engraving by Abraham Wolfgang Küfner, Nürnberg 1795, first published as the frontispiece to Ludwig Albrecht Schubart, *Gallerie ausgezeichneter Handlungen und Karaktere aus der Französischen Revolution* (Nürnberg 1796), vol. 1, reproduced by Rolf Reichardt, "Der Nachwelt Denkmahle unserer Zeiten aufzustellen: Zum kommemorativen Potential der "Tableaux historiques de la Révolution française" in Deutschland (1795–1819)," in *Bildgedächtnis eines welthistorischen Ereignisses: Die Tableaux historiques de la Révolution française*, ed. Christoph Danelzik-Brüggemann and Rolf Reichardt (Göttingen: Vandenhoeck & Ruprecht, 2001), 143–212, at 151; (3) "La Liberté Armée du Sceptre de la Raison foudroye L'ignorance et le Fanatisme" ("Liberty armed with the scepter of reason strikes ignorance and fanaticism"), engraving by Chapuy after Boizot, 1789, Musée Carnavalet, reproduced by Michel Vovelle, *La Révolution française: Images et recit, 1789–1799*, vol. 1: *De la prérévolution à octobre 1789* (Paris: Livre Club Diderot/Messidor, 1986), 300–301; and (4) "Trinité Conventionelle," 1792, Bibliothèque Nationale, Cabinet des Estampes, Qb 26 décembre 1792, reproduced in James A. Leith, "Les Étranges Métamorphoses du Triangle pendant la Révolution Française," in *Les Images de la Révolution Française (Actes du Colloque des 25–27 Oct. 1985 tenu en Sorbonne)*, ed. Michel Vovelle (Paris: Publications de la Sorbonne, 1988), 251–59, here figure 4, and in James Leith, "Images of the Sansculottes" in *Iconographie et image de la Révolution française: Actes du colloque tenu dans le cadre du 57e congrès de l'ACFAS les 15 et 16 mai 1989*, ed. Claudette Hould and James Leith (Montréal: Association Canadienne-Française pour l'avancement des sciences, 1990), 131–60, figure 10, at 143.

26. See the caricature "The deluge," 1798, attributed to Gottfried Mind, reproduced in *Zeichen der Freiheit: Das Bild der Republik in der Kunst des 16. bis 20. Jahrhunderts*, ed. Dario Gamboni and Georg German (Bern: Stämpfli & Cie, 1991), (exhibition catalogue Bernisches Historisches Museum and Kunstmuseum Bern), catalogue entry 453.

27. For further information, see the bibliography in Gudrun Gersmann and Hubertus Kohle, "Auf dem Weg ins 'juste milieu': Frankreich 1794–1799," in *Frankreich 1800: Gesellschaft, Kultur, Mentalitäten*, ed. Gudrun Gersmann and Hubertus Kohle (Stuttgart: Franz Steiner, 1980), 9–22, 15n39.

28. Reichardt, Prints (cit. n. 24), 225; Beltran/Carré, *La fée* (cit. n. 8), 23.

29. Compare Willem Hackmann, *Electricity from Glass: The History of the Frictional Electrical Machine 1600–1850* (Alphen aan den Rijn: Sijthoff & Noordhoff, 1978); J. L. Heilbron, *Electricity in the 17th and 18th Centuries. A Study of Early Modern Physics* (Berkeley and Los Angeles/London: University of California Press, 1979), plate 10.1; Oliver Hochadel, "Wo der Funke übersprang: Die soziokulturellen Milieus der Elektrisiermaschine in der deutschen Aufklärung," in *Instrument—Experiment: Historische Studien*, ed. Christoph Meinel (Berlin & Diepholz: Verlag für Geschichte der Naturwissenschaften und der Technik, 2000), 295–306; and Heiko Weber, *Die Elektrisiermaschinen im 18. Jahrhundert* (Ernst-Haeckel-Haus-Studien, vol. 7), (Berlin: Verlag für Wissenschaft und Bildung, 2007).

30. See Schaffer, "Consuming Flame" (cit. n. 7), 513 and plate 24.10.

31. James Delbourgo, *Electricity, Experiment and Enlightenment in Eighteenth-Century North America* (Columbia University PhD thesis, 2003), published as *A Most Amazing Scene of Wonders. Electricity and Enlightenment in Early America* (Cambridge, Mass. and London: Harvard University Press, 2006), particularly chapter 4, "Electrical Politics and Political Electricity," 119–64 (here p. 134 and 135). I would like to thank James Delbourgo very much for providing me with a copy of his then-unpublished thesis.

32. Michel Vovelle, *La Révolution française: Images et recit, 1789–1799*, vol. 2: *Octobre 1789—Septembre 1791* (Paris: Livre Club Diderot/Messidor, 1986), 85, upper right. Interestingly enough, the bolts emanate from a hand that holds a piece of paper that reads "Affiliation suprême," an allusion to the "être suprême," which had replaced the Christian God.

33. Reichardt, "Die Bild gewordene Revolution" (cit. n. 21), 172.

34. Michel Vovelle, *La Révolution française: Images et recit, 1789–1799*, vol. 4, *Juin 1793 à prairial an III (mai 1795)* (Paris: Livre Club Diderot/Messidor, 1986), 294.

35. Four examples that obviously try to reactivate the older Christian iconography are: (1) "Das Schaudervolle, die Kirch und das gantze Reich durch sein Gift zerstörende Unthier des Freygeistes" ("The Horrible Monster of Free Thinking that destroys the Church and the whole Reich with its poison"), 1791, anonymous etching in the Staatsbibliothek Berlin, shows Enlightenment as an apocalyptic dragon and as the source of all evil that can only be overcome by the allied forces, the three eagles, which throw lightning at the monster; (2) "Das unersättliche Thier der Nationalversammlung" ("The insatiable/voracious animal of the National Assembly"), 1792, attributed to Johann Martin Will, Kupferstichkabinett of the Veste Coburg; (3) "Germania verteidigt sich gegen Revolutionstruppen" ("Germania defends itself against revolutionary troops"), anonymous etching, 1794, Landesbibliothek Coburg; all three are reproduced in the exhibition catalogue *Freiheit—Gleichheit—Brüderlichkeit* (cit. n. 21), 417, 433, 434; and (4) "Der Corsische Giftbaum" ("The Poisonous Tree from Corsica"), around 1814, mentioned

in *Karikaturen* (Bern: Benteli, 1972), (exhibition catalogue Kunsthaus Zürich), catalogue entry L 25.

36. "O Frankreichs Lilie? Warum ließt du dich von Ungeheuern entblättern?" ("O Lily of France! Why did you allow the monsters to strip your leaves?"), etching by P. Scherz and G. Pokorny, 1793, today at the Heeresgeschichtliches Museum in Vienna, shows how the Revolution—that is, the snakes (with silhouettes of leaders of the French Revolution)—tear the Bourbon lily (with silhouettes of prominent members of the Royal family) apart. The other European kingdoms/crowns throw flames and lightning against the snakes, thus signifying the military coalition against France; reproduced in Freiheit—Gleichheit—Brüderlichkeit (cit. n. 21), 405.

37. A similar use of the lightning motif characterizes another work by Gillray: "The Valley of the Shadow of Death," 1808: Napoleon surrounded by adversaries after the Spanish war. See M. Dorothy George, *English Political Caricature 1793–1832: A Study of Opinion and Propaganda* (Oxford: Clarendon Press, 1959), plate 46 and p. 144: "A flaming Papal tiara darts thunderbolts: 'Dreadful Descent of ye Roman Meteor.'"

38. The colored engraving "Gerechte Klagen" ("Just Complaints," 1817) shows a "sacred" thunderstorm that attacks those marketeers who charge the poor exaggerated prices for food and other goods in the famine of 1817; see Reinhardt and Bichler, *Blitz und Funke* (cit. n. 10), 53.

39. Hubertus Fischer, *Wer löscht das Licht? Europäische Karikatur und Alltagswelt 1790–1890* [Schriften zur Karikatur und kritischen Grafik, vol. 2] (Stuttgart: Hatje, 1994), 60.

40. Ibid., 160–64.

41. Take, for instance, the oil painting "Der gerächte Neger" ("The Revenge taken for the Black") by Johann Heinrich Füssli, 1806/1807 (Hamburger Kunsthalle); in 1807, England decided to completely abandon the slave trade. Füssli shows the powers of nature to be "the voice with which he [the former slave] speaks." The freed black slave (egalité) supported by a girl (liberté) looks at the slave ship struck by lightning (thus setting the inmates free); see Peter Thurmann, in *Europa 1789: Aufklärung—Verklärung—Verfall*, ed. Werner Hofmann (Cologne: DuMont 1989) (exhibition catalogue Hamburger Kunsthalle), 327.

42. Reichardt, "Prints" (cit. n. 24), 225.

43. Robert Darnton, "Introduction," in *Revolution in Print* (cit. n. 24), XIII.

44. Michael Diers, *Schlagbilder. Zur politischen Ikonographie der Gegenwart* (Frankfurt am Main: Fischer—Taschenbuch—Verlag, 1997).

Part 3
THE LIGHTNING ROD AS COMMODITY IN NINETEENTH CENTURY AMERICA

LIGHTNING RODS AND THE COMMODIFICATION OF RISK IN NINETEENTH CENTURY AMERICA

Arwen Mohun

IN THE LATE SUMMER OF 1853, Herman Melville spent a few months living just outside the village of Pittsfield in the Berkshire Mountains of Massachusetts. Chronically short of money, he was gathering inspiration for short stories that could be sold to popular magazines. The following spring, drawing on his experiences in Pittsfield, he wrote "The Lightning-Rod Man," which appeared in the August 1854 issue of *Putnam's Monthly Magazine*.[1]

Told from the perspective of a nameless householder, Melville's brief narrative recounts the visit of a lightning rod salesman during a summer thunderstorm. Refusing his host's offer of a place near the fire, the visitor plants himself in the center of the room, holding tightly to "a polished copper rod, four feet long, lengthwise attached to a neat wooden staff, by the insertion into two balls of greenish glass, ringed with copper bands [insulators]. . . .The metal rod," Melville informs the reader, "terminated at the top tripodwise, in three keen tines, brightly gilt."[2] Jupiter Tonans, as the narrator calls him, then attempts to frighten his host into buying his wares. "I warn you, sir, quit the hearth. . . . Are you so horribly ignorant then as to not know, that by far the most dangerous part of a house during such a terrific tempest as this, is the fire-place?"[3] Both men then fall into an argument about the efficacy of rods and the relative danger of lightning. Finally, the narrator grows so frustrated by the salesman's evasive patter that he breaks the rod and kicks his visitor out into the storm, berating him with a speech about the hubris of testing God's will by employing technology.[4]

This short story has become fodder for two generations of twentieth century literary critics. Most of them agree that the salesman represents benighted, mean-spirited, intolerant, evangelical Protestantism and the narrator, the forces of rationality.[5] In contrast, many nineteenth century readers would have recognized the literal basis of Melville's story.[6] At the time of its publication, a growing number of lightning rod salesmen had begun to wander the villages and back roads of America pitching their wares to both the wary and the unsus-pecting. Melville himself supposedly encountered such a character during his Pittsfield stay.[7]

Lightning rod salesmen were a relatively new phenomenon in 1854. Al-though lightning rods had been around for nearly a century, they had just begun to change from a homemade device erected by knowledgeable farmers, me-chanics, blacksmiths, and others to a commodity made in a factory and sold and installed by salesmen or lightning rod companies (fig. 8.1). This transition co-incided with what American historians have called the "market revolution"—a period in which increasing numbers of Americans (including, significantly, small-town and rural Northerners) became more tied to a cash economy and the values of a market society.[8]

Why would such a technology become commodified? And how did both the technology itself and the social relations around it change as this happened? To answer these questions, it is important to understand something about the nature of the technology itself. Lightning rods (or more accurately, lightning protection systems of which the rod is the most visible and symbolic part) are a safety tech-nology (like seat belts, fire alarms, and radon detectors) that falls into a category of devices primarily useful for mediating risk. As commodities, safety technolo-gies can be quite difficult to sell. People are hesitant to invest their hard-earned money in a form of insurance they may never need. Unlike fire or life insurance (at least in their twentieth century forms), safety technologies can also fail or, if faulty or badly designed, can actually make an accident worse. Nineteenth cen-tury lightning rod systems seem to have been particularly prone to failure (or at least the failures attracted the attention of the newspapers). Melville's narrator, for example, brings up the fact that lightning had struck a local church steeple armed with a rod only the week before—a rod the lightning rod man had in fact installed.[9] Because of the problems of uncertainty and failure, people who sell safety technologies tend to rely on two strategies to bolster their message: creat-ing a heightened perception of risk so that potential customers can imagine themselves as potential victims (a strategy with great potential for abuse) or com-bining safety with some other value—aesthetic beauty, social status, and so on.

The success of various kinds of risk-mediating devices also depends on whether they can catch the imagination of an era. Most pre–twentieth century safety devices (guards around fireplaces, safety harness on horses, etc.) were me-chanical—it was obvious from looking at them how they work. Lightning rods were anomalous among the common pre–twentieth century safety devices be-cause they were based on a new set of scientific theories. It was not obvious to the scientifically untrained how they work. One needed some understanding of, or at least belief in, electricity to see a lightning rod as more than a metal spike decorating a building. Electricity, as a natural phenomenon and a subject of sci-

Figure 8.1.   Warranty issued by Orcutt's Patent Lightning Rod Company. Warshaw Collection, National Museum of American History, Washington, D.C.

entific discovery, was also a matter of intense popular interest in the eighteenth and nineteenth centuries. It is no accident that the growth of widespread popular interest in lightning rods coincides with what historians of science such as Sally Kohlstedt have described as the first great period in which interest in and exposure to scientific ideas became part of American culture.[10] Lightning rods nicely suited a world that was still largely rural, unmechanized, and therefore more aware of the risks associated with weather.[11]

In a broader sense, this is also a story about the cultural and technological making of the modern world. Lightning rods are the first historical embodiment of three distinctive characteristics of the modernity. Poised along the rooflines of rural America, they bear witness to the origins of what social theorists now identify as modern industrial "risk societies."[12] They are also part of the story of the creation of consumer cultures in which everything, including the fear of being hit by lightning, is subject to commodification. And they show how the spread of scientific theories reshaped how ordinary people understood and dealt with their physical environments.

The first lightning rods began appearing atop buildings in the 1750s. For much of the next one hundred years of their existence, lightning rods and accompanying grounding apparatus were one-of-a-kind systems, made by blacksmiths, mechanics, and handy homeowners. They were shaped from relatively generic forms of iron and wood, much as one might make horseshoes or door latches out of pieces of purchased metal. For instance, in the 1753 version of *Poor Richard's Almanack*, Franklin recommends making the rod out of the "rod-iron used by nailers."[13] By the early decades of the nineteenth century, some makers might employ factory-made wire or purchased glass insulators, but these systems remained essentially homemade.[14]

In Franklin's time, lightning rods were a phenomenon of cities and gentlemen's plantations in the South. Writing to Genevan philosophe Horace-Bénédict de Saussure in 1772, Franklin himself pictured rods as widely used on "private houses in every street of the principal Towns . . . on Churches, Public Buildings, Magazines of Powder, and Gentlemen's Seats in the Country."[15] One of Franklin's fans, Charles Woodmason, a South Carolina plantation owner, captured his peers' enthusiasm for the device in a "Poetical Epistle" he sent to Franklin in 1752:

> No fire I fear my dwelling shou'd invade,
> No bolt transfix me, in the dreadful shade;
> No falling steeple trembles from on high,
> No shivering organs now in fragments fly,
> Nature disarms, and teaches storms to spare . . . .[16]

This pattern of dissemination makes sense when one considers that in the United States until the 1820s, information about the new technology spread primarily through private correspondence among gentlemen interested in natural philosophy, a few publications such as the *Gentlemen's Magazine*, and probably most importantly, the presentations of itinerant lecture-demonstrators who, before the era of nineteenth century small-town lyceums, traveled a circuit of the great coastal cities. For example, Ebenezer Kinnersley, the most famous of these early popularizers, regularly lectured on electricity in Philadelphia, Boston, and New York between the 1750s and 1770s. His syllabus, published as a broadside, promised instructions on how to "secure Houses, Ships, &c." through the use of lightning rods.[17] However, the historical record does not indicate exactly who heard this message or what use they made of it. It is likely that Kinnersley and other speakers addressed a select audience who had the time, literacy, money, and proximity to attend.

Franklin and his correspondents viewed this device as a kind of quintessential Enlightenment technology—a boon to mankind born of the marriage of scientific theory and experimentation embodied in a relatively simple technology. However, that boon was apparent only to those who shared an Enlightenment mind-set and a belief that science really did describe the physical world and offer a means to control and direct it.

The economic character of America before the nineteenth century explosion of industrialization also suggests a limited audience for this new technology. Lightning rods were exceedingly rare in the rural eighteenth century north not only because of lack of exposure to scientific ideas but also because the average farmer had neither the cash for the raw material to make a system nor access to technical explanations about how to do it.

This pattern changed in the nineteenth century, thanks to what historian David Jaffee calls the "village Enlightenment": an ante-bellum process of the "democratization of knowledge" through the distribution of printed materials and the spread of educational institutions through the northeastern United States.[18] Beginning after 1815 and with increasing frequency in the 1830s, discussions of lightning and lightning rods appear in a wide variety of popular journals read by rural and small town northerners. These journals are quite diverse, ranging from the *The Genesee Farmer and Gardner's Journal*[19] to the *New York Religious Chronicle*. The articles talk in extensive detail about new theories of electricity. They assume that readers are familiar with electrical apparatus such as the Leyden jar. And they make arguments about why readers should install lightning rods, give directions for erecting rods, and provide explanations for failures.[20] This enthusiasm for and access to technical and scientific knowledge is apparent in Melville's story. Both protagonists couch their argument in highly technical terms—discussing, for example, the relative conductivity of various substances (wood, metal, and cloth) and the relative effects of the upstroke and downstroke of lightning.[21] Melville himself was both a product of and a participant in the "village enlightenment" process. So too were the Berkshire farmers and townspeople from Pittsfield with whom he mingled during that thundery late summer of 1853. They had ready access to a lending library, the Pittsfield Library Society (founded 1801), and flocked to the Lyceum to hear lectures from such luminaries as Ralph Waldo Emerson, Henry Ward Beecher, and Horace Greeley.[22]

Before the 1850s, these publications assumed that their readers would make and install their own lightning rods (if they had them at all) or would commission a blacksmith or local mechanic to do it under their supervision. For instance, in the summer of 1832, *The Genesee Farmer* published an article from a reader asking for directions for choosing and installing rods on his barn.[23] In response several readers wrote in not only prescribing the height of the rod, how to fasten it to the building, and how to ground it, but also providing the scientific rationale for each specific point.[24] The subtext of many of these articles is that readers should adopt this useful scientific and technical knowledge as a matter of self-help and a demonstration of prudence and morality. For instance, the editors of *The Genesee Farmer* repeatedly berated readers who failed to take precautions: "the farmer who neglects this mode of preserving his house or his barn can have little claim on the charity or commiseration of the public."[25] Farmers were urged to employ "the aids philosophy and science have provided."[26] Similar advice and exhortation appears in a wide variety of farmer's journals, religious papers, and *Scientific American*.[27]

This kind of do-it-yourself article rapidly evaporated after 1850 as these devices became commodified—that is, patented, manufactured in factories, and sold by specialized companies. One can almost watch the conversation about self-help through the acquisition of useful knowledge disappearing. This is perhaps most vividly exemplified in the pages of *Scientific American*, which moved from giving useful advice toward evaluating the virtues of various patent systems and commercially available products.[28] Instead of telling readers how to make useful devices for themselves, the magazine became a kind of armchair world's trade fair of devices others have made for the marketplace.

Rural participants in both the village Enlightenment and the market revolution were a tempting target for a new group of opportunists: inventors with dreams of exploiting patents, manufacturers, and eager salesmen. Unlike their urban counterparts, rural people were cautious entrants into consumer cultures, worried about the opinions of their neighbors as well as their limited supply of money.[29] Lightning rods, with their preexisting meanings of science, rationality, and prudence, and their conspicuous position on the tops of houses and barns nicely bridged older values of prudence and morality with the emerging market culture of status through consumption and display.

The devices these entrepreneurs peddled often bore little resemblance to what one author called "the blunt, rusty iron rods which are in so general use."[30] They manifested whorls, tridents, side appendages, glass balls, twisted rods, hollow rods, square rods, round rods, oval rods—in short, every shape the human imagination could conjure up. Part of the reason for this baroque ornamentation was the patent system. The first "lightning conductor" or "lightning rod" patents began appearing in the 1840s. Since everyone (including patent examiners) knew that Franklin had invented the lightning rod nearly one hundred years previously, patentees typically claimed some form of "improvement" on the original device. Added elements or novel shapes allowed patentees and manufacturers to distinguish their products from those of their rivals and from previous patented devices. Most, of course, claimed each added bend and whorl and bit of gilding was utilitarian, adding to the efficacy of the rod in guiding "electric fluid" safely into the ground, but of course it did not hurt that they were decorative and looked expensive (see fig. 8.2).

The elaborate designs of these rods also added value in a different way. Even if the rod was never struck by lightning, it was still worth owning for the decorative and status value it added to one's house and barn. To have a lightning rod at all made one kind of statement, but to have a rod with extra curlicues and gild-

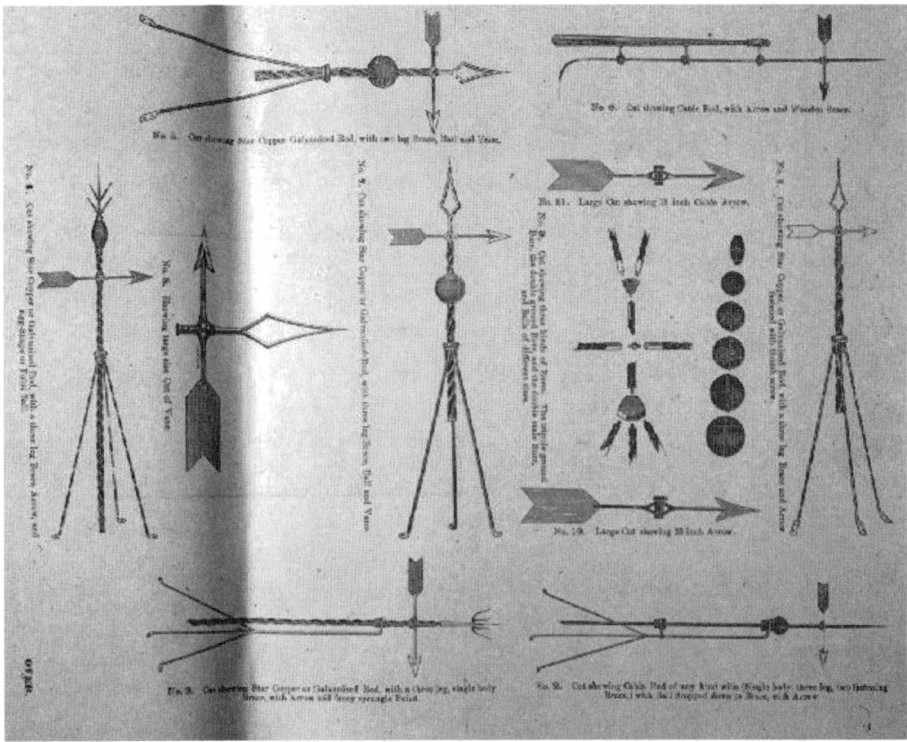

Figure 8.2.   Lightning Rod Catalogue, c. 1870. Warshaw Collection, National Museum of American History, Washington, D.C.

ing, harmonizing (if that is the right word) with increasingly elaborate Victorian architecture, was quite another.

Patents in and of themselves do not prove use since most patented devices never make it to the marketplace. But at least some of these early patentees rapidly moved to exploit their intellectual property. For instance, in 1851, G. W. Otis of Lynn, Massachusetts, patented a square rod with multiple points each tipped with gold (fig. 8.3). Soon thereafter he licensed the rod to the Lyon Manufacturing Company of New York—an event that was announced in *Scientific American*.[31]

The owner, Lucius Lyon, mounted an interesting marketing effort that piggybacked on both the institutions of the Enlightenment and an emerging market culture. Not only did he exploit the authority of the patent system (it must be good if it is patented) but he also solicited testimonials from ministers, professors of science, and heads of various schools. He entered the rod into prize competitions at various fairs including the New York Crystal Palace where it reputedly won a prize. Lyon was already the author of a treatise on lightning protection, which added further authority.[32]

The pamphlet also played on the aesthetic dimensions of Otis' rods. A testimonial from a "Prof. J. Ennis" made this dimension explicit. He said "as architectural

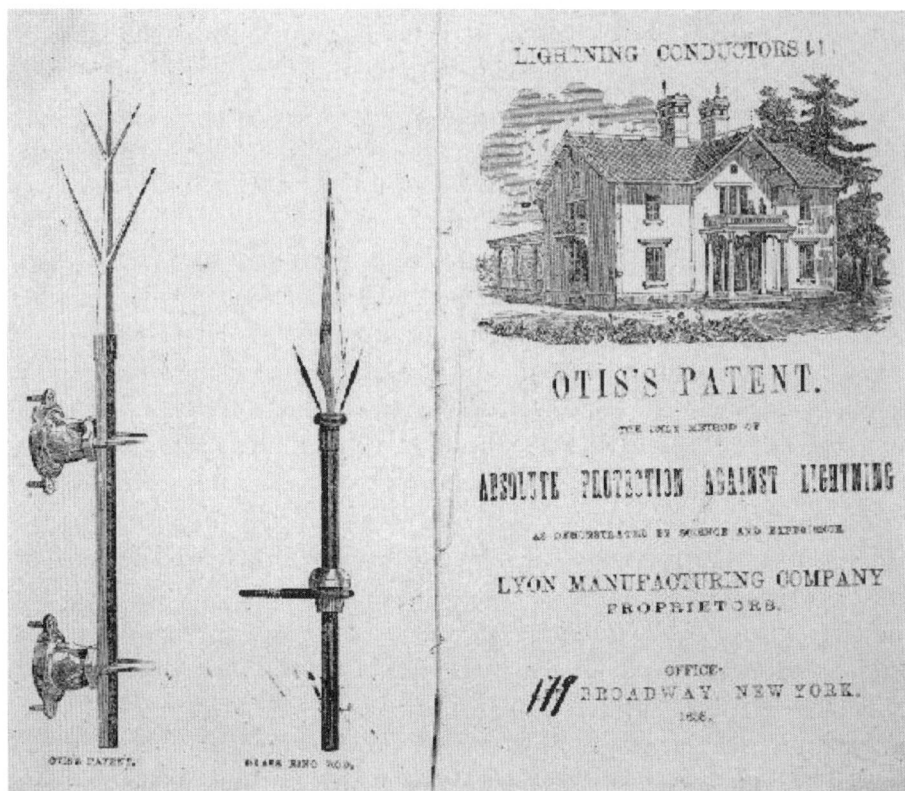

Figure 8.3.   Otis's Patent Lightning Rod Catalogue. Warshaw Collection, National Museum of American History, Washington, D.C.

taste is now a prominent feature in nearly all buildings, this new arrangement possesses a special recommendation. It adds decidedly to the beauty of the building, and thus serves the double purpose of protection and decoration."[33]

Undoubtedly, some farmers, mechanics, and clever people continued to make and erect their own lightning rods, and some lightning rod manufacturers sold their product through mail order or through the general store.[34] But most companies preferred the kind of salesmen described in Melville's story (although presumably with a better ability to get along with customers). These "lightning rod men" not only sold the system but also supervised its installation, which meant they needed a variety of skills and contacts.[35]

It is not entirely clear why lightning rod companies chose this form of marketing. In general, peddling was a very common form of sales in this period and was a cheap way for companies to sell over a wide geographic area.[36] Part of the reason may also have to do with the nature of the product. The enlightened farmer could erect his own rod. Lightning rod men knocked on the doors of people who had not yet been convinced that they needed lightning protection. Unless someone's house or barn had already been hit, it was unlikely that they would run down to the general store, and even then they might be more likely to abide by the folk wisdom that suggests lightning never strikes the same spot twice. Given the mixed press around lightning rods, sales pitches offered the opportunity to explain why the particular system in question would not fail.

To add value to their product and to slow competition from local mechanics, lightning rod companies strove to create at least the impression of a monopoly on expertise. Writing in 1879, crusader John Phin railed against the lack of information available to those who wished to make and install their own rods: "It is true that we have a few pamphlets and one small book; but they can hardly be dignified with the name of treatises on the subject since they are all written in the interest of some particular patent, and were never intended to give such information as would enable an ordinary, intelligent mechanic to erect a lightning rod for himself."[37]

Phin was also furious about the mystique surrounding patented devices. "Almost all the lightning-rods sold by itinerant vendors are patented; and it may therefore be worth while to remind our readers that *all essential requisites for perfect protection may be embodied in a rod which does not infringe on any patent.*"[38]

Because lightning rod men rapidly gained a very bad reputation as shysters, they could not necessarily embody the kind of authority that would sell their product.[39] Instead, they came armed with the kinds of pamphlets that Phin derided as well as warranties that gave the authority of the legal system to their sales pitches. These pamphlets are a formulaic compendium of various kinds of authority. They typically include a lengthy technical discussion of the nature of lightning and lightning conduction, testimonials from famous or authoritative men—typically scientists, heads of schools, and government officials—and quotes from books and articles about lightning.[40] At least some salesmen also carried elaborately illustrated books detailing the decorative elements that could be added to a system, thereby reinforcing the decorative side of the sales pitch (fig. 8.4).[41]

While the debates in farmers' journals make it relatively easy to find the voices of this generation of lightning rod users, the predominance of sales and prescriptive literature in the later period makes it more difficult to ascertain how potential consumers heard these sales pitches, why they bought or did not buy these products, and what their purchases meant to them. Surviving invoices from the second half of the nineteenth century suggest that most lightning protection systems cost between $65 and $200—5 to 10 percent of the cost of building a medium-priced house in the same period.[42] The relatively high cost of this safety technology suggests a substantial impact on family economies and raises the question of how such purchases were negotiated within families. Despite the seemingly male-gendered character of this technology, some evidence suggests that because lightning rods were an addition to houses and because they were

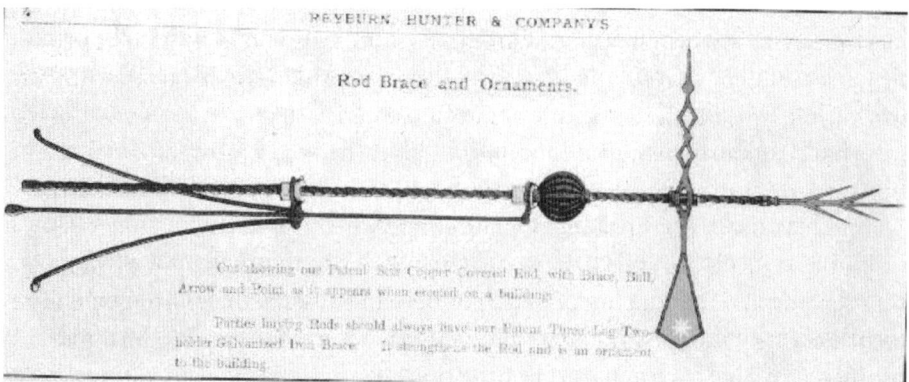

Figure 8.4.   Ornamental Rod, North American Lightning Rod Company Catalog. Chicago Historical Society.

used to protect families as well as property, the decision may have fallen partly to women. Widely circulated domestic manuals such as Catherine Beecher's *Housekeeper and Healthkeeper* (1873) included a section that specified what kind of lightning rod system to select and how to ensure it was properly installed.[43] Phin claimed that lightning rod men did prey on nervous women: "approaching a house in the absence of the men, they so terrify the ladies . . . that when the proprietor returns he gets no peace until a rod is ordered," but that kind of remark discounts the growing number of middle-class women with enough scientific knowledge to argue with salesmen, should they have the chance.[44] Moreover, the sales pitch was made within a space that was increasingly becoming both a site of commodity display and a culturally defined feminine space—which suggests it might have been gendered in a complex way no matter who was the audience.

Between the 1750s and the 1850s, the adoption and commodification of lightning rods in the United States passed through three distinctive stages. The first lightning rods were adopted by cultural elites whose ideas about the risk accompanying lightning and what to do about it had been reshaped by the Enlightenment. The rods they had erected were simple, utilitarian devices made by blacksmiths and other craftsmen. In the 1830s and 1840s, the spread of scientific ideas about electricity through popular magazines inspired a much broader audience, notably educated farmers, to erect rods. Manufacturers and inventors took note of this growing interest, and by the late 1840s, one could buy rods from specialized companies represented by itinerant salesmen. These rods were far more elaborate than their predecessors had been, both to allow patenting for novelty and to add value to the product as an architectural element.

Underlying this whole process are three stories: the metastory is about what some sociologists have called the emergence of risk-societies in the modern world. The two specific stories are about the spread of scientific ideas and the way they have reshaped risk perception and about the way industrialization and a market economy have allowed savvy entrepreneurs to grab hold of culturally resonant ideas about risk to create new products.

Lightning rods are still with us, but they no longer carry the powerful meanings that inspired eighteenth century gentleman to versify their virtues and nineteenth century farmers to debate the best way to erect them. Modern rods are subtle—architects do their best to hide them, and they are nearly culturally invisible as well. Twenty-first-century people do not think a lot about electricity. We rarely open our newspapers to read about lightning strikes, although clearly they still occur. When we do think about lightning, it is often in terms of surge protectors, not lightning rods. More often we worry about large systems failures, global warming, and epidemiological risks like skin cancer—these are the risks that speak to the cultural imagination of our own time.

Notes

1. Lea Bertani Vozar Newman, *A Reader's Guide to the Short Stories of Herman Melville* (Boston: G. K. Hall & Co., 1986), 269f.

2. Herman Melville, "The Lightning-Rod Man," *The Piazza Tales and Other Prose Pieces, 1839–1860* (Evanston, Ill.: Northwestern University Press, 1987), 119.

3. Ibid.

4. Ibid, 124.

5. For a summary of the various arguments, see Newman, *A Reader's Guide* (cit. n. 1), 269–82. Interpreters have tortured themselves to explain the internal inconsistencies in the story, particularly this last speech that seems to make the narrator (who is supposed to be the Enlightenment figure) into the spokesperson for a pre-Enlightenment religious fatalism. The religious interpretation does not fit with the powerful meaning lightning rods had in Melville's time as an Enlightenment technology. Hershel Parker's argument that this story is part of a genre of "Yankee peddler" stories makes more historical sense. See Parker, "Melville's Salesman Story," *Studies in Short Fiction* 1, no. 2 (1964): 154–58.

6. The first literary scholar to recognize the literal aspects of this story was Alan Moore Emery, "Melville on Science: 'The Lightning-Rod Man,'" *New England Quarterly* 56 (1983): 555–68.

7. Newman, *A Reader's Guide* (cit. n. 1), 270.

8. Charles Sellers, *The Market Revolution: Jacksonian America, 1815–1846* (New York: Oxford University Press, 1991) is the defining synthesis, but there is not much gender and consumption in it.

9. The narrator asks "of what use was your rod then?" True to the rhetoric of the time, the salesman responds: "Of life-and-death use." He blames the failure on his workman's failure to properly install the device (he failed to insulate it from other metal fittings on the roof). "Not my fault but his," he explains. Melville, "The Lightning-Rod Man," (cit. n. 2), 11. Discussions of failures begin with Franklin and are common across the nineteenth century. Improper installation is the most common explanation. See, for example, Benjamin Franklin to David Hume, January 21, 1762, *Papers of Benjamin Franklin*, vol. 10 (New Haven: Yale University Press, 1959), 19; and *Otis' Patent Lightning Conductors* (New York: Lyon's Manufacturing Company), 6f. Arthur R. Bostwick, citing the work of physicist Oliver Lodge at the end of the nineteenth century, refutes these arguments, stating instead that earlier failures were a result of incomplete understanding of the physics of lightning; "The Modern View of Lightning Rods," *The Literary Digest*, October 27, 1894, 764.

10. Sally Kohlstedt, "Parlors, Primers, and Public Schooling: Education for Science in Nineteenth-Century America," *Isis* 81 (1990): 424–45.

11. I went through the *Pittsfield Sun* for August and September 1853 (which the critical literature suggests Melville read) looking for ads for lightning rods. I did not find any, but I did find a number of accounts of buildings and people being hit by lightning. See, for example, *Pittsfield Sun*, August 17, 1853, and September 1, 1853.

12. The term was coined by Ulrich Beck. Beck's work originally appeared in 1986 in a German language edition, *Risikogesellschaft: Auf dem Weg in eine andere Moderne*, but came to be broadly read because of an English language translation, *Risk Society*, trans. M. Ritter (London: Sage, 1992). Anthony Giddens offers a complimentary approach in *Modernity and Self Identity: Self and Society in the Late-Modern Age* (Cambridge, Mass.: Polity Press, 1991). See also Mary Douglas and Aaron Wildasky, *Risk and Culture* (Berkeley: University of California Press, 1983).

13. *Poor Richard Improved: Being an Almanack and Ephemeris* (Philadelphia: B. Franklin and D. Hall, 1853), n.p.

14. The author of "Lightning Rods," *Genesee Farmer*, December 22, 1838, 401, hoped that readers would not be deterred from erecting rods because they could not obtain glass insulators. Homemade wooden ones would do just as well. In a transitional article, *Scientific American* touted the virtues of iron wire manufactured by Cooper and Hewitt "which answers for every purpose for lightning conductors." "Lightning Rods for Houses," *Scientific American* 7 (July 10, 1852): 344.

15. Benjamin Franklin to Horace-Bénédict de Saussure, October 8, 1772, *Papers of Benjamin Franklin* (cit. n. 9), vol. 19, 325.

16. Charles Woodmason, "Poetical Epistle" (1754), *Papers of Benjamin Franklin* (cit. n. 9), vol. 59.

17. J. A. Leo Lemay, *Ebenezer Kinnersley, Franklin's Friend* (Philadelphia: University of Pennsylvania Press, 1964), 73f.

18. David Jaffee, "The Village Enlightenment in New England, 1760–1820," *William and Mary Quarterly* 47 (1990): 327–46, at 327f. I would extend the range into the mid-Atlantic and place the timing somewhat later. In his study of Milwaukee newspapers between 1837 and 1846, Donald Zochert identified many articles about science,

mostly republished from East Coast newspapers and journals. See "Science and the Common Man," in *Science in America since 1820*, ed. Nathan Reingold (New York: Science History Publications, 1976), 7–32.

19. Cited hereafter as *The Genesee Farmer*.

20. See, for example, "Lightning Rods," *New York Religious Chronicle* 3 (July 9, 1825): 111, which discusses a paper on lightning rods published in Silliman's *Journal of Science*. An article in the *Christian Register* 10 (February 12, 1831): 28, explains the lightning rod construction theories of Londoner John Murray as described in a paper on "Atmospheric Electricity."

21. Melville, "Lightning-Rod Man" (cit. n. 2), 120–24, 122.

22. Richard D. Birdsall, *Berkshire County: A Cultural History* (New Haven: Yale University Press, 1959), 30, 154, 171.

23. "Lightning Rods," *The Genesee Farmer* 2 (August 25, 1832): 269.

24. Ibid., (September 8, 1832): 284f; ibid., (September 22, 1832): 299.

25. "Lightning Rods for Barns," *The Genesee Farmer* 3 (September 16, 1837): 290.

26. "Danger to Barns from Lightning," *The Genesee Farmer* 3 (September 23, 1837): 299.

27. I searched the APS Index to Early American Periodicals for "lightning rods" and found nearly forty articles for the period between 1820 and 1850. See, for example, "Proper Construction of Lightning Rods," *The Cultivator* 4, new series (May 1847): 161; and "A Word about Lightning Rods," *Scientific American* 3 (August 26, 1848): 387. The articles in religious papers are particularly interesting. They are explained in part by Herbert Hovenkamp's thesis that science and religion were not necessarily viewed as antithetical by these pre-Darwinian enlightened Protestants. See Herbert Hovenkamp, *Science and Religion in America, 1800–1860* (Philadelphia: University of Pennsylvania Press, 1978).

28. See "Lightning Rods for Houses" 7 (July 10, 1852): 344; and "More about Lightning Rods," *Scientific American* 8 (July 9, 1853): 341, which features the Otis rod discussed later.

29. Although her primary focus is women, Catherine Kelly provides a useful summary of these tensions. See Catherine E. Kelly, *In the New England Fashion: Reshaping Women's Lives in the Nineteenth Century* (Ithaca, N.Y.: Cornell University Press, 1999), especially pp. 9f, and ch. 8.

30. "Lightning Rods," *Scientific American* 3 (April 15, 1848): 237.

31. "More About Lightning Rods," *Scientific American* 8 (July 9, 1853): 341.

32. *Otis' Patent Lightning Conductors* (cit. n. 9), 23; Lucius Lyon, *A Treatise on Lightning Conductors* (New York: G. P. Putnam, 1853). Interestingly, the literary critic Emery, "Melville on Science" (cit. n. 6), mistook Lyon's treatise for a legitimate scientific work. I do not know whether it means anything that Melville and Lyon had the same publisher.

33. *Otis' Patent Lightning Conductors* (cit. n. 9), 24.

34. H. F. Morrow, for example, set up a shop in Chester, Pennsylvania, to sell and install rods. After a brief while, he gave his rod the name "Morrow Corrugated Rod," implying that it was a patented rod. See [no title], *Delaware County American*, May 24, 1865; "Copper Lightning Rod," May 19, 1865; "Busy," July 7, 1865. One of Lyon's direct

competitors, the Quimby Company, installed most of their rods within a fifty-mile radius of Manhattan using their own mechanics but would ship systems and directions to more distant locations. They seem to have found a market in Caribbean and Central American sugar plantations. See *Circular of A. M. Quimby & Son, Dealers in Quimby's Improved Lightning Rods for Houses and Vessels* (New York: Baker, Godwin & Co., 1854).

35. "My workman was heedless. In fitting the rod at top to the steeple, he allowed a part of the metal to graze the tin sheeting. Hence the accident. Not my fault, but his." Melville, "The Lightning-Rod Man" (cit. n. 2), 120.

36. Timothy B. Spears, *One Hundred Years on the Road: The Traveling Salesman in American Culture* (New Haven: Yale University Press, 1995).

37. John Phin, *Plain Directions for the Construction and Erection of Lightning Rods* (New York: Industrial Publishing Company, 1879), v.

38. Ibid., italics in the original.

39. Otis' pamphlet also contains what could be read as a frank admission of the problem of how these devices were sold. "Intelligent local agents are being appointed in every town and county in the United States, to avoid the impositions of irresponsible lightning-rod peddlers, whose flagrant misrepresentations have defrauded thousands in different parts of our land, besides increasing the exposure of their families and property to the fury of the lightning blast." *Otis' Patent Lightning Conductors* (cit. n. 9), 32.

40. In addition to the Otis pamphlet, see C. J. Hubbell, *The Hubbell Patent System* (Washington, D.C.: C. W. Brown, Printer, 1885); H. G. Denniston, *Lightning Rods: Their Adoption and Value when Properly Constructed and Scientifically Applied* (New York: The Williams Printing Company, 1889); *Circular of A. M. Quimby & Son, Dealers in Quimby's Improved Lightning Rods* (New York: Baker, Godwin & Co., 1854).

41. *North American Lightning Rod Company Catalog* (Philadelphia: Rand-McNally, n.d.), 6f.

42. Invoices in the Warshaw Collection files suggest that customers paid about twenty cents per foot between the 1860s and 1890s, buying several hundred dollars worth of rod, connections, and grounds for a house and outbuildings. See W. H. Demorest, "Invoice" (October 1, 1866); New York Star Lightning Co. "Invoice" (September 13, 1884); Atlantic Lightning Rod Company (1893) Lightning Rods—Box 1, Warshaw Collection, NMAH. The lightning-rod man in Melville's story wants a dollar a foot, which seems to me dramatic license although prices may have dropped over the century as production became cheaper and more efficient. For house prices, see Cooperative Building Plan Association, *Complete Collection of Shoppel's Modern Houses* (New York: Co-operative Building Plan Association, c. 1885).

43. Catherine Beecher, *Miss Beecher's Housekeeper and Healthkeeper* (New York: Harper and Brothers, Publishers, 1873), 369. See also *The Book of the Household; or Family Dictionary of Everything Connected with Housekeeping and Domestic Medicine* (London and New York: The London Printing and Publishing Company, ca. 1870), 345.

44. Phin, *Plain Directions*, (cit. n. 37), 27.

Earth Grounds and Heavenly Spires

Lightning Rod Men, Patent Inventors, and Telegraphers

Elizabeth Cavicchi

Introduction

THE NEARER LIGHTNING FLASHES, the more vulnerable we feel. That feeling came to many residents of the open spaces of nineteenth century America as their churches, barns, and prairie homesteads became frequent targets of lightning that accompanied the region's destructive thunderstorms. Like Melville's "Lightning Rod Man," salesmen of the time reaped commercial advantage by exhorting fear and promising the rod as a means to control nature's fury.[1] No clear understanding of how the rod functioned was provided or sought by most rod installers, makers, or their public. Rods went up in good weather and—if everyone was lucky—were never put to a real test. This separation from the actual phenomena allowed for installations that were inadequate to deal with a real strike; even the patented contraptions often turned out to be swindles.

A different outlook took hold among another nineteenth century community whose operations were daily affected by lightning: the telegraphers. Telegraphic signals were weak and could not compete with lightning whenever it disturbed the lines. Getting the signal through to its destination involved observing lightning's behaviors and improvising ways to safely divert it. Telegraphers could not ignore lightning or shortchange the ground connection. This interactive experience developed telegraphers' understandings of paths lightning took in their wiring and showed up by contrast the groundless myths and frauds of the sales-oriented rod profession.

This contrast is especially apparent in how the two extreme ends of lightning protectors were treated in home rod installations and at telegraphic stations. The upper visible ends of many marketed devices were decorative, boasting superfluous metal points on fluted hollow poles. The lower end stopped short, seldom going far enough underground to reach a reliable reservoir for a bolt's excess charge. In telegraphy, there was a reverse in emphasis. Lightning arresters put high up on telegraph lines were evident only to the trained—and continually checking—operator's eye. Ground excavations were substantive and essential to maintain if the signal was to go through. Thus by looking at the rod's ends, we gain an entry to disparate cultures and understandings that arose around lightning protection in nineteenth century America.

The Story of One "Lightning Rod Man"

In the mid-nineteenth century, philosophical proprietors anxious to protect their own buildings were not the only ones putting up lightning rods. The notion of making profit by outfitting others' houses arose to men who, often lacking homes of their own, were free to roam about, taking on any unprotected residence along their route as a potential sale. One of these peddlers was thirty-year-old James Wylie (d. 1866?) who left the Vermont farm where he was raised to go out and see the world. One November day in 1852, he wrote home from Nashville, Tennessee: "Brother Augustus, I take the present opportunity to inform you that I am a traveling in the south and a selling Lightning rods and find it a good business and like traveling in this country very well the people are kind and hospitable to strangers I am a driving a span of horses that cost 225 dollars and a good wagon and like my business very well."[2]

James was then on route to the Atlanta, Georgia, base of an outfit run by a certain Mr. Ladd (possibly a family acquaintance). Six teams of men and horses had started off traveling together, but they were about to part ways, some continuing on with Ladd to Georgia, others going to Alabama.

The following April, James was still in good health and spirits when he chided his brother for not writing: "I . . . did not know but you ware all dead or forgot this poor old bachelor . . . but I am still knocking about and climbing a houses and putting the genuine frame rods on them."[3] A month later, impatience began to show in James' correspondence. Mr. Ladd had taken off for Ohio, stranding James in Atlanta with responsibility for shutting up shop: "Mr. Ladd wants me to stay until the rods are all sold and the business closed up and I think it will take until July to do that and maybe longer but he thinks of coming back if he does I shall come home as soon as he gets back."[4]

There was no quick release from the job, or return to beloved Vermont for the now-homesick James. That summer stretched on and on as its heat scorched the corn and made outdoor work intolerable: "the hands have all quit but me and I shan't stay much longer . . . the weather is very warm and it is hard work and money won't tempt me to follow this business much longer unless the weather becomes cooler."[5]

In August, conscientious James was off in Tennessee, still under the direction of Mr. Ladd's letters: "Mr. Ladd . . . wants me to stay awhile longer and sel the horses and a few more rods—we have three horses yet to get rid of and about one thousand feet of rods but if the horses ware sold I should start for home next week but they are not and so I don't know when I shal start."[6]

James' mind clothed Vermont in a rosy hue. Months before, he had been agog over the belles who "dress so neet" and play heavenly piano music while taking snuff.[7] Now he urged "tel the girls not to all get maried before I get back to Vermont"; Vermont girls were ahead of those southern "whoshers and corn crackers."[8]

The next spring, James wrote a prescient epistle, which was also the last of those preserved from this stage in his life. One sunny day, to get to where he had to do a repair, he drove his horses fifteen miles to a train stop, then went twenty-five miles more by train. After finishing the repair, the next train was not due for hours, so James started walking along the tracks. Fatigued, he took respite by sitting in a store. On getting up, he felt "it began to grow dark the next that I know three men had hold of me holding me up and putting camphor on my head . . . have had a severe head ache ever since that fainting spel."[9] A day or more passed before he got back to the horses. The incident is telling, both of daily rigors in work and transport and of James' health. A decade later, James succumbed in a similar way: "James Wylie . . . died from a sunstroke in a harvest field near this place (Fayette Iowa)."[10]

James' letters disclose how lightning was far more distant from a rod-outfitter's concerns than the orders of an ever-absent boss, Mr. Ladd. James understood farming and horses; he was keenly observant of local crops "caries and plumbs . . . and new Irish potatoes," and seasons—noting a "backward spring" and a dry summer.[11] Yet he had no acquaintance with electricity and nothing in his letters suggests that his employment involved him with it, or in working out the paths lightning might take in striking a house. James spent his days atop houses and driving horses—he never mentioned excavating to ground the rods. His eventual interest lay not at either end of the rod but only in dispensing with his obligations to the business and in dreaming about home—and about getting rich by mining the Tennessee hills for copper.

The Rods, Tips, and Grounds of Patent Inventors

James' commitment to his absentee boss—and to repair a damaged fixture— suggests that he would not intentionally cheat anyone. However, most iterant peddlers were not so scrupulous: "Next to the substitution of sawdust packages for counterfeit money . . . the business of putting up lightning rods is a favorite field for the operations of the swindling fraternity."[12] An early twentieth century U.S. government report linked that fraud to the treatment of the rods' two ends: "Rods of every description were . . . erected at excessive cost. . . . The object was to make a great showing with a minimum of material and labour; to accomplish this the conductors in many instances were discontinued a few inches below the surface of the ground."[13]

These elaborate contraptions—outclassing the straight iron bars peddled by James—were products of another new profession, the "lightning rod makers" of whom the 1850 U.S. Census registered a total of 13; a decade later their numbers had risen to 164.[14] Many makers obtained patents for their inventions. The original patents now provide a resource for depicting nineteenth century lightning protection.[15] Each patent included diagrams, design and construction specifications, and explanation. A three-dimensional model was also required; a few are preserved at the Smithsonian Institution. Some patents were sold or licensed to other firms, thus distributing the invention across different territories.[16] One example is that of David Munson, the most prolific patenter of lightning rod apparatus, holding at least sixteen patents covering items made at his Indianapolis establishment.[17] His patent for tubular rods from sheet copper was bought by Lockart & Co., of Pittsburgh, and then by Reyburn, Hunter & Co., of Philadelphia, who marketed Munson tubes along with their own.[18]

The rod itself might not be a stock item; it could be made of different metals and take various forms. From Franklin's time to that of James Wylie, most rods were lengths of iron bar or chain that was braced or linked together. Experienced experimenter and Smithsonian director Joseph Henry concurred with the 1824 French Academy recommendation for a three-quarter-inch diameter solid iron bar.[19] However, as copper's high conductivity became known through telegraphy, many rod outfitters switched to copper, boasting of its superiority in their ads. Yet copper did not then provide the benefit we might suppose for it: the low-grade copper used in rods was impure, poorly conducting, brittle—yet still costly—and thus not a clear preference to iron.[20] The replacement of iron by copper typically brought with it a change from solid rods to hollow tubes. Although this was a cost-saving measure on the part of the maker, it was promoted as an advantage.

Another partial understanding about electricity supported this claim. Just as static electricity stays only on the surface of metals, lightning was supposed to run only down the rods' surfaces.[21] John Phin, a critic of the lightning protection trade, regarded this as a fallacy trumped up to sell flimsy sheet metal finery. He argued that static electricity was not a suitable model for the behavior of lightning discharges: "while static electricity . . . always disposes itself on the surface of bodies . . . electricity in motion pervades the entire substance of the bar . . .

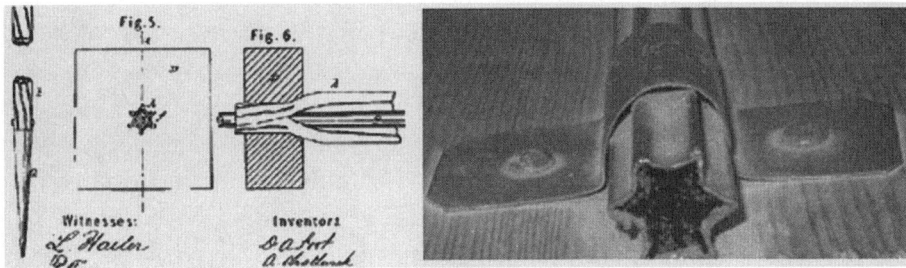

Figure 9.1.   *Left*: The patent diagram (93609) of Foot and Chadwick's lightning rod with a star cross section. The rightmost diagram shows how copper sheeting was pulled through the star die shown in the middle. *Right*: The patent model of Foot and Chadwick's star rod is catalog number 326310 in the Smithsonian collection.

consequently the power of such a bar to convey electricity is measured by the quantity of metal that it contains, and not by the extent of surface."[22]

However, the surface area argument won firm adherents among patenters — and some support from Joseph Henry's research.[23] Prominent maker David Munson used this argument to explain his system of "perfect protection" from lightning. He rolled copper sheet into fluted hollow tubes, which he claimed worked by "presenting a greater amount of surface upon which the electric fluid can act . . . thereby insuring its promptness of action . . . and thereby removing all danger from lightning."[24] He said these elegant tubes doubled as gutters to conduct both rainwater and lightning into the ground.[25]

Other patents capitalized on the ease of manipulating sheet metal by advancing hardware that was whimsical in appearance and justification. Strips of sheet copper were fed between the hand-cranked rollers of Stearns' portable rod-maker; out they came as twisted and corrugated hollow rods.[26] Rather than rolling sheet into tubes, Lyon fitted rods together by interleaving long copper strips so their cross section (in the short dimension) looked like the spokes of a wheel. This rod's outer edges were serrated "to receive or break up the force of . . . electricity in its passage to or from the earth"; the barbs were there to catch and impale electricity.[27] Minnesotans Foot and Chadwick combined the Munson tube with the pointy edges of Lyon's rod by drawing copper tubing through a toothed die. The drawn hollow rod with its star-shaped cross section was reputed to "furnish in a small compass a great extent of surface for the electric current" (figure 9.1).[28]

Ohioan Kleckner took the common concern for maximizing surface area to an extreme by perforating his hollow star rod and the trident of its upper tip with rows of evenly spaced holes (figure 9.2). These drill holes were seen as doubling the surface available to lightning by acting as means by which "the [electric] fluid may pass from the exterior to the interior of the rod, and vice versa."[29]

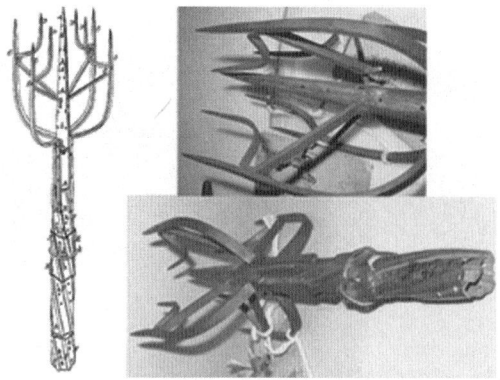

Figure 9.2.   *Left*: Kleckner's patent drawing (162828) of a twisted hollow rod with trident top perforated with drill holes throughout. *Right*: Photos of Kleckner's patent model (catalog number 308577) in the Smithsonian collection.

As the rod went skyward, its upper tip was often a separate attachment. Care was lavished upon these fixtures, whose beauty and metalworking craft render them still desirable today.[30] Stressing simplicity, Joseph Henry steered the public away from succumbing to such embellishment in his public advice on the subject. He advocated terminating the rod with a *"single point* tipped with platinum," an item which any ordinary instrument-maker could produce. He warned that ornate tips and thin coatings risked meltdown under the strong currents of an actual strike, and that extra prongs might provoke errant sideways electrical discharges.[31] This was the experience of Cincinnati rod-maker James Spratt who, having "frequently observed instances of destruction of the points by lightning, . . . was led to the plan of forming my points of a number of metals incased one within the other."[32]

Although Spratt let observation of the melted tip inform him in redoing its structure, it was unusual for a maker to put experience with lightning strikes to use. Many patents and brochures featured multiple-pointed "crowns" and spears that were expressions of design rather than findings based on electrical research. They justified these bristles of points and sharp edges by adapting Franklin's stance on points: makers contended that more metallic points would attract more lightning into the conductor, away from the house.[33] Testimonials appended to sales brochures announced "cheerful" satisfaction with points: "I have examined . . . the Patent Insulated Lightning Rod . . . [is] constructed on sound scientific principles . . . the multiplicity of points at the top . . . will not all be likely to be fused and blunted."[34]

Some owners described a rod's successful handling of a strike—but without the backup of hard evidence: "I think I saw the lightning pass down the rod on my kitchen chimney; it passed off with a very vivid flash, without, however, doing any damage—the rod was probably the means of saving my house."[35]

Among the names listed in endorsement of rods, that of "Samuel Morse, Inventor of the Magnetic Telegraph" appeared on more than one company's promotion.[36] Perhaps as tenuous a link as that influential name underlay the notion

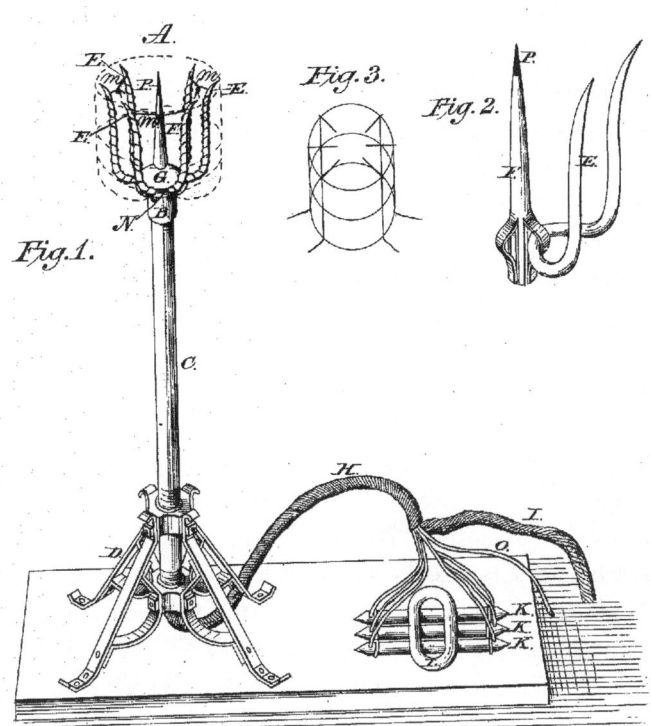

Figure 9.3.   The trident roof fixture (on left) of Bryan's patent 160151 terminates underground with an "earth battery" depicted on the right. The trident is made of pointed horseshoe magnets, covered in a cage to keep out birds.

that magnetized points on rooftops could function as safe attractors of lightning. Philadelphian Bryan patented such a "Magnetic Lightning-Rod" whose trident top was assembled from three upended steel horseshoe magnets, the better to magnetically draw lightning into the rod (figure 9.3).[37] Even ordinary peddlers took up the cause, making for "a great FLOURISH of TRUMPETS about *magnetized* points" while lacking any scientific basis for their claims.[38]

Patenters and salesmen championed excess in the rod's upper termination, but its underground lower end was often entirely neglected, as was likely the case with James Wylie's installations. Joseph Henry checked up on rodded houses that were damaged by lightning and found their rods stopped in packed, dry earth. This inadequate grounding had left lightning free to wreak havoc in coursing toward conductive masses, such as the gas mains, which were becoming more common by the midcentury.[39] These direct observations influenced Henry's view on what constituted a sufficient grounding. In the 1840s, Henry felt it would do to submerge the rod in an underground trench filled with charcoal.[40] Three decades later, he considered it paramount to solder the rod's end to nearby gas or water mains; otherwise the lightning would break violently through to the mains.[41] The example of a New Haven church bolstered others' arguments for including gas mains in grounding systems. This church was repeatedly damaged when lightning discharges passed from its rod's base through a two-foot brick wall to reach the city gas mains.[42]

Gas and water mains provided such extensive underground networks that they exposed, by contrast, how ineffectual all the other installations were. In rural areas where mains were not available, it was difficult to provide an underground network sufficient to discharge the lightning without undertaking substantial excavation. In such cases, one author advised that the rod's end be embedded in several charcoal-filled trenches leading to all nearby bodies of water.[43] Charcoal was advantageous both by the extended surface contact it made with the surrounding ground, and by its tendency to retain water and thus constitute a moist, more conductive path for the discharge.

Calamities, such as that of the New Haven church, aroused investigative interest from some installers.[44] In Boston, inspecting disaster sites was routine for both young William Orcott in the 1840s and Fred Harris four decades later.[45] Invariably the fault lay in rods' inadequate grounding.[46] Harris believed that observing stricken buildings was the only way to build up relevant experiences of lightning's behavior:

> Careful examination will show that there were either poor connections in the rod, or a very slight connection with the earth. . . . A conductor that is inserted one or two feet into the ground by pushing an iron bar down until it strikes a stone is scarcely any better than a rod that is cut off . . . above the ground. . . . People are satisfied if the end . . . is out of sight . . . [but] the rods are of no practical use. . . . My experience has taught me . . . the earth terminal should be most particularly attended to. . . . A conductor cannot be properly arranged by a person without experience. . . . Details that would be overlooked by a novice are sometimes the most important.[47]

Providing adequate, functional lightning protection was not simply a matter of putting up metal rods (as James Wylie had done); it involved careful analysis of a building's structure and its surroundings, informed by experiences of how lightning shatters anything blocking its route to ground. Shortchanges in the rod's ground connection was a ploy not restricted to peddling "lightning rod men." Even the established manufacturer, David Munson, treated the ground with some abandon. His manual for rod installation went to great lengths propounding the superiority of tubular copper rods, but about the ground it merely prescribed that "the rod should extend far enough below the surface . . . to reach a permanently moist stratum . . . say four feet. The hole may be made . . . with an iron bar."[48]

The adverse effects of this minimalist attitude are disclosed in a letter written by a local watchmaker to John Wise, a notable critic of lightning rods. Lightning had rampaged his house, darting to the gas pipe:

> Mr. Munson hearing of the circumstance immediately came up to see me about it. After making an examination and probing around where the rod entered the ground finally concluded that it was caused by a want of dampness in the ground, although this occurred after two days of rain. In order to impress me . . . he . . . dug a hole around the rod, and then dumped in [char]coal and poured on it . . . water, then covered it up and telling me that it would never occur again, and if it did he would pay me five hundred dollars in cash. He furthermore said, if I would say nothing about it he would patronize me hugely in my business.[49]

Munson's perfunctory response completely ignored the issue of the gas line, but what seemed to have angered the homeowner was Munson's greater concern for his business image than for genuinely resolving the problem.

While most rod designers expended their imagination on rooftop finery, a few were inventive about what they buried underground. The arguments given to explain these contraptions drew on ideas about electricity that were often weighted more with imagery than testable demonstration. A coil of copper ribbon was one common method of grounding rods, but the potential benefit of its large surface area was obviated by coiling it so tightly that little true earth contact was made (figure 9.4, left).[50] Similarly, the points of J. B. Burleigh's two-foot-diameter "Equilibrium Disk"—meant to be planted six feet underground—were alleged to "increase many times the inductive power" of the rod (figure 9.4, right).[51] Although a committee of the Franklin Institute (where Burleigh was a prominent member) sanctioned the equilibrium disk, they condemned its design for presuming that points underground would disperse electricity just as in air and for the insufficiency of its metal surface to adequately contact the earth.[52]

Even more bizarre was James Bryan's system for bringing about equilibrium between the electricities of air and earth by means of trident magnets posted on rooftops to collect atmospheric electricity and direct it by cable to an underground "earth magnetic battery." This battery supposedly restored electrical balance by uniting atmospheric electricity with galvanic currents extracted from the earth (see figure 9.3). Bryan's underground battery modeled an induction coil in form, not function. Its core of steel magnets was overwound first with a primary coil terminating in galvanic plates, and then with a secondary coil whose ends went up the lightning cable to the roof.[53] The whole was embedded in sulphur, said to intensify the current being educed from the earth. Deriding Byran's assembly as "one of the most ridiculous and unscientific combination of conductors . . . ever

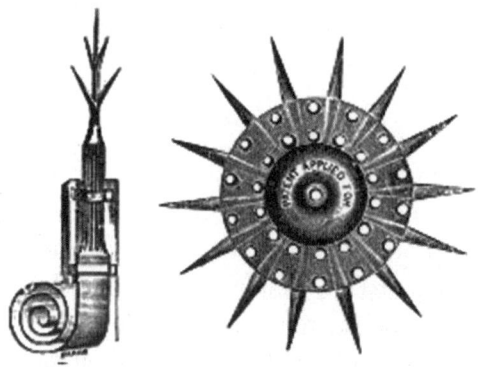

Figure 9.4.   *Left*: Kissell & Blickensderfer's 1867 brochure showed the rod's
trident top and the coil of copper ribbon
that was buried underground as a
ground. *Right*: Burleigh's "Equilibrium
Disk" was buried underground as a
ground for lightning rods; its points were
supposed to discharge the electricity.
Journal of the Franklin Institute 92
(1871): 4.

suggested for lightning protection," fellow Pennsylvanian Spang countered with
his own "reliable system" consisting of a perforated iron pipe buried vertically to
catch rainwater—and lightning—from gutters and direct both water and charge
into the earth.[54]

The lightning rod's ends were distinguished in many ways: up and down; vis-
ibly prominent and out of sight; a spire crowned with fine points or a bar
pounded in the dirt—perhaps into a maze of charcoal trenches. In the instance
when lightning did strike the top, it might not take the rod's prescribed path to
the bottom but might instead divert to some unseen reserve of water or metal pip-
ing. Since lightning strikes were not an everyday occurrence, itinerant lightning
rod men, such as James Wylie, had no direct feedback on their installations.
Patent inventors were more stable and identifiable than peddlers, yet they too
reaped profits by inflating costs while neglecting function.[55] The public did not
realize that nothing in the process of getting a patent certified that the invention
fulfilled its reputed claims. There was no accountability, no actual test. Phin, a
critic, railed: "Nine tenths of all the inventions protected by patents are perfectly
worthless. . . . All the various devices of hollow rods, twisted rods, hacked rods . . .
are perfectly childish."[56]

Developing understandings of how to direct lightning safely to ground took
critical experience and time. Despite the claims so often made, there was no in-
vincible patent protection from lightning—yet there were ways of learning to become more critical. Homeowners, like Munson's client, learned they had been
"had" when their rods failed. From the damage a stroke left behind, conscientious inspectors such as Joseph Henry and Fred Harris learned to infer the breaks
in rod continuity and ground features that had stopped or redirected the stroke.
Eventually, through a crossover of expertise with telegraphy, it became possible
to interactively test an implanted rod's continuity and grounding by including it
in a circuit with a battery and galvanometer.[57] Such galvanometric testing, while
not foolproof, enabled knowledgeable installers to interactively evaluate rods

and their surroundings—an interaction whose absence had given rise to such extensive fraud.

Telegraphers' Lightning Arresters

From the mid-nineteenth century on, telegraphy extended across increasingly long distances in America, steadily involving more people in its operation and maintenance. The United States census of 1850 records 544 telegraph operators, indicating that telegraphy was a small skilled profession, on the order of that of bankers (552), gas fitters (564), or undertakers (495), and far more numerous than those identified as lightning rod manufacturers (13). Telegraphers increased in number by nearly a factor of four in successive decades: 1,956 in 1860; 8,500 in 1870; and 22,809 (including 1,131 women) in 1880.[58]

These telegraphers worked a wiring that differed materially from most power lines and circuitry of today. One wire only—not two—carried the electric signal between any pair of stations; the circuit was completed by taps made at each station to the common earth. Such inclusion of earth in electrical paths was integral to eighteenth century friction machines, where high tension electricity sparked as it discharged to anything grounded. One early (1828) American telegraph used this form of electricity, keying it from a friction machine, through one wire to a recorder and then to ground.[59] However, the first telegraphs operating on galvanic sources were strung with multiple signal wires and one common "return" wire.[60]

The possibility of dispensing with the return (or "home") wire emerged in 1837 during Karl August Steinheil's experimenting to construct a twelve-mile telegraph near Munich where he was an instrument-maker and conservator.[61] His trials soon made evident what the lightning rod installers long disregarded: a reliable earth connection depended on a large area of surface contact between earth and the electrical system: "Quite recently, I made the discovery that the ground may be employed as one-half of the connecting chain. . . . It is necessary that at the two places where the metal conductor is in connection with the [earth], the former should present very large surfaces of contact. . . . Not only do we by this means save half the conducting wire, but we can even reduce the resistance of the ground below what that of the wire would be, as has been fully established by experiments."[62]

In the United States, Samuel Morse's early efforts independently yielded the same observation. Morse's business colleague Amos Kendall even experimented with running the telegraphic circuit off the ground—with no other battery—by embedding a copper ground plate at one end and a zinc plate at the other. He described this in a private letter, saying "I have all along said the ground was to become our battery."[63]

The ground return was built into subsequent telegraphic circuits, giving a common natural reference for all stations, no matter how distant. The interactive relations arising among earth, air, and telegraphic wiring, figured in how telegraphic circuits were designed and used. For example, local soil conditions affected the grounding provided by the three-foot-square metal plate typically implanted horizontally in it;[64] in very dry soils, where "earth cannot be found," a relative ground could be made by putting in several plates at different tensions. Nearby stations had to coordinate the materials used in the ground plates; if these were made of dissimilar metals, a permanent galvanic current would flow between them, undermining the telegraphic signal. And it took special care to insulate the wiring carrying the telegraphed signal from the ground—any direct contact introduced an inadvertent ground: "Dead earth in American phraseology . . . occurs when the line at any point touches the earth. . . . It practically divides the line in two . . . each terminal station working on its own battery to the fault.[65]

For example, the bare iron wire used for most overhead telegraph wires leaked electricity at the support posts, especially in wet weather. Although underground or submarine cables were coated in an insulator such as gutta-percha, moisture eventually degraded that insulation, making for further faults. How the ground return worked was puzzling. It was popularly pictured as equivalent to a wire: "the earth is practically one vast conductor."[66] But that image did not stand up to close inspection. Unlike ordinary wires, the earth ground showed no resistance to electricity—and yet if a tray of soil was inserted into a circuit, it was far more resistive than any wire. Increasing a wire's diameter reduced its resistance, but applying this analogy to earth would imply that the earth plates needed to be enormous. Further, the earth return was unlike ordinary conduction in that it did not exhibit the chemical decomposition products that accompanied galvanic currents. Observing that the analogy between earth and a resistive wire did not hold up to scrutiny, a leading British telegrapher offered an alternative understanding: "For myself, I have long seen the confusion in which this question involves us, and have been unable to admit the existence of [earth conduction]. . . . [We] are thus reduced to . . . rejecting altogether the idea of conduction . . . and of regarding the globe merely as a vast reservoir of electricity.[67]

Confusing as it was to understand, the ground figured in everyday telegraphic operations. Routine maintenance involved setting up tests to distinguish and track down leaks, wiring breaks, and inadequate grounds,[68] as advised in one manual: "the first business in the morning is to examine the batteries, test the lines, and ascertain if the connecting lines are all in working order . . . note [this] in the [office] journal."[69]

Telegraphers were in such continual interaction with ground, and each other, that their exchanges became encoded in shorthand—such as "G" for ground. The parodies by which telegraphic insiders poked fun at public inexperience with electricity resembled the critical responses to lightning rod men.[70] For example, a traveling telegrapher stops by an outpost telegraph office. In one tale, playing on the local's naiveté, this passing telegrapher claims things will work better if the ground line is disconnected and replaced with a line ending in a pail of water kept under the table. In another, the traveling expert spies a broken wire that turns out to be the ground, whose break had gone unnoticed for months since it was accidentally cut.[71] Telegraphic experience engendered a practical knowledge that was essential in ferreting out such faults, yet inscrutable to the uninitiated.

The environmentally scaled circuits of telegraphy encompassed both earth and sky. Just as earth could hamper the signal from below, atmospheric electricity—and lightning—disrupted telegraphic circuits from above. Wires spanning the big spaces of the American landscape were particularly afflicted: "Lines traversing several hundred miles, north and south, were subjected to repeated and almost constant interruptions. The adjustment of the apparatus had to be changed from moment to moment . . . very destructive . . . sometimes totally destroying [apparatus] . . . and at other times it has temporarily rendered ineffective the electromagnet."[72] This problem showed up immediately with the first trial telegraph installations. Morse attempted to redirect the high voltage atmospheric electricity to ground by sending line current through a conducting ball with ground lines nearly grazing its surface, but this proved inadequate.[73]

A telegrapher's pleas moved Joseph Henry to observe for himself what happened in a telegraph office while an electrical storm raged outside. The electrical ferocity borne by the wires astonished even the dean of American science. In that Philadelphia office, the incoming telegraphic signal wire was merely an inch away from the ground wire. Henry described how every few minutes there passed between these two wires "a series of sparks in rapid succession . . . simultaneous with a flash in the heavens. . . . The effect became so powerful that the superintendent, alarmed for the safety of the building, connected the [signal] wire with the city gas pipes, and thus transmitted the current silently to the ground. I was surprised at the quantity and intensity of the current."[74]

Henry advised running ground lines up the individual telegraph poles, mounted to not quite touch the signal wire, just like the wire arrangement in the Philadelphia office. However, there was no simple solution to the problem, and lightning continued to plague American telegraphy, giving rise to a diverse range of inventions. To protect their circuits, telegraphers devised "lightning arresters." All these arresters operated similarly. A single wire, suspended outdoors on poles, provided the electrical link between one telegraph station and the next. The signal's transmission was along this wire. The arrester was placed on it, just before the wire went inside the office with its delicate instruments. In the arrester, the signal current passed through a wire or surface, which was almost in contact with a ground line. At the low voltages of ordinary transmissions, the signal current would pass through the arrester, into the office equipment where the operator decoded it. But when an electrical storm raged outside, disturbances of high voltage and high current were induced on the long suspended lines, even in cases where no lightning was in sight. These disturbances overwhelmed the signal and endangered office equipment. The arrester circumvented this calamity by providing a quick route to ground that could be taken only by high voltage electricity, leaving the signal unaffected. The disturbance would tend to spark through the arrester's short air gap to get to ground, rather than traverse the high resistance electromagnet coils intervening between the arrester and the office's own ground wire. Sometimes a short length of very thin wire was inserted between the arrester and the office to behave like a fuse: under high currents (such as a direct lightning strike to the lines), it burned and broke the circuit. No device was infallible; in a severe storm, it was best to do as the Philadelphia superintendent had and disconnect the office from the outside line.

A multitude of arrester designs emerged along the American telegraph lines. Few were patented; individual improvisation was more prevalent. In a crunch, almost anything handy might do, even a cup of water. Wires coming from office equipment could be immersed in the water so their ends nearly touched a ground line, allowing high voltages to jump through the water gap to ground.[75] Or, instead of the cup, a water-filled flowerpot was fitted with a thin gauge wire that went through the water and out the hole in the pot's bottom, and carried the signal current. Any excess electricity was expected to dissipate in the water, or burn and break the fine wire, thus protecting the office equipment. A related device sent the signal onto a central line of fine wire hanging in the center of a grounded brass tube weighted at its lower end (figure 9.5).

Electricity of high voltage and current would jump from the fine wire to the ground, breaking the wire and letting the weight fall to open the circuit.[76] In this case, the line had to be repaired after every break, whereas the electromagnet-activated spring switch of another arrester opened when line current in the coils was high and sprang back on its own to reestablish the telegraphic circuit.[77] Royal House, a well-known telegraphic inventor, patented a similar arrester as part of his extensive system for postal telegraphy.[78]

An alternative design used side-by-side metal plates; one was at ground, the other at line potential. These plates were separated by a tiny air gap across which high voltage electricity could dart. The plates' adjacent edges were serrated with sharp points, to assist electricity's jump to ground, an analogy to the lightning rod tips (figure 9.6, left).[79] However, lightning's terrific currents could literally fuse the two plates' edges together. Fusion protected the equipment but also sent the signal to ground, which lost its information. In a later design, the ground plate also had a vertical post with a fine insulated wire carrying the signal coiled

Figure 9.5.   In this lightning arrester by Royal House, the signal wire on the left attaches to the ornamental brass support and enters a fine wire that passes through a hole in the cylindrical weight and connects to the office equipment. The fine wire supports the weight by a silk thread so that when high voltages pass on the fine wire, the wire breaks and the weight falls and transmits the discharge to ground. Turnbull, *Electro-Magnetic Telegraph*, 188.

around it. Very high voltages would simply dart across the gap between the line and ground plates while lesser disturbances migrated from the coil to the inner ground post (figure 9.6, right).[80]

A design that was widely adopted in mid-century telegraphy expanded the surface area held in common between line and ground by positioning the plates in parallel, one above the other. To emulate the sharp points of lightning rods, a wire bristle resembling a wire hairbrush made up the facing surface; its thousands of wire tips nearly grazed the ground plate—and sometimes dropped out (figure 9.7, left). The wire surface was hard to maintain; its teeth dropped out and

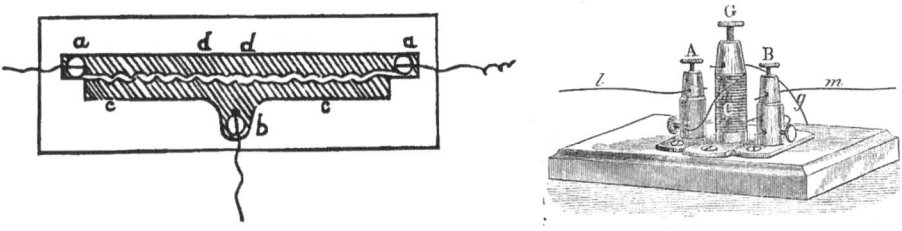

Figure 9.6.   *Left*: In telegrapher Charles S. Bulkley's arrester, the signal enters the top brass plate from the left; the office wire leaves it at the right. The lower serrated plate is grounded; it leads high voltages away from the office. *Right*: Line wire enters this arrester at post A, then coils around the grounded middle post G with a thin insulative sheet separating it from ground, and goes on the office connection at post B. Lightning pierces the insulation and discharges to ground via post G. *Left*, Turnbull, *Electro-Magnetic Telegraph*, 187; *right*, *The Telegrapher*, May 25, 1865, 100.

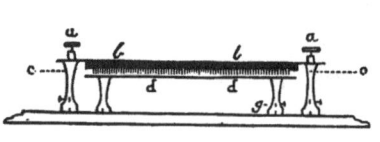 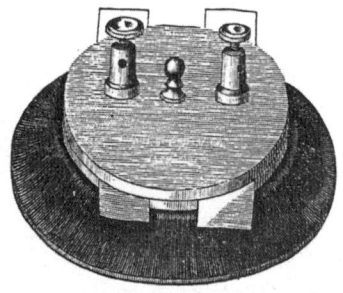

Figure 9.7.   *Left*: A lightning arrester where the top plate is in the form of a wire brush with multiple teeth just grazing the top of the grounded lower plate dd. *Right*: A lightning arrester produced by Boston instrument-maker Thomas Hall. The signal and office lines enter the posts on the top plate. The lower plate is grounded and separated from the top plate by a thin insulating tissue which breaks down under the high discharges of lightning. *Left*: Turnbull, *Electro-Magnetic Telegraph*, 187. *Right*: Shaffner, *Telegraph Manual*, 571.

the separation distance warped. This instrument evolved into a sandwich where a nonconducting tissue, such as paper, silk, or gutta-percha, intervened between the ground and line plates, keeping them a fixed distance apart (figure 9.7, right). The tissue prevented ordinary current from escaping the line while high voltages pierced through to ground.[81] But the erratic electricity of summer afternoons would not drain off the line, rendering "the wires almost entirely useless for hours . . . sometimes for days."[82]

No device could effectively insulate telegraphy from the vagaries of weather's electricity. Yet it was just this continual close contact between signals, and what telegraphers called the "fantastic tricks . . . [of the] Spirit of the Storms,"[83] that deepened telegraphers' understanding of lightning. As compared to the lightning rod men who rigged houses and then left without realistically testing their rods, telegraphers were always dealing with what environmental electricity did to their instruments and signals. In response, telegraphers adjusted circuits, repaired wiring and ground connections, and invented new instruments to protect their circuits—and then observed what happened next. This interactive exchange between instrument modification and observing the phenomena developed their experience. Telegraphers, as a group, were learning to handle electricity through interacting with it—a way of learning inaccessible to most lightning rod men and patent inventors.

Learning by Working with Lightning's Fire

Two themes informed the work of all the inventors and installers, one psychological, which addressed people's fear of lightning, and the other physical, which sought to safely dissipate lightning's discharges. Where people's hold on fear was greater—involving personal property and lives—their attention to electrical matters weakened, leaving them vulnerable to both deception and harm. Where people observed lightning's effects and were provoked to investigate it, those fears receded, enabling them to work and learn.

Telegraphers and rod inventors pulled differently on these two strands, the psychological and the physical strand. Telegraphers experienced atmospheric electricity as integral with the daily running of their apparatus. They could not ignore high voltage electricity; it came with each passing storm. If a line had no protective arrester, or if it failed, they had to improvise, and through improvising they learned about materials, lightning's terrific discharges, and the paths it took to ground. Because telegraphers were always signaling each other, they shared ideas and revised devices as a community, learning together. By contrast, the lightning rod men were not present when lightning hit. Patent inventors could get feedback on the adequacy of their work only by taking the time to investigate and dissect a striken house. Because few did this with the intent to remedy the rod's failings, their patented fixtures often became fanciful reflections of how lightning was believed to behave, rather than successive efforts at diverting lightning to a safe ground. Rather than coalescing as a community, their efforts splintered competitively.

Evidence of this story is recorded in the various forms taken by the lightning conductor's ends. For lightning rod men, from peddlers such as James Wylie to patent-holding businessmen such as David Munson, it was only the upper, visible end that mattered. Spiky rooftop spires often terminated as a rod forced into dry dirt. But telegraphers did not need to erect tall rods to receive lightning; it alighted of its own on their wires. Instead of the rod's top, telegraphers were concerned with its bottom. They worked at maintaining a good ground and devising instruments to bleed high voltage electricity off the signal lines and direct it to ground.

History is a doubled mirror, looking into both where we were and where we might be. These stories of professions and instruments gone by can still speak to us today and can suggest how the ways we experience natural phenomena provoke different inventive responses. Where the experience rests on accounts lacking direct tests with nature, all anyone can make of it will be a sham easily struck down by an actual bolt. But when experiences offer many ways of interacting with natural effects, any inventions exerted to handle those effects will undergo continual testing and growth.[84] These stories also reveal how community enables people to deepen their learning. Where community is diverse and interactive, everyone benefits; their work creates something real. But where people's work is without substance and isolated, there is also no basis for community; rivalries fashion fraud.[85]

The psychological and physical aspects also underlie a tension between work that is shaped by its market and work that evolves as an ongoing process of research. Lightning rod men and makers operated on the stage of the market; they benefited more from gauging what people would buy and from cultivating persuasive tactics than by studying electricity's strange behaviors. Conversely, telegraphers' primary commitment was to sustain and work with the electrical signal; this entailed experimenting directly with it and inventing apparatus that would immediately be put to a practical test. Yet there was some crossing over between the differing pulls of market and research on these professions. The effectual grounding of telegraphers found advocacy and, in some cases, a role in the lightning protection of buildings: "The rod should . . . have a large plate or bar of metal at the base, according to the arrangement of the electric circuits at all the telegraph stations."[86] And although telegraphers were daily involved in the intricacies of weak signals, faulty connections and noxious battery fluids, they were also affected by such market agencies as monopolies and patent rivalries among the many incipient American telegraph companies.[87]

Lightning is one of the most tremendous actions of nature that ordinary people witness directly. Its destructive power and paths are well beyond our control, and yet it does not forever freeze us in terror; we can respond to its behavior and our fears. In these responses, the worlds of human doings and markets come together with nature's ways and wonders. We can hope that strategies and understandings, such as worked out by lightning rod men and telegraphers, will arise as people today deal with the perennial occurrences of lightning—and with the newer dangers of our present world.

Notes

I thank the former Dibner Institute for the History of Science and Technology and the Edgerton Center, MIT, for support and resources during my development of this study. Dibner Fellows, including Alain Bernard, Abigail Lustig, Christopher Smeek, George Smith, Chen Pang Yang, responded to the work in progress. Judith Nelson and Howard Kennett searched out relevant materials in the former Burndy Library (now subsumed into the Huntington Library, San Marino, California). Robert Post encouraged my evolving ideas and added further suggestions. Roger Sherman searched for nineteenth century lightning rods and lightning arresters in the instrument collections of the National Museum of American History, Smithsonian Institution; Ronald Brashear located materials in the Dibner Library. Alan Janus and Mark Kahn facilitated my research of John Wise materials at the Archives Division, National Air and Space Museum, Smithsonian Institution. Carolyn Cooper shared methods for researching patents. I thank the Bakken Museum and David Rhees for supporting my participation in the conference "Taming the Electrical Fire." Peter Heering, David Rhees, and Oliver Hochadel offered comments, reflections, and focus in responding to this manuscript. Alva Couch insightfully extended my interpretations. The teaching of Philip Morrison, which deepened my sensitivity to inventive and observant responses to nature, is here continued in his memory.

1. Herman Melville, "The Lightning-Rod Man" (1853) in *The Piazza Tales and Other Prose Pieces 1839–1860* (Evanston, Ill.: Northwestern University Press, 1987). Melville's short story is available online in full, for example, at http://www.online-literature.com/melville/2056/.

2. James Wylie, November 14, 1852, Nashville, Tennessee; Augustus Wylie papers, New England Genealogical Society Library, Boston, Mass.

3. Ibid., April (1853?), Cave Springs, Georgia.

4. Ibid., May 29, 1853, Atlanta, Georgia.

5. Ibid., July 10, 1853, Calhoun, Georgia.

6. Ibid., August 9, 1853, Charleston, Tennessee.

7. James Wylie, letter of April 1853.

8. James Wylie, letter of August 9, 1853. James never married.

9. James Wylie, May 29, 1854(?), Loudon, Tennessee; see also the August 9, 1853 letter. However, the following year, James wrote home from Iowa, where he had bought land for framing and timber.

10. A. W. Callender, letter of March 16, 1901, Fayette, Iowa; Augustus Wylie papers, New England Genealogical Society Library, Boston, Mass.

11. James Wylie, letters of May 29, 1853, Atlanta, Georgia, and July 10, 1853, Calhoun, Georgia, Augustus Wylie papers, New England Genealogical Society Library, Boston, Mass.

12. John Phin, *Plain Directions for the Construction and Erection of Lightning-Rods* (New York: Handicraft Pub. Co, 1871), 27.

13. O. S. Peters, *Protection of Life and Property against Lightning*, Technologic Papers of the Bureau of Standards (Washington, D.C.: U.S. Government Printing Office, 1916), 8.

14. J. D. B. DeBow, *Statistical View of the United States . . . Being a Compendium of the Seventh Census (1850)*, vol. 5 (1854; repr., New York: Gordon and Breach Science Pub., 1970), 127; *Eighth Census of the United States (1860)* (New York: Norman Ross Pub., 1990), 666–67. Lightning rod men such as James were lumped among the ranks of over ten thousand peddlers tabulated.

15. All U.S. patents are now searchable online from the website http://www.uspto.gov/patft/index.html. Patents issued before 1976 are available in image format only, and can be accessed only by patent number. For determining patent numbers of inventions, see M. D. Leggett, *Subject Matter Index of Patents Issued from the United States Patent Office from 1790 to 1873* (Washington, D.C.: 1874).

16. For background on nineteenth century patenting, see Carolyn C. Cooper, *Shaping Invention: Thomas Blanchard's Machinery and Patent Management in Nineteenth-Century America* (New York: Columbia University Press, 1991).

17. The "Improvement in Lightning-Rods" U.S. patents of David Munson, Indianapolis, Indiana, include August 5, 1856, 15491; November 1, 1864, 22880; February

11, 1868, 74406; November 17, 1868, 84210; February 1, 1870, 99461; March 8, 1870, 100549; January 3, 1871, 110778; September 19, 1871, 119043; and July 23, 1872, 129675–7.

18. Circular of 1877, North American Lightning Rod Company of Reyburn, Hunter & Co., Philadelphia, Pennsylvania, held in Warshaw Collection, Archives Center, National Museum of American History, Smithsonian Institution, Washington, D.C.

19. Joseph Henry to John Torrey, August 2, 1838, *The Papers of Joseph Henry*, vol. 4 (Washington, D.C.: Smithsonian Institution, 1981), 81–83; Joseph Henry to M. W. Jacobus, July 19, 1843, in ibid., 372; "Directions for Constructing Lightning Rods," *Smithsonian Miscellaneous Collections*, 10, 1871, 1.

20. Phin, *Plain Directions* (cit. n. 12), 11; I. N. Miller, "Lightning and Lightning Rods," *American Electrical Society Journal* 1 (1875–77): 36–48. The best copper available was that mined at Lake Superior. Miller reports that copper from other sources could be more resistive than iron. Chicago rod company Kissell & Blickensderfer advertised their use of "pure COLD ROLLED Lake Superior Copper" in their pamphlet "The Best System of Protection from Thunder Storms" (Chicago: Church, Goodman & Donnelley, 1867).

21. For critiques of this assumption in practice see Phin, *Plain Directions* (cit. n. 12), 11; Miller, "Lightning" (cit. n. 20), 40; John Phin, "The Form of Lightning-Rods," *Popular Science Monthly* 7 (1875): 399–402; and H. W. Spang, *A Practical Treatise on Lightning Protection* (Philadelphia: Claxton, Remson & Haffelfinger, 1877), 121.

22. Phin, *Plain Directions* (cit. n. 12), 11; see also Phin, "Form of Lightning-Rods" (cit. n. 21).

23. Joseph Henry, "On the Protection of Houses from Lightning," *Proceedings of the American Philosophical Society* 4 (1845): 179–80; reprinted in *The Scientific Writings of Joseph Henry*, vol. 1 (Washington, D.C.: Smithsonian Institution, 1886), 231f.

24. Munson patent, August 5, 1856, 15491.

25. David Munson, *A Chapter on Thunder and Lightning: Their Causes and Effects Together with a Description of David Munson's Copper Tubular Lightning Rod with Spiral Flanges as Perfect Protection against the Disasters of Electricity* (Indianapolis: Indianapolis Journal Co., 1858), 16. Also see Patent of February 11, 1868, 74406.

26. Charles Stearns, Lowell, Massachusetts, U.S. Patent of September 20, 1859, 25534.

27. Amos Lyon, Worcester, Massachusetts, U.S. Patent of July 11, 1854, 11261.

28. David A. Foot and Avery Chadwick, Winona, Minnesota, U.S. Patent of August 10, 1869, 93609. A sample of the Star Copper Lightning Rod, manufactured by Hooker and Rich of Gainesville, Georgia in 1872, is catalog number 326310 in the Smithsonian's National Museum of American History collection.

29. John Kleckner, Canton, Ohio, U.S. Patent, January 26, 1874, 162828. Kleckner's patent model is catalog number 308577 in the National Museum of American History, Smithsonian Institution, Washington, D.C.

30. While the brochures of lightning rod firms typically included illustrations of the many-pointed tips, their texts argued for the functionality and efficacy of the devices without giving primacy to their decorative qualities. Today, that very quality of decoration and

craft imbues lightning rod artifacts with value for collectors. Illustrations of lighting rod points and paraphernalia, along with estimated market prices, are provided by Russell Barnes, *The Lightning Rod Collectibles Price Guide* (Austin, Tex.: R. Barnes, 1997).

31. Joseph Henry to James Rodney, March 25, 1846, *The Papers of Joseph Henry*, vol. 6, 1992, 404, where he mentions McAllister of Philadelphia, and Pike of New York, as suppliers of rod tips; "Directions for Constructing Lightning-Rods," *Smithsonian Miscellaneous Collections* 10 (1871): 1f.

32. James Spratt, Cincinnati, Ohio, U.S. Patent, May 4, 1852, 8930.

33. See George W. Otis, Lynn, Massachusetts, U.S. Patent of July 21, 1868, 8025; S. J. Mitchell, St. Louis, Missouri, U.S. Patent, May 23, 1865, 47846. For critiques of the view that lightning rod points attract lightning, see "Lightning Conductors and Attractors," *Scientific American* 1 (1859): 367; H. W. Sprang's critique in *A Practical Treatise on Lightning Protection* (Philadelphia: Claxton, Remsen & Haffelfinger 1877), 92, and Phin's critique, *Plain Directions* (cit. n. 12), 21. Illustrations of nineteenth century lightning rod points, as now identified by collectors, are provided in Russell Barnes, *The Lightning Rod Collectibles Price Guide* (Austin Tex.: R. Barnes, 1997).

34. "Otis's Patent. The only method of Absolute Protection Against Lightning" (New York: Lyon Manufacturing Co., 1858), 28.

35. Reyburn, Hunter & Co., *Lightning Rod Works*, Philadelphia, 1858, 8, Warsaw Collection, Archive Center, National Museum of American History, Smithsonian Institution, Washington, D.C.

36. Samuel Morse is listed as endorsing both the Pennsylvania Lightning Rod Co., Warsaw Collection, Archive Center, National Museum of American History, Smithsonian Institution, Washington, D.C., and the Cleveland Lightning Rod Co., formerly at Burndy Library, Cambridge, Mass.; now in the collection of The Huntington Library, Art Collections, and Botanical Gardens, San Marino, Calif.

37. James C. Bryan, Philadelphia, U.S. Patent of February 23, 1875, 160151.

38. "Otis's Patent"(cit. n. 34), 21.

39. Joseph Henry, "Record of Experiments," entries of July 16, 1841 (printed in *Papers of Joseph Henry*, vol. 5, 73–74), and May 18, 1858 (unpublished manuscript in Smithsonian Archives). Joseph Henry's drawing in this entry shows that lightning hit a house's rod at the tip, left the rod to pass along the roof's coping, then followed a gutter and then leapt to a gas pipe, leaving a destructive path. Henry published his 1858 observations in "Meteorology in its Connection with Agriculture," Agricultural Report of Commissioner of Patents, 1859, 461–524, reprinted in *The Scientific Writings of Joseph Henry*, vol. 2 (Washington D.C.: Smithsonian, 1886), 376–78. Municipal gas lighting is referenced in Martin V. Melosi, *Coping with Abundance: Energy and Environment in Industrial America* (Philadelphia: Temple University Press, 1985), 58.

40. Joseph Henry to James Rodney, March 25, 1846 (printed in *Papers of Joseph Henry*, vol. 6, 404).

41. Joseph Henry, "Directions for Constructing Lightning-Rods," *Smithsonian Miscellaneous Collections* 10 (1871): 1f.

42. Benjamin Sillimen Jr. documented the strikes on the church in *Principles of Physics* (Philadelphia: Bliss, 1870), 662. Other references to the story appear in David

Brooks, "Lighting and Lightning Rods," *Journal of the Franklin Institute* 94 (1872): 132–43; "Ground Connection of Lightning," *Popular Science Monthly* 1872: 120–21; John Mott, "Lightning and Lightning Rods," *Journal of the Franklin Institute* 96 (1873): 118–27; H. W. Spang, *Practical Treatise* (cit. n. 21), 103f.

43. Miller, "Lightning and Lightning Rods" (cit. n. 20), 46.

44. For a list of 505 buildings (with and without rods) damaged by lightning from 1665 to 1883, see Arthur Parnell, *The Action of Lightning Strokes in Regard to the Metals and Chimneys of Buildings* (London: 9 Conduit St, 1884). Examples of houses protected by a patent lightning rod are used as testimonials in the *Circular of William A. Orcutt, Manufacturer of Orcutt's Patent Lightning Rods* (Boston: John M. Hewes, 1857).

45. *Circular of William A. Orcutt*, 31; *Lightning* broadside of Orcutt, c. 1841, The Huntington Library, San Marino, Calif.

46. Fred E. Harris, *On Lightning Conductors* (Boston: J. E. Farwell, 1885). Harris' diagram on page 15 shows lightning hitting a rodded house, following the conductors down the sides, and then jumping to the cows standing in moist earth in the basement. The rod was not properly grounded; the cows were killed.

47. Harris, *Lightning Conductors* (cit. n.46), 3, 19–20.

48. Munson, *Chapter on Thunder and Lightning* (cit. n. 25), 17.

49. Mr. C. A. Ferguson to John Wise, quoted in John Wise manuscript "Thunderbolts and Lightning Rods," text of lecture given at the Franklin Institute, John and Charles Wise Ballooning Collection (Acc. 2001–0002), Archives Division, National Air and Space Museum, Smithsonian Institution, Washington, D.C.

50. Kissell & Blickensderfer, *The Best System of Protection from Thunderstorms* (Chicago: Goodman & Donnelly, 1867). See Miller's critique, "Lightning and Lightning Rods" (cit. n. 20), 46.

51. "The Equilibrium Disk," *Journal of the Franklin Institute* 92 (1871): 4.

52. Phin, *Plain Directions* (cit. n. 12), 24 and Spang, *Practical Treatise* (cit. n. 21), 131f.

53. James C. Bryan, Philadelphia, U.S. Patents of February 23, 1875, 160151–2. See Spang's critique, *Practical Treatise* (cit. n. 21), 132–34.

54. Spang, *Practical Treatise* (cit. n. 21), 134–37; also Henry W. Spang, Reading, Pennsylvania, U.S. Patents of September 7, 1875, 167415, and October 23, 1877, 196493.

55. Similar fraudulent behavior among traveling medicine salesmen of the time is described in Ann Anderson, *Snake Oil, Hustlers, Hambones: The American Medicine Show* (Jefferson, N.C.: McFarland & Company, 2000). One source quoted there reports that for the person who used medical galvanic belts, "all the wearer ever got out of his belt was a dream—and a blister" (35).

56. Phin, *Plain Directions* (cit. n. 12), iv.

57. Spang, *Practical Treatise* (cit. n. 21), 112–15.

58. DeBow, *Statistical View of the United States* (cit. n. 14), 128. *Eighth Census of the United States (1860)*, 676f.; *Ninth Census of the United States (1870)*, Francis A. Walker, 1872, 706; *Tenth Census of the United States (1880)*, 746.

59. Ken Beauchamp, *History of Telegraphy* (London: Institution of Electrical Engineers, 2001), 22.

60. Ibid., 32.

61. Alto Brachner, "C. A. Steinheil, a Munich Instrument Maker," *Bulletin of the Scientific Instrument Society* 12 (1987): 3–7.

62. Karl August Steinheil, "Upon Telegraphic Communication, Especially by Means of Galvanism," *Annals of Electricity, Magnetism, and Chemistry* 3 (April 1839): 509–20.

63. For Morse's use of the ground return, see Lewis Coe, *The Telegraph: A History of Morse's Invention and Its Predecessors in the United States* (Jefferson, N.C.: McFarland & Co, 1993), 71. Amos Kendall's letter of April 2, 1846, Washington, D.C., to Thomas Clark, Treasurer, New York, is held in the Papers of Amos Kendall, Library of Congress Manuscript Collection, Washington, D.C.

64. The grounding plate is described in a footnote in J. E. Smith, *Manual of Telegraphy Designed for Beginners* (Chicago: Western Electric, 1865), 41.

65. Thomas Lockwood, *Electricity, Magnetism and Electric Telegraphy: A Practical Guide and Hand-book* (New York: D. Van Nostrand, 1883), 231.

66. *Manual of Telegraphy and Description of Instruments* (New York: J. H. Bunnell and Co., 1889), 9.

67. Charles V. Walker, *Electric Telegraph Manipulation* (London: George Knight and Son, 1850), 37.

68. For the problems in making and testing a ground, see R. S. Culley, *A Handbook of Practical Telegraphy* (London: Longman, 1863), 54–58, 103–17; William Henry Preece and James Sivewright, *Telegraphy* (New York: D. Appleton, 1876), 251–81; Smith, *Manual of Telegraphy* (cit. n. 64), 41–47; Lockwood, *Electricity, Magnetism and Electric Telegraphy* (cit. n. 65), 195–96, 228–30.

69. Tal Shaffner, *Telegraph Manual: History and Description of the Semaphoric, Electric and Magnetic Telegraphs of Europe, Asia, Africa and America* (New York: Pudney & Russell, 1859), 752.

70. See the tales in *Anecdotes of the Electric Telegraph* (London: William Tegg and Co., 1848), and William John Johnston, comp., *Lightning Flashes and Electric Dashes: A Volume of Choice Telegraphic Literature, Humor, Fun, Wit & Wisdom, Contributed to by all the Principal Writers in the Ranks of Telegraphic Literature as Well as Several Well-known Outsiders* (New York: W. J. Johnston, 1877).

71. "Telegraphic Humor," *The Telegrapher*, October 15, 1866, 41. A genderized variant of this story is expressed in the poem "Out of Adjustment" in *Lightning Flashes and Electric Dashes* (cit. n. 70), 62, where a passing telegrapher applies his expertise not to mock the local female telegraph operator but to fix a fault and thereby steal a kiss from her.

> For I was a knight of the telegraph key,
> And knowing that currents when terribly weak
> Beget a fell anguish, a dire misery
> That pen can't portray nor human voice speak,
> My heart urged me forward; go in there I must,
> And do what I can to offer relief . . .

72. Shaffner, *Telegraph Manual* (cit. n. 69).

73. Laurence Turnbull, *The Electro-Magnetic Telegraph: With an Historical Account of its Rise, Progress and Present Condition* (Philadelphia: A Hart, Carey, Hart, 1853), 182. Turnbull's book is online at Google books http://books.google.com/books?vid=HARVARD32044056200454&printsec=titlepage.

74. Joseph Henry, "On the Relation of Telegraph Lines to Lightning," June 19, 1846, *Proceedings of the American Philosophical Society*, iv, 260–68; reprinted in *The Scientific Writings of Joseph Henry*, vol. 1 (Washington, D.C.: Smithsonian Institution, 1886), 251. See also "Meteorology in its Connection with Agriculture," Agricultural Report of Commissioner of Patents, 1859, 461–524, reprinted in *The Scientific Writings of Joseph Henry*, vol. 2 (Washington, D.C.: Smithsonian Institution, 1886), 380–84.

75. Smith, *Manual of Telegraphy* (cit. n. 64), 47; Turnbull, *Electro-Magnetic Telegraph* (cit. n. 73), 188.

76. Turnbull, *Electro-Magnetic Telegraph* (cit. n. 73), 188. The arrester with the weighted line was used by all stations operating with designed equipment by Royal House.

77. Shaffner, *Telegraph Manual* (cit. n. 69), 567–69.

78. Royal E. House, Binghamton, New York, U.S. Patent of February 20, 1874, 180090.

79. The "Bulkley protector," which had two brass plates, is described in Turnbull, *Electro-Magnetic Telegraph* (cit. n. 73), 187; a later two-line version is illustrated in Spang, *Practical Treatise* (cit. n. 21), 171. Because any office intermediate on a telegraph line would have two incoming wires, its arrester was fitted with two-line plates; see Spang, *Practical Treatise* (cit. n. 21), 170–71. A serrated grounded plate placed near the telegraph line where it is held on a pole was patented by J. L. Finn, Elyria, Ohio, on March 24, 1858, U.S. Patent 75889. The telegraphic line itself provided the serrations in a patent device that ran the telegraph line in a long zigzag with each pointed zag almost grazing the ground; J. N. Gamewell, U.S. Patent of August 7, 1855, 13389.

80. "A New Lightning Arrester," *The Telegrapher*, May 29, 1865, 100.

81. The wire card or "Carey's protector" is described in Turnbull, *Electro-Magnetic Telegraph* (cit. n. 73), 187; both it and the parallel plates are discussed in Shaffner, *Telegraph Manual* (cit. n. 69), 570f.

82. "A New Lightning Arrester," *The Telegrapher*, May 29, 1865, 100.

83. *American Telegraphic Magazine* 1, 280.

84. In an essay citing recent analyses of urban structures at stress with natural elements and written with the hindsight of Hurricane Katrina (2005), Ari Kelman reflects on what history shows about the limitations, failings, and lessons of even our best efforts to control nature; "Nature Bats Last: Some Recent Works on Technology and Urban Disaster," *Technology and Culture* 47 (2006): 391–402. While modern technological constructions benefit from extensive engineering tests and experience, the colossal failures of some are traced to organizations driven by production and management concerns where data analyses on physical, engineering, and safety issues are minimized and not admitted to the decision-making process and culture; for examples, see Eda Kranakis, "Fixing the Blame: Organizational Culture and the Quebec Bridge Collapse," *Technology and Culture* 45 (2004): 487–518; Philip Tompkins, *Apollo, Challenger, Columbia: The Decline*

of the Space Program: A Study in Organizational Communication (Los Angeles: Roxbury, 2005). The role of feedback and interaction in technological development is discussed in Henry Petroski, *Success through Failure: The Paradox of Design* (Princeton, N.J.: Princeton University Press, 2006); and Davis Baird, *Thing Knowledge: A Philosophy of Scientific Instruments* (Berkeley: University of California Press, 2004). In applying this kind of argument to education, the learners' interactive experiences with nature (or other subject matter) are crucial to developing workable understandings whereas formulaic instruction is analogous to the ungrounded rod; see John Dewey, *Democracy and Education* (1916; repr. New York: Free Press, 1944); Frances Hawkins, *The Logic of Action* (Boulder: Mountain View Center, 1969); Eleanor Duckworth, *"The Having of Wonderful Ideas" and Other Essays on Teaching and Learning* (1987; repr. New York: Teachers College Press, 2006).

85. Experiences and developments of diversity in communities are analyzed by Shelly Harrell and Meg Bond, "Listening to Diversity Stories: Principles for Practice in Community Research and Action," *American Journal of Community Psychology* 37 (2006): 365–76; specific narrative examples of diversity and community are provided in the same journal's volume 37 special issue.

86. "Lightning Conductors," *Scientific American* 1 (1859): 305–6. Spang, *Practical Treatise* (cit. n. 21), 129, used a galvanometer to evaluate the groundings of his installations. He measured the earth ground resistance of a telegrapher's earth terminal (50 ohms), and found that it was much lower (and thus more effective in conducting discharges) than that of a typical lightning rod termination (70–1600 ohms). These measurements offer a check on the rod's grounding without requiring the hazard of a lightning strike.

87. David Paul Hochfelder, "Taming the Lightning: American Telegraphy as a Revolutionary Technology, 1832–1860" (PhD diss., Case Western Reserve University, 1999); Robert Luther Thompson, *Wiring a Continent: The History of the Telegraph Industry in the United States: 1832–1866* (Princeton, N.J.: Princeton University Press, 1947).

Part 4
LOOKING AT LIGHTNING: MODELS, INSTRUMENTS, AND MODERN RESEARCH

THE LIGHTNING ROD
A CASE STUDY OF EIGHTEENTH CENTURY MODEL EXPERIMENTS

Willem D. Hackmann

Experiments to Isolate and to Imitate Electrical Phenomena

For almost two hundred years the term "electricity" denoted static electrical phenomena. To William Gilbert in 1600, it meant electrical attraction to distinguish it from magnetic attraction. More properties were added as these were discovered by increasingly sophisticated electrical devices coupled to a theoretical framework that was also evolving—electric repulsion (1629), glow (1705), transmission or conduction (1720), spark and shock/pain (1730), and heat (1740)—so that by the 1740s, an electrical phenomenon had to possess *all* these properties to be considered *true* electricity. The discoveries of these properties involved two basic types of experimental procedures: (a) experiments to discover or isolate the properties of specific natural phenomena, such as whether the electric spark is "hot," and (b) laboratory experiments with models. Underlying both strategies was an intuitive belief in the fundamental regularity of natural processes. This became the raison d'être for these two research strategies: (a) the use of the analogous argument in the framing of physical theories, and (b) the model experiment in which natural processes were "re-created" in the laboratory by means of small models. Model experiments were used when strict inductive procedures could not be applied, or when no direct experimental intervention was possible.

An early example of a model experiment in the history of electricity is the sulfur ball apparatus of Otto von Guericke, constructed in the 1660s to demonstrate that the nature of gravitation was not magnetic—as had been argued by Gilbert—but electrical, caused, according to von Guericke, by the friction of the air on the rotating earth. The sulfur ball represented the Earth, which was made up mainly of "sulfurous particles," and the hand with which it was rubbed while being rotated represented the atmosphere. According to an anonymous reviewer in the *Philosophical Transactions* (probably Robert Boyle or Robert Hooke), with this globe the "Impulsive, Attractive, Expulsive and other vertues of the Earth . . . may be ocularly exhibited."[1] This model was intended to confirm a broad hypothesis about the nature of gravitation, but model experiments could be much less ambitious. In the realm of electricity, a frequently discussed model experiment was devised by Henry Cavendish in 1776. Cavendish used a model fish to demonstrate that the apparent dissimilarity between the so-called animal electricity and "common [laboratory] electricity" was caused by differences in the relationship between various electrical factors, such as intensity, capacity, and resistance in the laboratory and in the fish.[2]

Similarly, the analogous argument underpinned experiments with models in the laboratory to elucidate the behavior and causes of lightning, earthquakes, and other natural phenomena. The use and development of such models in a specific topic such as lightning and the lightning rod illustrate the intricate relationship that evolved during the eighteenth century between theory, scientific observations, and laboratory experiments.[3]

Atmospheric Electricity and Model Experiments

Atmospheric electricity was an immensely popular subject in the second half of the eighteenth century. The most common underlying theoretical framework was Franklin's concept of a single electric fluid. Most atmospheric phenomena such as lightning; the aurora borealis[4]; whirlwinds; the formation of clouds, hail, and rainstorms; and earthquakes were thought to be caused by the restoration of imbalances in the amount of electric fluid in different regions of the atmosphere. It is instructive to examine how the various theories that were developed relate to laboratory model experiments. This also highlights a key aspect of eighteenth century experimental science — the importance placed on model experiments in cases where the natural phenomena were too large or too complex to be studied in the laboratory directly. The technique of experimenting with scaled-down laboratory models has been extremely fruitful in the history of science in general and electricity in particular. It could also result in triviality, as in (with the benefit of hindsight) Joseph Priestley's model experiment in the late 1760s, which sought to demonstrate the electrical nature of earthquakes. He passed a powerful electrical discharge from a Leyden battery through a wet plank floating in a basin of water. This caused wooden blocks placed on the plank to tumble. The plank represented the earth and the blocks, buildings. The lateral force of the discharge caused the plank to move and the blocks to fall over. Priestley equated this force with that produced by electric discharges coursing through the earth's crust.[5] This leads to the intriguing question: what made some model experiments more "successful" than others?

The Formation of Clouds: Electrical Theories and Models

One of the most important atmospheric phenomena thought to have a bearing on natural electrical phenomena—in particular on the occurrence of rain, hail, and thunderstorms—was the formation of clouds. Laboratory experiments in 1742 on the evaporation of water caused John Theophilus Desaguliers to conclude that the air was electrified positively, and that evaporation and the continuing vapour state of water in the atmosphere were due to water particles receiving "electric virtue" from the air, causing these particles to repel each other. Any subsequent condensation was owing to the loss of electricity, allowing the fine water particles to conglomerate. This view was held by Franklin and in modified form by most natural philosophers, including Volta and Horace Bénédict de Saussure.[6]

In 1753 Franklin determined that clouds were usually charged negatively, but sometimes also positively.[7] This would remain a thorny issue during the eighteenth century. The study of atmospheric electricity resulted in the development of sensitive electrometers by de Saussure, among others, in the latter part of the eighteenth century. Several mechanisms were proposed to account for the variations of the electricity of the atmosphere and of the clouds.[8]

It was argued that once the clouds had reached a substantial charge they would discharge in a lightning strike, and this, too, was demonstrated by model experiments. A particularly striking model experiment was devised by the Dutch natural philosopher Martinus van Marum in 1787. It was an extension of a similar experiment performed by William Henley in London some years previously.[9] Van Marum made two "artificial clouds" from large bladders filled with hydrogen gas. One was charged positively and the other negatively. This caused the bladders to rise toward the ceiling and to approach each other slowly. When a short distance apart, a spark jumped across the gap and both clouds immediately began their descent. According to Van Marum, this model experiment demonstrated the observed behavior of clouds before and during a thunderstorm. To make the demonstration more spectacular for his audience, he arranged for a third balloon filled with a mixture of hydrogen and air to be placed between the two clouds so that it exploded with a resounding bang when the spark occurred.[10]

Another arrangement with which Van Marum demonstrated the same phenomenon was based on the "plate of air condenser" devised by Aepinus in 1759. Van Marum probably got the idea for this model experiment from Priestley.[11] Two large, smooth circular boards were suspended from the prime conductor of his generator. The top board had its lower face and side covered with tinfoil while the bottom board had its upper face coated with varnish and bronze powder and its side with tinfoil. This board was connected to the earthed negative conductor of the electrical machine. At the center of each board was a large copper sphere to act as a spark gap between the condenser plates. When the electrical machine was set in motion, streams of sparks were observed to flow from the upper to the lower sphere, and from there along the bronzed surface "like numerous lightning bolts" to the earthed tinfoil side. Van Marum argued that there was a close correspondence between the bronzed surface of the board and a cloud, as either one was only a partial conductor of electricity. According to Van Marum, this and other model experiments he had performed demonstrated that the lightning discharge was more complex than had been assumed by Franklin and, shortly afterward, by the Franklinist Beccaria because it did not simply consist of a positively electrified cloud discharging its excess electricity onto a negative cloud or onto the earth. Another important factor was the mechanism of electrical induction, which Lord Mahon had argued was the cause of the "return stroke" during a thunderstorm.[12]

Model Lightning Rods

The most dramatic aspect of atmospheric electricity was the thunderstorm. Its study led to the first practical technological benefit of electrostatics—the lightning rod.[13] In 1708 Samuel Wall alluded to the similarity between lightning and the electric glow, and to the extremely small sparks that could be produced at that time. The analogy between the two sets of phenomena became increasingly obvious as electrical laboratory apparatus increased in power. A dramatic breakthrough occurred with the invention of the Leyden jar in 1746. This device, connected to the new generation of electrical machines, produced sparks that had all the appearance of miniature lightning bolts, and on a small scale, even some of the destructive power. During the next twenty years or so, the main research on lightning rods took place in the laboratory, primarily by means of models.

Franklin's proposed original "sentry-box experiment" to test the hypothesis that lightning is an electrical discharge between clouds, or between clouds and the earth, was described in 1749 and communicated to Peter Collison the following year.[14] Franklin proposed his sentry-box experiment after describing a series of laboratory experiments in which charged insulated conductors were discharged by a nearby pointed conductor, from which he concluded the potential efficacy of the lightning rod.[15]

In his 1749 paper, Franklin described two model experiments mimicking electrified clouds and the action of pointed rods. In the first, a pair of large brass scales was hung from the crossbeam of a balance by insulating cords of silk. The beam was suspended from the ceiling by a packthread so that the bottom of the scales was about one foot from the floor; as the weight of the scales caused the packthread to untwist, the scales moved in a circular motion, passing over an iron punch placed upright on the floor. As one of the electrified scales moved across the punch it would discharge with a snap, but when he terminated the punch by a sharply pointed needle, the scale would be discharged silently. In the second experiment, an electrified pasteboard tube behaved in a similar fashion when approached by the punch and then by the punch terminating in a needle.

Franklin was well aware of the difference between grounded rods used to protect property and ungrounded test rods used for experiments with lightning in atmospheric laboratories, but he underestimated the danger of experimenting with ungrounded conductors, possibly because of his experience with models in the laboratory—hence the shock of the scientific community when Professor Richmann was killed in his St. Petersburg atmospheric laboratory in 1753, even if Priestley wrote that he wished to have been "the first martyr of the electrical sciences."[16]

Two questions dominated the development of the lightning rod in the second half of the eighteenth century. First, does the rod neutralize or attract and conduct away to earth the electric fluid of the thundercloud? Second—and what would become of particular interest in England—should the rod terminate in a knob or in a sharp point? Franklin was unsure of the answer to the first question. In his laboratory experiments he observed a glow discharge emanating from the tip of his miniature rod when the charge of the model cloud was neutralized (lost its charge). He observed that with an insulated rod, this process stopped after a few minutes but, conversely, would last as long as the cloud was charged if the rod was earthed. He also noticed that the glow discharge was enhanced when the model rod ended in a sharp point. Hence, he advocated that the real rod should be terminated likewise.

The main fears were whether the rod would actually attract the lightning stroke, a prospect that filled people with dismay, and whether this was less likely to occur if the rod ended in a ball rather than in a sharp point. The court painter and amateur electrical experimenter Benjamin Wilson in 1777 devised a portable model centered on the Leyden jar to demonstrate the safety of a ball-shaped or blunt termination over a sharp one [fig. 10.1]. He attached to the outer coating of a large Leyden jar a brass fork with one tine ending in a sharp point and the other in a ball. The fork faced a large brass ball on a separate support. The inner coating of the Leyden jar was connected to a long curved brass rod terminating in a small ball (A) while the support for the large brass ball had a similar curved rod terminating in a small ball (B). The sleeves ending in the ball and the point could be moved toward the large ball on the other support until the discharge from the Leyden jar would strike either the ball or the point, or both. According to Wilson, in this experiment the point drew a spark three times longer than the ball conductor; therefore a lightning rod terminating in a ball was less dangerous.[17]

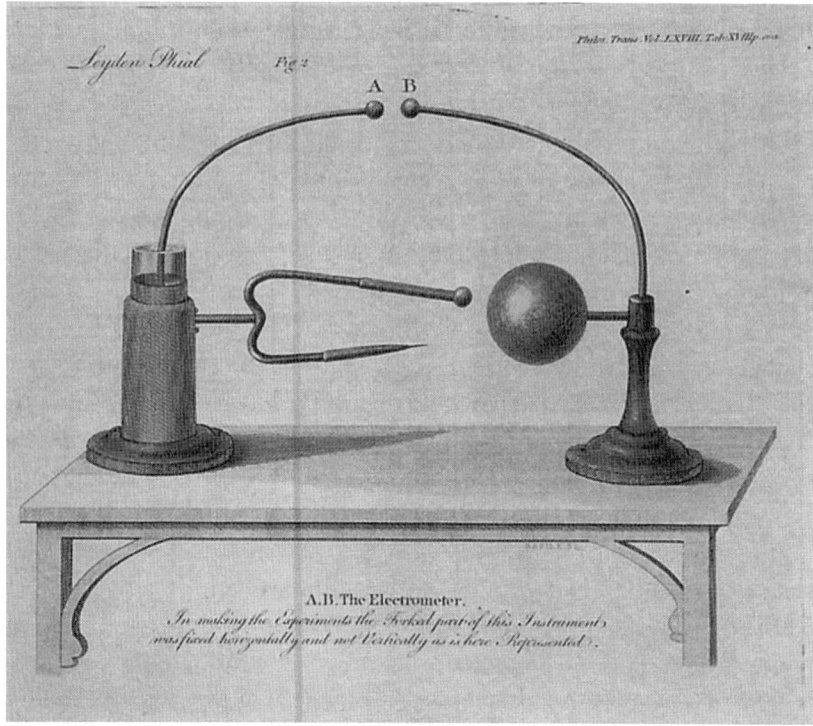

Figure 10.1.   Benjamin Wilson's point-ball apparatus to model lightning strikes. *Philosophical Transactions of the Royal Society* 68 (1778): pl. XVIII, fig. 2.

The most impressive model experiment to try to settle this matter was performed by Wilson at the "Pantheon" in London, in the presence of his patron, George III, in 1777.[18] This was in response to debates on how to protect the Ordnance Board's powder magazines after it moved from Greenwich to Purfleet, and is discussed in detail in this volume by Rod Home. What is important in our context is that the ensuing debates centered on what were considered the "right" interpretation of the lightning rod model experiments, which in truth were inconclusive. We now know that Wilson was correct when he claimed that points could not discharge distant thunder clouds, but Franklin's followers were also right when responding that the higher the rod protruded above the building the more effective its protection.

After the Purfleet magazine was rather embarrassingly struck by lightning in May 1777 in spite of being protected with Franklin's rods, Wilson seized the initiative and with backing from the court constructed a dramatic model in which he tried to be as realistic as possible.[19] His electrical machine was of conventional size, but its conductor was extremely large, similar to Franklin's made of pasteboard tubes but coated with tinfoil, 155 feet long and about 16 inches in diameter, and suspended by silk cords from the theater's ceiling. The discharge could be augmented by a second conductor consisting of 3,900 yards of copper wire. This arrangement, although less efficient than a good-sized Leyden battery, certainly had more dramatic appeal. It represented an electrified cloud, under which he placed a model house, representing the Purfleet gunpowder magazine, complete with lightning rod. Because he could not move his artificial cloud, he could instead move his model beneath it by pushing it along grooves. After demonstrating his model experiment to the royal court, the Ordnance Board, and members of the Royal Society, he entertained the paying public with a daily show during the autumn of 1777.[20]

Rod Home has described Wilson's conclusions of his (model) experiments in detail, which led Wilson to advocate blunt over sharp-pointed rods. However, most natural philosophers did not accept Wilson's reasoning and continued to advocate Franklin's pointed rod.[21] It makes little difference, of course, whether the real lightning rod terminates in a small ball or point.[22] However, it did result in an interesting controversy concerning the validity of Wilson's model spearheaded by the Franklinists: would it be better to model the electrical behavior of clouds with small conductors or Leyden jars than with Wilson's huge cumbersome pasteboard cloud, which was much more difficult to control in a laboratory environment? In more abstract terms, how analogous should a model based on analogy be? William Swift imitated Wilson with a much smaller scale model consisting of a pair of prime conductors, one positively and the other one negatively charged. He added an insulated vessel of water supported above the model powder house to imitate rainfall.[23] Others, such as the skilled instrument-maker Edward Nairne, argued that his small-scale model could control the key factors much more precisely than Wilson's large and cumbersome arrangement.[24]

To avoid misunderstandings, it should be pointed out that not all laboratory experiments were based on models. In this category were experiments to determine the most effective metal for the lightning rod. Especially noteworthy in this respect were the experiments performed by Martinus van Marum with the instrument-maker John Cuthbertson on the conducting power of different metals. They determined that copper was the best conductor of electricity and produced the least "lateral" discharges ("side flashes") of all the metals tested. Copper, offering the least resistance, would therefore be the most suitable metal for lightning rods if cost were no object.[25]

Laboratory Models to Didactic Instruments

By the late 1780s the concept of the lightning rod was well established in the scientific community, although it would take another hundred years before appreciation of the complex mechanism of the lightning flash and its action on the

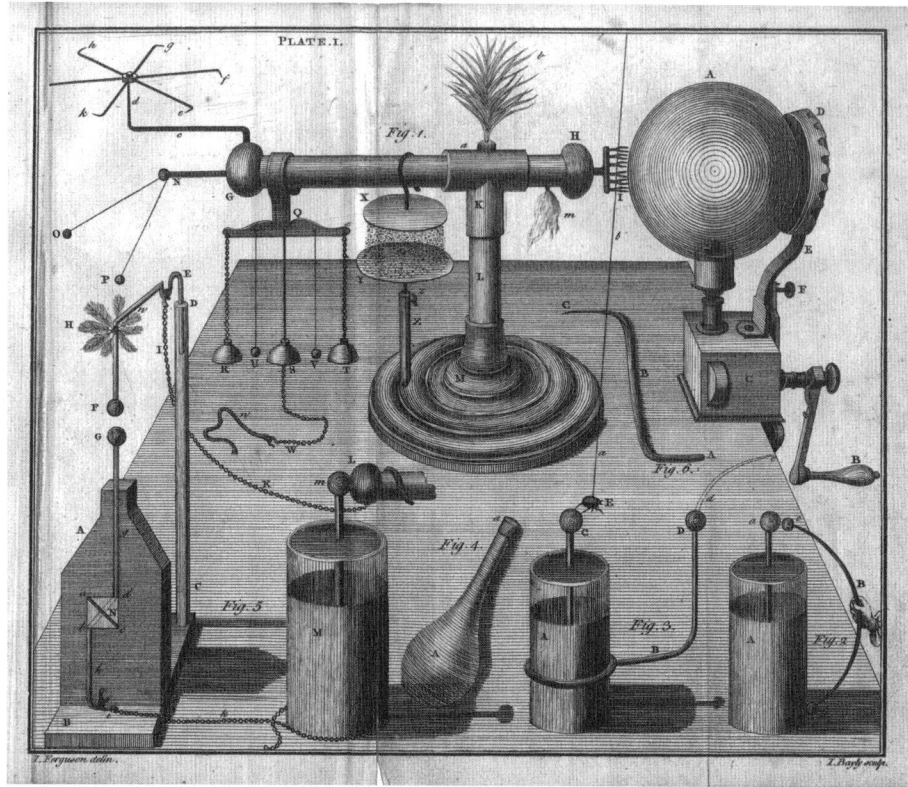

Figure 10.2.   Portable globe machine by Edward Nairne with electric kit, including the model thunder façade. James Ferguson, *An Introduction to Electricity* (London, 1770), Pl. I.

lightning rod would advance further.[26] Knowledge about the properties of electrical phenomena evolved through manipulations in the laboratory; from these, Franklin and others (especially his followers) derived the models to investigate the behavior of lightning and lightning rods. From these prototypes evolved the didactic instruments that became a delightful feature of eighteenth century (and later) natural philosophy textbooks and of lecture demonstrations. Didactic model lightning instruments are a good example of this process from laboratory prototype to what Bachelard has called *un théorème réifié.*[27]

A typical early electrical kit with demonstration apparatus (fig. 10.2) was marketed by the London maker Edward Nairne and publicized by the itinerant lecturer James Ferguson in 1770.[28] The mahogany model is of the gable end of a house with lightning rod, the brass wire terminating in a ball above the chimney. Above it is another brass ball and wire forming a spark gap and supported by a glass tube. When electrified by a charged Leyden jar in circuit, the downy feathers expanded like a thundercloud and shrank when the discharge occurred. A spark jumped with a sharp snap between the two balls, but when the one at the chimney end was replaced by a sharp point, the charge was drawn off silently instead. When the model rod was interrupted by rotating a small square piece of mahogany set in the gable to which was attached a small section of the rod, the spark jumping across this gap blew this portion out of the façade, reenacting the behavior of an interrupted lightning rod in real life. Thus, this model demonstrated the basic phenomena associated with the lightning rod: the effects of blunt endings and points, and the damage that could be caused by a discontinuous conductor. According to James Ferguson, this demonstration model was devised by Dr. James Lind of Edinburgh.[29]

In a more sophisticated version devised a little later, the gable end formed part of a mahogany house with hinged walls, enclosing a spark-gap arrangement set in a small brass cylinder on a wooden pedestal. In this was placed some gunpowder. The spark from a Leyden jar ignited the gunpowder and the little house collapsed with a great bang. The earliest description of the English version of this model was by George Adams in the mid 1780s, but a simpler French version in which the explosion was caused by an explosive cylinder consisting of a tube of rolled paper enclosing a spark gap and gunpowder was described by the Abbé Nollet in the mid seventies. This model was purchased with spare explosive cylinders. One such model in Florence demonstrates by its vaulted roof that it was inspired by the contemporary powder magazines.[30] In yet another variation, a spark gap in a model house of tin ignited alcohol and caused a merry blaze.

Several models marketed by the instrument-makers re-created real-life events. The electric obelisk that collapsed when struck by a miniature lightning bolt reenacted a real obelisk damaged in this manner in London in 1774. A unique surviving model (fig. 10.3) in the Oxford Museum of the History of Science was made by W. & S. Jones of London to replicate an actual event described in the 1750s. The occupants of a house received severe shocks from the metal patterns in their wallpaper as their house was struck. In the model, the figures are placed in front of small spark gaps made by wires in the walls of the house. They were knocked over by the sparks when a Leyden jar was discharged through wires.

Such demonstrations also became popular in late eighteenth century textbooks on "rational recreations" for budding natural philosophers. Typical examples are the works by Guyot in France, William Hooper in England, and Weber and Langenbucher in Germany.[31] Among Weber's recreational electrical toys is a church with contemporary lightning rod, while Langenbucher's more sophisticated model (fig. 10.4) also has a ship with rod and a suspended "cloud" delicately balanced on a pivot.

A similar cloud can be seen in Kirchhof's popular textbook, in principle similar to the model cloud described by Franklin in his paper of 1749.[32] The psychological intent of such apparatus becomes very clear from the frontispiece in Langenbucher's book. The model church depicts a real natural event in miniature. It is

Figure 10.3.   Unique model house, c. 1820, with paper figures by W. & S. Jones of London. In the Museum of the History of Science, Oxford. Inv. No. 75608.

even more obvious in the engraving from a book by Beck published in 1787 (fig. 10.5): the "act of God" is the almost hidden hand of the operator in the act of discharging the Leyden jar.[33] Such demonstration devices were designed to teach truths about nature, and the gestures of the manipulator that accompanied them were supposed to be invisible.[34]

Discussion

In natural philosophy, model experiments were especially popular in a subject like atmospheric electricity in which little direct experimental intervention could be made. It was not possible to experiment with real clouds (electrified

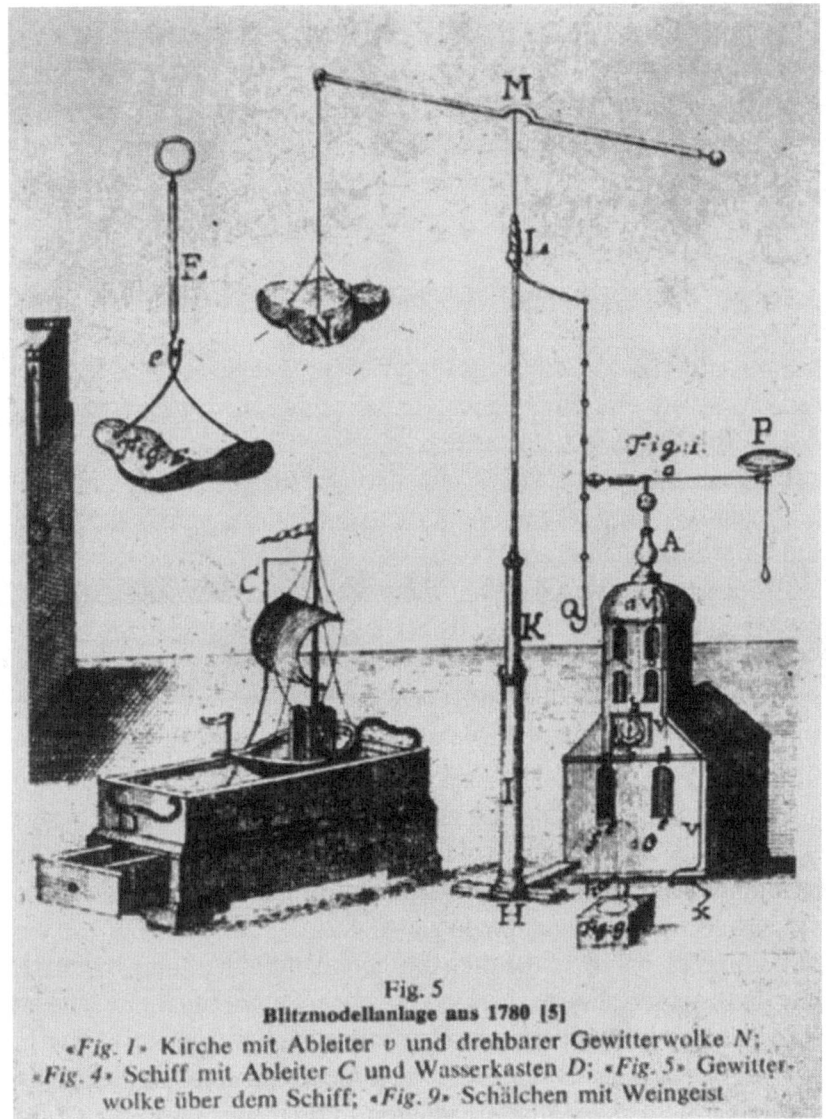

Fig. 5
Blitzmodellanlage aus 1780 [5]
«*Fig. 1*» Kirche mit Ableiter *v* und drehbarer Gewitterwolke *N*;
«*Fig. 4*» Schiff mit Ableiter *C* und Wasserkasten *D*; «*Fig. 5*» Gewitter-
wolke über dem Schiff; «*Fig. 9*» Schälchen mit Weingeist

Figure 10.4.   Model church and ship experiments with moveable cloud. Langenbucher, *Beschreibung einer beträchtlich verbesserten Elektrisiermaschine.*

or not) in the laboratory and with lightning strikes, but these large-scale phenomena could be manipulated with models in the laboratory.[35] Franklin's hypothesis (1749) by means of the analogous argument that artificially produced electricity and lightning were essentially similar led him to re-create lightning in his laboratory with models, and he then proved his analogy by means of the electric kite experiment. He determined that the natural electricity obtained in this way exhibited the same properties as the species of electricity generated by

Figure 10.5.   Real church and model church struck by lightning. Beck, *Kurzer Entwurf der Lehre vonder Elektricität.*

his electrical machine. Both passed the same electrical tests first determined in the laboratory: they charged Leyden jars, produced sparks, gave shocks, and had the properties of attraction and repulsion.

Franklin's explanation of the probable action of the lightning rod was based entirely on his observations of how pointed conductors behaved in the laboratory. Such grounded conductors, he observed, attracted electrical charge from a body at a considerable distance, but it was through experience with actual lightning rods that he realized that, in addition to their supposed property of preventing a stroke, they could also conduct a stroke safely into the ground.[36]

The history of the eighteenth century lightning rod illustrates that in common with contemporary experimental science, it was made up of a mixture of instrumental observations and research with conceptual models of brass and wood. Similar techniques were developed in mechanics, or what today we would call civil engineering, in particular in England. Inevitably, distinctions become blurred as past masters in this art, such as James Watt and John Smeaton, were instrument-makers turned civil engineers. In the 1760s Watt's seminal improvements were made on a model steam engine, while, shortly before, Smeaton developed his trial models to estimate the friction and efficiency of mill wheels and test the most effective hydrodynamic shape of ship hulls.[37]

Here one might be able to postulate a difference in attitude toward the veracity of applying such models between academic natural philosophers such as George Atwood and nonacademic experimentalists such as Franklin, Watt, and Smeaton. Atwood used his "fall machine" in an academic setting to demonstrate important aspects of Newtonian mechanics but argued that models like those of Smeaton could never challenge the foundation of Newtonian mechanics because of the interference of friction in real-world machinery.[38] Similarly, Atwood had constructed for him a model bridge of brass and wood while he was a member in 1798 of a Parliamentary committee to test Thomas Telford's concept of a large iron bridge across the Thames to allow larger ships into the London docks. Atwood concluded that Newtonian mechanics could explain all the properties of complex stone and iron arches. He legitimized his model by invoking rational mathematical analysis while another contemporary potential bridge builder, Tom Paine, claimed that his designs were fundamentally natural models based on the laws of spiders' webs.[39] The debate here was between academic practices and the empirical procedures of the artisan, but the analogous foundation of such models was not in doubt. The mechanics of Newton had been taught by models since the late seventeenth century and were first formalized in academic pedagogy by the great Dutch academic university teacher Jacob van 's Gravesande, who claimed that he was honored to be following in the footsteps of Newton.[40] The impact of such practices on experimental modeling in electricity in general and on the lightning rod in particular requires further study.

In the case of the lightning rod, experiments based on analogy and models were fruitful, but they could also be misleading. The success of models was judged by how well they represented (simulated) the natural phenomenon (or set of phenomena) under consideration. Indeed, the success of replication was their sole criteria since there is no logical means of testing their validity.[41] Thus, logically, there is no reason why the phenomena re-created in the laboratory should be the same as the ones in nature—not least because of the simple practical problem of differences in scale.[42] In mechanics the puzzle of scale—postulating large structures from small models—was already tackled in Galileo's *Two New Sciences* (1638), and formulated in a best-selling mechanics textbook by William Emerson in 1758.[43]

In natural philosophy, there could be only weak or strong models depending on how closely they were tied to the observed phenomena.[44] In this sense Priestley's electrical earthquake model was weaker than Franklin's lightning rod model, for although both were based on theories tied to observations, earthquakes had not yet been observed so thoroughly and their theories were considered more hypothetical and generated more controversy. The underlying conceptual frameworks, too, influenced the models and the direction taken by experiments with models. In the case of the eighteenth century lightning rod, the underlying concept of its modus operandi was that its action was analogous to a rain pipe through which the electric fluid was conveyed safely from the heavens to earth.[45] The study of model experiments indicates that specific models generally went into decline not because they became less successful in reproducing the phenomena within their own framework but because their conceptual framework was superseded by a more successful one.[46]

The inspiration for such models was by no means solely scientific in the narrow, modern sense, but also cultural. Von Guericke's electric terella, for instance, was inspired not by research in electricity but by his views of cosmology. It had been Priestly who with hindsight described von Guericke's investigations in the more narrow framework of electrical phenomenology. In the eighteenth century, these models represent very specific views of nature in which the phenomena are reduced to a few simple elements or factors of operation.

Von Guericke made his electric terella of sulfur because he argued that the earth was mainly composed of sulfurous matter. It is hard to believe that Volta and Wilson actually thought that clouds were composed of pasteboard and tinfoil or, in the case of Van Marum, of bronze powder and varnish, and yet these were the substances with which they mimicked nature. The importance for these modelers was to arrange ingredients in such a way that they could produce the specific phenomena observed. One would consider there to be an epistemological difference between observing a glow in a partially evacuated vessel and comparing this with the aurora and with simulating the imperfect conducting properties of a thunder cloud by coruscations darting over bronze powder as done by Van Marum. It is a great pity that there is so little discussion of these experiments in eighteenth century literature.

From our perspective, the main problem with this type of research was that the experiments were interpreted in terms of macroscopic observable behavior and not synthesized to idealized behavior (forces) between microscopic particles. This approach proved to be successful at the macrotechnological level of designing lightning rods. The main eighteenth century investigators of the lightning rod made no clear-cut distinction between conceptual models in the laboratory and the real world phenomena experienced by the senses. Cracks began to appear in this approach in the 1760s. Thus Priestley stated that he accepted Franklin's single-fluid theory of electrical action over the two-fluid theory because it explained electrical phenomena more simply, but he admitted that neither theory could be proven.[47] The shift in attitude became more definite in the late eighteenth century, spearheaded by the new generation of mathematical physicists, such as Cavendish in England and Coulomb in France, who had a fundamentally different attitude toward the macroscopic explanatory models, regarding them as a means to assist with the mathematization of nature rather than as descriptions of reality.[48]

In the study of atmospheric electricity, as in other branches of eighteenth century natural philosophy, a complex link was established between instrumentation, theory, and the observer, which in the construction of the lightning rod was intended to dispel fear and ignorance.

Notes

1. Anonymous review in *Philosophical Transactions of the Royal Society* 7 (1672): 5103–5; and Willem Hackmann, *Electricity from Glass: The History of the Frictional Electrical Machine 1600–1850* (Alphen aan den Rijn: Sijthoff and Noordhoff, 1978), 25.

2. H. Cavendish, "An Account of Some Attempts to Imitate the Effects of the Torpedo by Electricity," *Philosophical Transactions of the Royal Society* 66 (1776): 196–225. Cavendish reached this conclusion by rearranging well-known electrical laboratory apparatus such as charged Leyden jars, brass chains, and electroscopes; Willem Hackmann, "The Relationship between Concept and Instrument Design in Eighteenth Century Experimental Science," *Annals of Science* 36 (1979): 205–24; and Hackmann, "Scientific Instruments, Models of Brass and Aids to Discovery," in *The Uses of Experiment*, ed. David Gooding, Trevor Pinch and Simon Schaffer (Cambridge: Cambridge University Press, 1989), 31–65, esp. 55f.; see also Simon Schaffer, "Fish and Ships: Models in the Age of Reason," in *Models: The Third Dimension of Science*, ed. Nick Hopwood and Soraya de Chadarevian (Stanford, Calif.: Stanford University Press, 2004), 71–105. Newly discovered electrical phenomena were considered electrical only if they passed what had become the traditional tests, such as charging a Leyden jar and deflecting a gold-leaf electroscope. For another case study, see the identification of electricity produced by the friction of steam from steam engines by Lord Armstrong and Faraday in Hackmann, "Electricity from Steam: Armstrong's Hydroelectric Machine in the 1840s," in *Making Instruments Count. Essays on Historical Scientific Instruments Presented to Gerard L'Estrange Turner*, ed. R. G. W. Anderson, J. A. Bennett, and W. F. Ryan (Aldeshot: Variorum, 1993), 147–73. What is interesting in this example is that once it had been established that the prime cause of this unusual phenomenon was the friction of the water/steam particles, this was readily accepted, as it did not challenge the existing "mind set" concerning the behavior of electricity. The electrostatic mind-set was altered dramatically by Volta's discovery of the electrochemical battery (1800) and Faraday's work on electromagnetic induction (1831).

3. For a more detailed discussion on analogies and model experiments, see Hackmann, "Scientific Instruments" (cit. n. 2), 45–48, and the references cited therein. Peter Galison and Alexi Assmus, "Artificial Clouds, Real Particles," in *The Uses of Experiment*, ed. David Gooding, Trevor Pinch, and Simon Schaffer (Cambridge: Cambridge University Press, 1989), 227, refer to the Victorian tradition of "mimetic experimentation" and state (231) that a number of late nineteenth century morphological scientists began to use the laboratory to reproduce natural occurrences such as cyclones or glaciers. In 1892 the geologist E. Reyer wrote that researchers had given up either because quantitative experiments

seemed impossible or because experiments had been unable to imitate (*nachbilden*) natural conditions. Now, by reproducing these phenomena—at least partially—much could be learned, thus heralding the beginning of a new experimental physical geology. Model-making was already one of the main techniques of seventeenth and eighteenth century experimental philosophy, and in my study of the Dutch natural philosopher Martinus van Marum (1750–1837), I called it "imitative experiments," based on van Marum's Dutch term *nabootsen*. See Hackmann, "Relationship between Concept" (cit. n. 2), 220–23, based on my unpublished Belfast University M.A. thesis (1970), "The Electrical Researches of Martinus van Marum (1750–1837)," 225–26.

4. Willem Hackmann, "Instrument and Reality: The Case of Terrestrial Magnetism and the Northern Lights (Aurora Borealis)," *Philosophy and Technology. Supplement to Philosophy* 38 (1995): 29–51.

5. In 1785 Priestley wrote to the Dutch natural philosopher Martinus van Marum encouraging him to improve on this experiment with his large twin-plate electrical machine, which he did the following year. Van Marum, "Eerste vervolg der proefneemingen, gedaan met Teyler's electrizeer-machine," *Verh. Teyler's IIde Gen.* 4 (1787); see Hackmann, *Electrical Researches* (cit. n. 3), 239f.

6. John Theophilus Desaguliers, "Some Conjectures Concerning Electricity, and the Rise of Vapour," *Philosophical Transactions of the Royal Society* 42 (1742): 140–43; and H. B. de Saussure, "Lettre," *Observations sur la physique* 25 (1784): 290–91. William Gilbert in *De magnete* (New York: Dover Press, 1958), 91, had already observed the attraction of water to rubbed amber. See also Willem Hackmann, "Instruments and Experiments: The Case of Atmospheric Electricity in Eighteenth-century Holland," *Tijdschrift voor de Geschiedenis des Geneeskunde, Natuurwetenschappen, Wiskunde en Techniek (GeWiNa)* 10 (1987): 190–207.

7. Franklin's letter to Collinson in September 1753; see I. B. Cohen, ed., *Benjamin Franklin's Experiments: A New Edition of Franklin's Experiments and Observations on Electricity* (Cambridge, Mass.: Harvard University Press, 1941), 225–52; see also I. B. Cohen, *Franklin and Newton* (Philadelphia, 1956).

8. Thus, in the 1780s de Saussure discovered with his specially designed atmospheric electrometers diurnal variations in the amount of electricity in the atmosphere, even during fine weather. He observed that these variations were greater in winter than in summer. His observations agreed with Volta's. In calm weather, the atmosphere was positive, although during stormy weather it was negative. De Saussure and Volta concluded (in accordance with Franklin's single fluid hydrostatic model of electrical action) that clouds were normally electrified positively and when highly charged gave up some of their excess electricity to the surrounding atmosphere. Both argued that a certain amount of electric fluid (positive charge) in addition to heat was necessary to change water into its vapour state. Thus, when clouds were formed by the heat of the sun, they obtained a certain quantity of electric fluid from the earth, which was left negatively charged. See H. de Saussure, *Voyages dans les Alpes*, vol. 2 (Neuchatel, 1786), 202–78. In *New Experiments on Electricity Wherein the Causes of Thunder and Lightning as well as the Constant State of Positive or Negative Electricity in the Air or Clouds Are Explained* (Derby, 1789), VIII, 103, the author, Abraham Bennet (the inventor of a sensitive gold-leaf electroscope), states that

during a period of fifteen years, the Italian natural philosopher, Giambattista Beccaria had never observed a serene atmosphere to be electrified negatively except on four occasions when his readings could have been affected by distant clouds, but this ran counter to the conclusion reached by the Dutch natural philosopher Cornelis Rudolphus Theodorus (later Baron) van Krayenhoff, who in 1783 observed that clouds were usually charged negatively. His explanation was based on Franklin's "insulated cup and chain experiment," a device that demonstrated the relationship between surface area (capacitance) and charge. According to this analogy, Van Krayenhoff argued that when the globules of water broke down into smaller particles of vapour, the surface area of the cloud became larger, thereby increasing its capacity to take up electricity. Conversely, the positive charge on a cloud was the result of a sudden decrease in its area caused by the condensation of the vapour back into water globules. Van Krayenhoff demonstrated this with his sponge analogy: when the sponge's area was decreased by squeezing it, the superfluous water ran out. According to this analogy rain, too, was an electrical phenomenon: the water vapour of clouds that lost charge would condense into rain. Electrical imbalances of clouds were caused by the heat of the sun, which resulted in non-uniform rates of evaporation. This produced pockets of differing amounts of electricity in the atmosphere since air was a bad conductor. The general circulation towards a state of electrical equilibrium in the atmosphere caused the formation of rain, hail, thunderstorms, and even (some thought) whirlwinds, waterspouts, and earthquakes. See C. R. T. van Krayenhoff, *Proefeener electrische natuurkunde in 't Fransch van den abt Jacquet* (Leiden, 1783), 155, 213. These considerations prompted Volta to ask Van Marum to investigate atmospheric electricity with his huge twin-plate electrical machine. Van Marum's experiments were on the whole negative. He could not increase the rate of evaporation of alcohol or water by electricity, nor find that electrified air held more water vapour, or that the atmosphere could be "rarified" by electricity. His attempts to discover whether barometric pressure was affected by electricity were inconclusive. See J. Bosscha, *La correspondance de A. Volta et M. van Marum* (Leiden, 1905), especially Van Marum's letter of August 31, 1788 (letter IX), 36–42, and Volta's of July 23, 1789, and of March 28, 1792 (letters XI and XII), 46–63.

9. William Henley used an electrified bladder coated with leaf copper to demonstrate that it could be silently discharged by a point conductor without moving closer to the bladder; *Philosophical Transactions of the Royal Society* 64 (1774): 133–52.

10. Van Marum, "Eerste vervolg" (cit. n. 5), 219–25. For a fuller discussion of Van Marum's experiments on atmospheric electricity and lightning rods, see Hackmann, "Electrical Researches" (cit. n. 3), 224–73.

11. J. Priestley, *The History and Present State of Electricity* (London, 1769), 232; Van Marum, "Eerste vervolg" (cit. n. 5), 229–31. He could have found a similar arrangement in the Dutch edition of T. Cavallo, *Volledige verhandeling over de electriciteit*, trans. J. T. Rossijn (Utrecht, 1780), 223–24.

12. Charles Mahon, *Principles of Electricity* (London, 1779). Another instance of the action of the return stroke is mentioned in Lord Mahon's "Remarks on Mr. Brydone's Account of a remarkable Thunder storm in Scotland," *Philosophical Transactions of the Royal Society* 77 (1787): 130–50. See also John Tyndall, *Lessons in Electricity at the Royal Institution, 1875–76* (London, 1876), 102–8.

13. Although medical applications were described in the 1740s, see the references in Willem Hackmann, "The Medical Electrical Machines of John Wesley and John Read," in *Musa Musaei: Studies on Scientific Instruments and Collections in Honour of Mara Miniati*, ed. Marco Beretta, Paolo Galluzzi, and Carlo Triarico (Florence: Leo S. Olschki, 2003), 261–77, and Hackmann, *John Wesley and His Electrical Machine* (London: John Wesley's House and The Museum of Methodism, 2003), 35; I am stressing here that the lightning rod was the first technological innovation based on electrical researches.

14. See the discussion by E. Philip Krider, "Benjamin Franklin and Lightning Rods," *Physics Today* 2006: 42–48.

15. I. B. Cohen, *Benjamin Franklin's Science* (Cambridge, Mass.: Harvard University Press, 1990), 70–108, and the footnotes on 228–41. As Cohen has pointed out, the only question in Franklin's mind that demanded experimental proof was whether clouds that contained lightning were electrified. If that were demonstrated to be the case there would be no doubt in Franklin's mind that lightning rods would work and draw off the electrical fire before the clouds could discharge in a bolt of lightning. If the lightning flash was merely a bigger spark discharge than that obtained in the laboratory, then there was no reason to suppose, he argued, that the change in scale would affect the action of pointed conductors in discharging charged bodies, whether small metal objects in the laboratory or gigantic clouds.

16. In experiments with the ungrounded rods (and with the electric kites), the usual laboratory practice was followed with the operator insulated on cakes of resin or on a stool with glass legs. This can be seen in some of the engravings of such experiments. As it happened, Richmann had not been insulated in this way when the accident occurred. In any case, not all experiments could be done with an ungrounded rod by the operator. For instance, to charge a Leyden jar from the electrified rod, the operator would have to be grounded (earthed). It is quite a miracle that more operators were not killed because it could have made little difference whether the operator was insulated or not. On the modern explanation of the electrical action of grounded and ungrounded lightning rods, see B. F. J. Schonland, *The Flight of Thunderbolts* (Oxford: Clarendon Press, 1964), 19–21, and see note 26.

17. B. Wilson, *Philosophical Transactions of the Royal Society* 68 (1778): 999–1011, Pl. XVII, fig. 3.

18. B. Wilson, *An Account of Experiments Made at the Pantheon on the Nature and Use of Conductors* (London, 1778). The impressive engraving of this experiment is reproduced in Rod Home's chapter in this volume.

19. John Boddington, et al., "Sundry Papers Relative to the Accident from Lightning at Purfleet," *Philosophical Transactions of the Royal Society* 68 (1778): 232–44.

20. T. Mitchell, "The Politics of Experiment in the Eighteenth Century: The Pursuit of Audience in the Manipulation of Consensus in the Debate over Lightning Rods," *Eighteenth-Century Studies* 31 (1998): 307–31.

21. G. J. Symons, ed., *Report on the Lightning Rod Conference* (London, 1882), 77.

22. But of course it does make a difference with the models because the ratios between the sizes of the ball and point to the artificial cloud is so much smaller, as are the charges; but see the chapter by Moore et al.

23. William Swift, "Account of Some Experiments in Electricity," *Philosophical Transactions of the Royal Society* 69 (1779): 454–61.

24. Edward Nairne, "Experiments on Electricity Being an Attempt to Shew the Advantage of Elevated Pointed Conductors," *Philosophical Transactions of the Royal Society* 68 (1778): 823–60.

25. Van Marum, "Eerste vervolg" (cit. n. 5), 165–69.

26. Lightning was described in the contemporary framework of the fluid (corpuscular) theory, and the action of the lightning rod in terms of what Sir Oliver Lodge called in 1892 the "rain pipe analogy." The rod conveyed the electric fluid from the heavens to the earth, or vice versa. Thus, if the resistance could be kept at a minimum by using a highly conducting metal, and by efficient grounding, there would be little possibility for the side flashes that caused most of the accidents and fires. As Lodge pointed out, this analogy would work perfectly well if lightning behaved as a steady continuous current, but this is not the case. In reality, the lightning flash consisted of a series of surges, and these rapid oscillations caused conditions undreamt of until the late nineteenth century, and not appearing in the small laboratory scale models used in the eighteenth century— in other words, before the work of Kelvin and others on "electrodynamic capacity" or "self-induction" and "electrical inertia" or "impedance." The rain pipe analogy was described by Sir Oliver Lodge in his *Lightning Conductors and Lightning Guards* (London, 1892), chap. 28. This analogy was aptly demonstrated by John Cuthbertson's statement that the electric fluid passed through a pointed rod like water through a pipe, in vol. 2 of his *Algemeene eigenschappen van de electriciteit, onderrichting van de werktuigen en het neemen van proeven in dezelve* (Amsterdam, 1782), "Experiment LXXX," 165. The oscillatory nature of the discharge was unwittingly hinted at by Priestley without realizing its significance; see F. W. Gibbs, *Joseph Priestley* (London, 1965), 47.

27. For a discussion, see S. Gaukroger, "Bachelard and the Problem of Epistomological Analysis," *Studies in the History and Philosophy of Science* 7 (1976): 189–244; and Hackmann, "Scientific Instruments" (cit. n. 2), 32.

28. James Ferguson, *An Introduction to Electricity* (London, 1770), Pl. I.

29. For a fuller description of these devices, which were variously called "thunder houses" or "thunder façades," see Willem Hackmann, *Catalogue of Pneumatical, Magnetical and Electrical Instruments in the Museo di Storia della Scienza, Florence* (Florence: Giunti, 1995), 98, 138–44.

30. This type was called a "Powder House." See items 166 and 167 in Hackmann, *Catalogue* (cit. n. 29), 142, for references.

31. E. G. Guyot, *Nouvelles Récréations Physiques et Mathématiques: Nouvelle Édition Corrigée et Considérablement Augmentée*, vols. 2 and 4 (Paris, 1773, 1775); W. Hooper, *Rational Recreations*, vol. 3 (London, 1774); J. Weber, *Abhandlungen von dem Luftelektrophor*, 2d ed. (Ulm, 1779); and J. Langenbucher, *Beschreibung einer beträchtlich verbesserten Elektrisiermaschine, samt vielen Versuchen und einer ganz neuen Lehre vom Laden der Verstärkung* (Augsburg, 1780).

32. N. A. J. Kirchhof, *Beschreibung einer Zurüstung, welche die anziehende Kraft der Erde gegen die Gewitterwolke und die Nützlichkeit des Blitzableiters sinnlich beweisen* (Hamburg-Berlin, 1781).

33. D. Beck, *Kurzer Entwurf der Lehre von der Elektricität* (Salzburg, 1787).

34. S. Schaffer, "Machine Philosophy: Demonstration Devices in Georgian Mechanics," *Osiris* 9 (1995): 159.

35. As we noted in the text, there was a great deal of controversy concerning the electrical polarity of clouds based on observations, which led to the model cloud experiments by Van Marum and others (see cit. n. 8).

36. B. J. F. Schonland, "Lightning Protection—200th Anniversary of the 'Philadelphia Experiment,'" *Journal of the Franklin Institute* 253 (1952): 375–504, especially section II, 384.

37. A. Q. Morton, "Men and Machines in Mid-18th Century London," *Transactions of the Newcomen Society* 64 (1993): 47–56; T. S. Reynolds, "Scientific Influences on Technology: The Case of the Overshot Waterwheel, 1752–1754," *Technology and Culture* 20 (1979): 270–95; and Larry Stewart, "A Meaning of Machines: Modern Utility and the Eighteenth-Century British Public," *Journal of Modern History* 70 (1998): 259–94.

38. Schaffer, "Machine Philosophy" (cit. n. 34), 157–82.

39. Schaffer, "Fish and Ships" (cit. n. 2); E. L. Kemp, "Thomas Paine and his 'pontifical' matters," *Transactions of the Newcomen Society* 49 (1977–78): 21–40; and John Keane, *Paine: A Political Life* (London: Bloomsbury, 1995).

40. W. J. 's Gravesande, *Mathematical Elements of Natural Philosophy, Confirmed by Experiments: or, an Introduction to Sir Isaac Newton's Philosophy*, 2 vols. (London, 1720–21), introduction.

41. In the case of model experiments, their success depends on how close the parameters of the real phenomenon can be modeled or paralleled in the laboratory. Cavendish's model of the electric fish was far more successful than Priestley's electric earthquake model. This essay has only dealt with the empiricism of modeling, not with such important practicalities as scaling. Scaling is a complex issue, as it is not only simply a matter of size. Thus, in the case of the lightning rod various scaling factors come into play, such as (1) the ratios of the size of the point or ball to the surface area of the "artificial cloud"; (2) the differences in the energy between the laboratory spark and the lightning flash; (3) the differences in the electromagnetic force fields generated; and (4) the oscillatory nature of the respective discharges (see note 26 for some of these unobservable complexities). Furthermore, as described by Schonland (*Flight of Thunderbolts*, 92–99; cit. n. 16), the appearance of the electric spark and lightning flash is also deceptive. The light seen is of the ionized gases of the return stroke traveling in the channel created by the initial lightning leader traveling downward from the clouds. Thus, the light caused by the ionized gases in an electric discharge travels in the opposite direction from that observed by Franklin and his followers—that is, from the negative to the positive conductor, with the result that the electron is given the negative and not the positive sign. See note 21 for some of these unobservable complexities. On a recent discussion on replication, see H. M. Collins, *Changing Order: Replication and Induction in Scientific Practice* (London/Beverly Hills: Sage, 1985), 73–74.

42. J. L. Heilbron, *Elements of Early Modern Physics* (Berkeley: University of California Press, 1982), 194–99.

43. Schaffer, *Fish and Ships* (cit. n. 2); and S. B. Hamilton, "Historical Development of Structural Theory," *Proceedings of the Institute of Civil Engineers* 3 (1952): 374–419.

44. A common strategy adopted in the eighteenth century by an opponent of a particular model was to argue that the modeler was involved in a circular argument. According to this, the same concepts used to make the model were then "proofed" by the model if it behaved as predicted or re-created the phenomenon. It is difficult to see how analogies can fully avoid circular arguments.

45. For instance, the phenomenon of induction (Lord Mahon's return stroke) was almost totally ignored in the context of the lighting flash and the behavior of the lightning rod; see also note 21. It also accounts for Van Marum's interest in investigating the conductivity of various metals of which the lightning rod could be manufactured (see note 25) and the discussion in note 41.

46. See the succession of aurora borealis models described in Hackmann, "Instrument and Reality" (cit. n. 4).

47. Priestley, *History and Present State* (cit. n. 11), 1775 edition, vol. 2, 41–52.

48. For further discussion, see Hackmann, "Instruments and Experiments" (cit. n. 6), 58–60; and Hackmann, "Instrument and Reality" (cit. n. 4), 47–51.

Prometheus' Tools
Instruments and Apparatus Used in Atmospheric Electricity Research and Experiments

Paolo Brenni

Introduction

Lightning and thunderstorms, with their wonderful and terrifying display of energy, are certainly one of the most impressive phenomena of nature. For centuries people associated them with the power and rage of gods. Civilizations in the Mediterranean and northern Europe, in the Middle and Far East, and in America had their own gods responsible for thunder and lightning. They often represented them surrounded by flames or holding lightning. Philosophers speculated about their nature and tried to explain them in terms of conflict between elements, air charged with fire, inflammable vapors and telluric fumes exploding in the atmosphere. In the Middle Ages, several theologians claimed that sinners were particularly exposed to the effects of lightning, which represented supreme and divine punishment. Several saints, such as Saint Barbara, were invoked as protectors against thunderstorms and lightning strikes.

During the Scientific Revolution, authors such as Descartes tried to explain the mechanism of thunderstorms, but their theories often revisited those of ancient philosophers. Only at the beginning of the eighteenth century was lightning seen to be related to sparks from the first electrostatic machines.[1] With the invention of Leyden jars and with the introduction of improved and larger friction machines, the idea that sparks and lightning shared a common nature became more established. Benjamin Franklin's well-known theories and experiments as well as Dalibard's experiments in France marked the birth of modern studies on atmospheric electricity.

In my essay I will retrace the history of the researches related to atmospheric electricity between the mid-eighteenth century and the first decades of the twentieth century. In the first part I will consider the apparatus and techniques used for studying the "fair weather" electricity. In the second part I will describe the first scientific observations on lightning and the early measurements of their physical parameters. Finally I will illustrate some of the most important projects and experiments conceived to harness atmospheric electricity.[2]

Measurement of "Fair Weather Electricity"

The mechanisms underlying atmospheric electricity are extremely complex and even today only partially understood. Atmospheric electricity is connected with charge separation involving thermodynamic phenomena, radiation ionization, and collision ionization effects. The ubiquitous electrification of the atmosphere is known as "fair weather" electricity. The atmosphere normally has a positive charge while the Earth is negative. In the simplest terms, we can say that thunderstorms act as batteries keeping the Earth negatively charged while the atmosphere is positively charged. During fair weather there is also a difference of potential between the surface of the Earth and the various layers of atmosphere, a fact that has been known since the discovery of the electric nature of lightning. It is also known today that normally this electric field is of the order of about one hundred volts per meter; but this value varies enormously during thunderstorms and can be influenced by an incredible number of meteorological, physical, and geographical factors. The measurement of this electric field and the determination of its temporal and spatial variations was, at least until the end of the nineteenth century, the main concern of scientists interested in atmospheric electricity. In 1860 William Thomson (later Lord Kelvin) described the basic task in terms of instruments: "Apparatus for the observation of atmospheric electricity has essentially two functions to perform; to electrify a body with some of the natural electricity, or with electricity produced by its influence; and to measure the electrification thus obtained.[3]

A vertical conductor in the atmospheric field becomes electrified by influence: the negative charges migrate to the upper part of it while the positive ones accumulate at the lower one. But if the top of the conductor is pointed, the negative electric charges disperse and the conductor becomes positive. In modern terms, the points act as "potential equalizers." These simple electrostatic phenomena formed the basis of most nineteenth century research in the field of atmospheric electricity.[4] After the pioneering experiences of Dalibard in France and Franklin in British America, there was a series of interesting observations done concerning the "electricity of the air." The French naturalist Louis Guillaume Le Monnier noticed that even during fair weather, a lightning rod showed signs of electricity. In fact, the bottom of the rod was able to attract some light powder. He also noticed that the electrification ceased during the night. His experiments were repeated and largely extended by the Italian natural philosopher

Giovanni Battista Beccaria, one of the most important electricians of the eighteenth century.[5] Beccaria, who devoted an important part of his work to atmospheric electricity, collected a great number of observations using simple pith ball electrometers connected to "exploring" iron wires, kites, and even rockets. In the following years several natural philosophers and inventors proposed improved electroscopes.[6] Horace Bénédict de Saussure, the famous Geneva natural philosopher, invented various types of instruments for exploring atmospheric electricity. First he used one of his electroscopes (generally a double silver-wire pendulum electroscope) with a fishing rod. Then, for reaching a significant height, he connected the electrode of the electroscope to a small lead attached to a thin metallic wire (fig 11.1a). The lead was thrown vertically (about 15 meters) until the stretched wire detached itself from the electrode, leaving the electroscope charged.[7] A third electrometer had a vertical rod of about 50 cm (fig. 11.1b). When the instrument was suddenly elevated from the ground to an approximate height of 1.5 meters, it immediately detected the presence of an electric charge. During this period, another electrician, Tiberius Cavallo, made extensive observations with his apparatus and electrometers based on an instrument developed by John Canton. All these observations, as well as other measurements made at this time, showed the diurnal as well as a yearly variation of atmospheric electricity, and a strong influence of meteorological conditions.

Around 1787 the Italian scientist Alessandro Volta and the British scientist Abraham Bennett (famous for his gold-leaf electroscope) independently discovered a successful method for detecting and measuring atmospheric electricity. They observed that the flame of a candle, connected with an electroscope, revealed atmospheric charges much easier and faster. The flame was a good potential equalizer and worked much better than the pointed conductors used in previous experiments. Volta placed the flame in a small lantern on the top of a long insulating rod, making the apparatus much more efficient, and the electroscope rapidly showed the presence of electricity in the air (fig. 11.2a). Volta also experimented with burning sulfur wicks and candles. Later in the nineteenth century, scientists used gas lamps and burning papers impregnated with lead nitrate. In the late 1850s William Thomson introduced a different type of potential equalizer, the "water drop collector," replacing a flame or pointed conductor. He fixed a well-insulated water reservoir at the place where the potential had to be measured and was connected with the usual electrometer. Droplets of water that flowed from a thin pipe inserted into the base of the reservoir carried with them electric charges until the potential of the reservoir was equal to the surrounding atmosphere. This system, which proved to be very good, was useful in windy conditions when the flames did not work. These collectors were subsequently improved and, together with the flame collectors, continued to be used well into the twentieth century. At the beginning of the twentieth century the ionizing proprieties of radioactive substances were studied, and scientists started using metallic collectors covered with radium or polonium salts.

The changes in potential equalizers ran parallel with improvements in electrometric instruments. I shall only mention the most significant ones. Until the beginning of the nineteenth century the most popular electrometers were double pendulum apparatus such as Saussure's or Volta's, or Bohnenberger's single-leaf instrument. The latter had a single gold leaf suspended between the poles of two Zamboni dry cells and could indicate not only the presence but also the polarity of a charge. In the 1830s Jean Athanase Peltier introduced a very sensitive apparatus that became popular in France (fig. 11.1c). His electrometer had a light, movable metallic needle with a small magnet suspended to a frame connected to quite a large spherical electrode fixed at the top of the instrument. A second insulated needle was fixed parallel to the first needle. The instrument was oriented along the Earth's magnetic meridian so that the two needles were parallel. For measuring, the electrometer was brought to different heights where the atmospheric field induced charges on the sphere. The electrostatic repulsion acting against the magnetic restoring force deviated the movable needle. The deviation was measured on a graduated scale. In 1846 the German scientist Elard Romershausen proposed an improved apparatus with a multiple-point collector and a measuring instrument combining a kind of Bohnenberger electrometer and a torsion balance. In 1853 the German physicists Johann Friedrich Dellmann (fig. 11.1e) and, later, Rudolph Kohlrausch used improved and complicated versions of Coulomb's torsion balance. In the late 1850s, the Italian Luigi Palmieri proposed another electrometer having a needle similar to Peltier's (fig. 11.1d). But in the former instrument, the movable needle was not pivoted like Peltier's electrometer. Instead it was attached to a bifilar, silk-thread suspension that exerted the force that counterbalanced the electrostatic repulsion.

In the 1850s and 1860s William Thomson introduced several sophisticated electrostatic measuring instruments, among them, the divided ring electrometer, the portable electrometer, and the quadrant electrometer, which were widely used for atmospheric electricity. The quadrant electrometer, which is an improved form of the ring version, has a light figure-eight-shaped needle suspended between four couples of metallic plates (fig.11.1f).[8] The charged needle was under the influence of charged plates. A small mirror attached to it allowed one to read the deflection using a reflected beam of light. This instrument was subsequently modified by Eleuthère Mascart (fig. 11.1g), Edouard Branly, and others. The portable electrometer, on the other hand, exploited the electrostatic attraction of two charged plates in the same manner as a condenser. Since 1861 an instrument of this latter kind, connected with a "water dropper," was used in the Kew Observatory for photographically recording the variation of atmospheric potential. In France, Mascart was the first to introduce recording instruments that

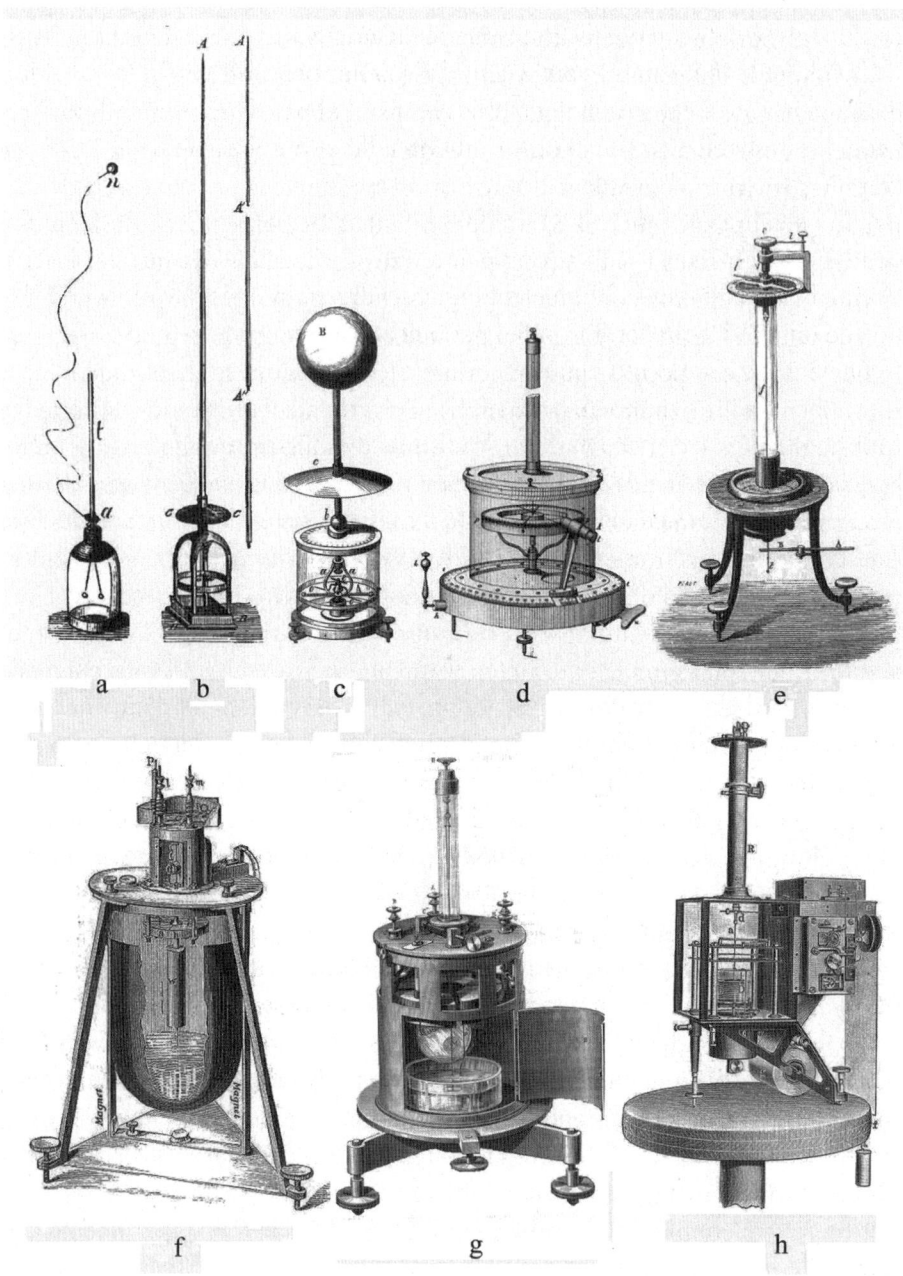

Figure 11.1.   Various types of electrometers used for measuring atmospheric electricity: (a) and (b) Saussure; (c) Peltier; (d) Palmieri; (e) Dellmann; (f) Thomson; (g) Mascart; and (h) Benndorff.

could register the variation of atmospheric electricity in a continuous fashion. In 1879 he used an apparatus composed of a quadrant electrometer with a writing pen, and in 1881 he made photographic recording apparatus in which a luminous beam was reflected by the mirror of the needle and fell on a movable photographic plate (fig. 11.2c). In Germany the most common recording instrument was the electrometer described in 1908 by Hans Benndorff. It was a modified quadrant electrometer with a long pointer attached to the needle (fig. 11.1h). Owing to a frame moved by an electromagnet driven by a clockwork switch, the needle touched the recording paper periodically, leaving on it a dotted line.

During the nineteenth century, several meteorological stations and physical laboratories were equipped with special "electrical observatories." Palmieri's Mount Vesuvius observatory of the 1860s provides a good example (fig. 11.2b). Special cabins were equipped with an exploring mast surmounted with a pointed electrode that could be elevated into the air using a system of ropes and pulleys. The top of the mast was connected with a measuring instrument. This kind of equipment represented state of the art technology for studying atmospheric electricity. But electrometry was also carried on systematically during scientific expeditions in the most remote parts of the world and in the air. Measurements with collectors and electrometers were a priority for courageous scientists and balloonists who explored the atmosphere in the nineteenth and early twentieth centuries.

Around 1850 tens of thousands of observations were collected in several European and American observatories. Nevertheless, in spite of the large number of observations, the data of various sources often showed remarkable discrepancies. As far as the development of instrumentation for measuring fair weather atmospheric electricity was concerned, one could argue that there was a continuous struggle for improving apparatus and observational techniques up to the end of the nineteenth century. In fact, many of these devices had originated in the mid-eighteenth century. It is true that one could also study diurnal and annual variation of the electric state of the air using galvanometers whose terminals were connected with a vertical conductor and the ground. But even the most sensitive galvanometers of the time could not clearly detect these variations. Currents of varying intensity and direction were detected during rain, snow, fog, stormy weather, and lightning stokes. But in these cases, the indications of galvanometers were so sudden and erratic that they were not very useful. In 1858 when Auguste de la Rive surveyed a century of investigations related to atmospheric electricity in his famous *Traité d'électricité théorique et appliqué*, he admitted that "quelle en soit la cause, une si grande divergence entre les résultants obtenus par des observateurs si distingués . . . nous montre combien il y a encore d'incertitude dans la détermination bien précise de l'électricité atmosphérique. Cette incertitude ne tient pas seulement à l'imperfection des instruments, mais bien aussi à la nature du phénomène qui est éminemment complexe."[9]

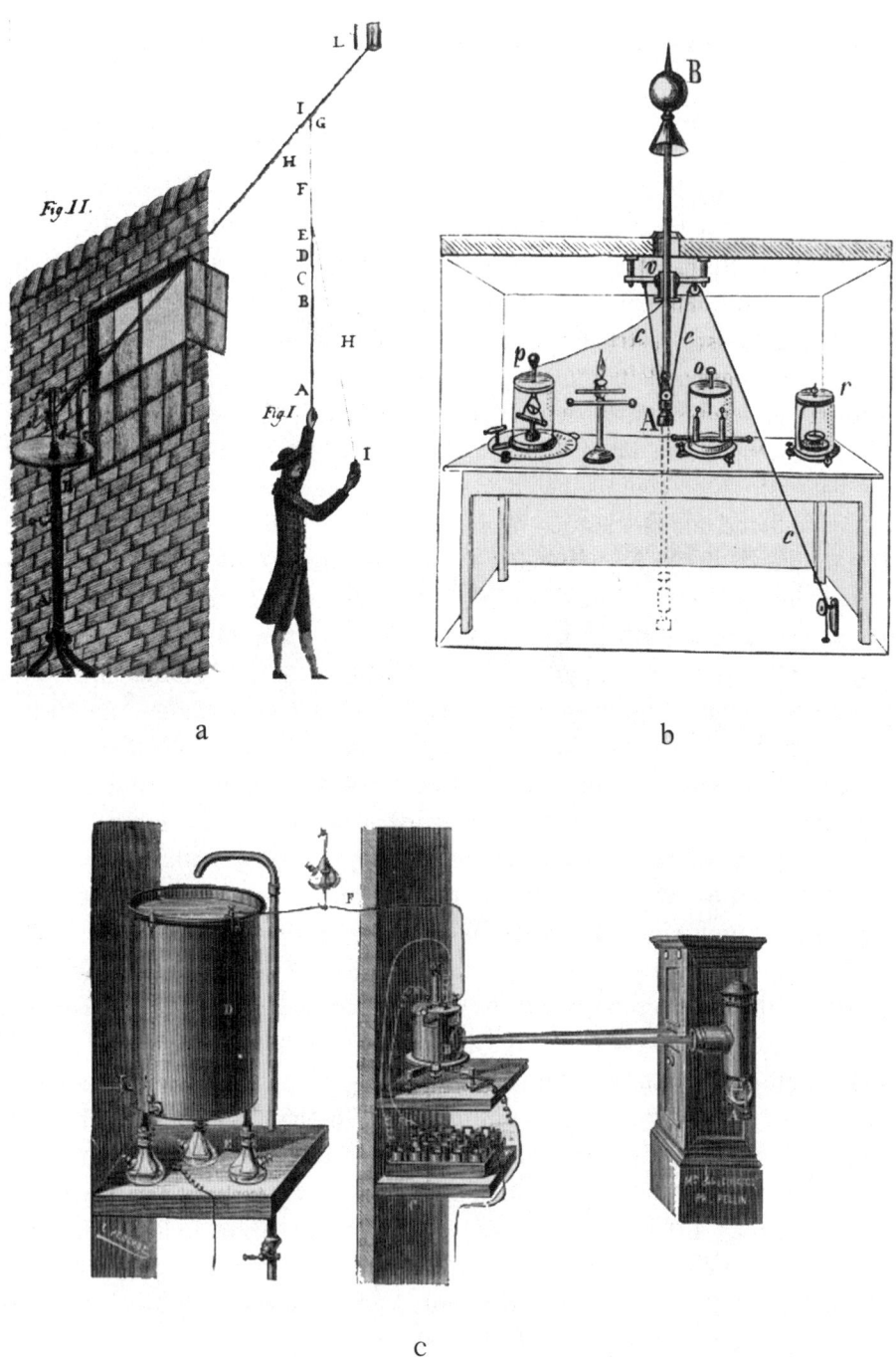

Figure 11.2. Detection and measurement of atmospheric electricity: (a) Volta's "lantern atmospheric electrometer" using a flame collector; (b) Palmieri's electric observatory; and (c) Mascart's apparatus with quadrant electrometer, water drop collector, and photographic recorder.

Only in the last years of the nineteenth century did a series of discoveries (such as ionization) open new perspectives in the study of atmospheric electricity.[10] The "German school" played a fundamental role in this research, represented by physicists Julius Elster and Hans Geitel, who were strongly influenced by the research of Viennese scientist Franz Exner. Exner and his followers tried to measure precisely the absolute value of the potential fall of the atmospheric field between the ground and an elevated point. He pointed out how many past measurements were often useless because they were obtained with instruments placed near buildings, walls, and terraces whose presence strongly modified the electric field. While "classical measurements" continued to be refined, phenomena related to ionization and conductibility of the air, solar radiation, electricity of precipitation (rain, snow, hail, etc.), and the radioactivity of air and ground began to be systematically investigated with a generation of new apparatus.[11] The dispersion of electricity in the air—first determined by Charles Augustin Coulomb in 1787 and in 1850 by the physicist Carlo Matteucci, who used an improved Coulomb's torsion balance—was extensively studied by Elster and Geitel, who demonstrated the presence of ions in the atmosphere (fig. 11.3c). They invented a dispersion apparatus that allowed one to measure the time necessary for unloading an insulated and charged conductor. In 1901 Arthur Ebert described an "ion counter" (fig. 11.3b), and in 1905 H. Gredien proposed a special condenser aspirator for measuring the conductibility of the air (fig. 11.3a). It was a cylindrical condenser (connected with an electrometer) in which it was possible to produce a current of air. These instruments and many others were connected with improved electrometers. In 1887 Exner invented a simple and compact aluminum-leaf electrometer, which was later modified by Elster and Geitel. It was used well into the twentieth century. But other measuring instruments, such as Wulf's single- or double-wire electrometer, Lindemann's electrometer (which was a combination of a quadrant and a wire), and Perucca's electrometer were widely used as well.

In spite of improved instrumentation, the increasing number of observational data, and the discovery of new phenomena that certainly played a fundamental role in the mechanisms of atmospheric electricity, the situation was far from clear. At the end of the nineteenth century Chauveau counted more than thirty different "serious theories" related to atmospheric electricity.[12] In 1900, Exner presented a paper to the congress of physics held during the Paris Exhibition, where he stated the following: "Il est impossible d'exposer ici toutes les théories, souvent très problématiques, qui ont été proposé sur l'électricité atmosphérique. . . . Je tiens à dire tout de suite qu'aucune des hypothèses actuelles ne rend compte de tous les phénomènes de l'électricité atmosphérique, de sorte qu'il nous semble possible qu'une série de causes différentes intervienne dans leur production."[13] Exner's words echoed the statement made by De la Rive about half a century before.

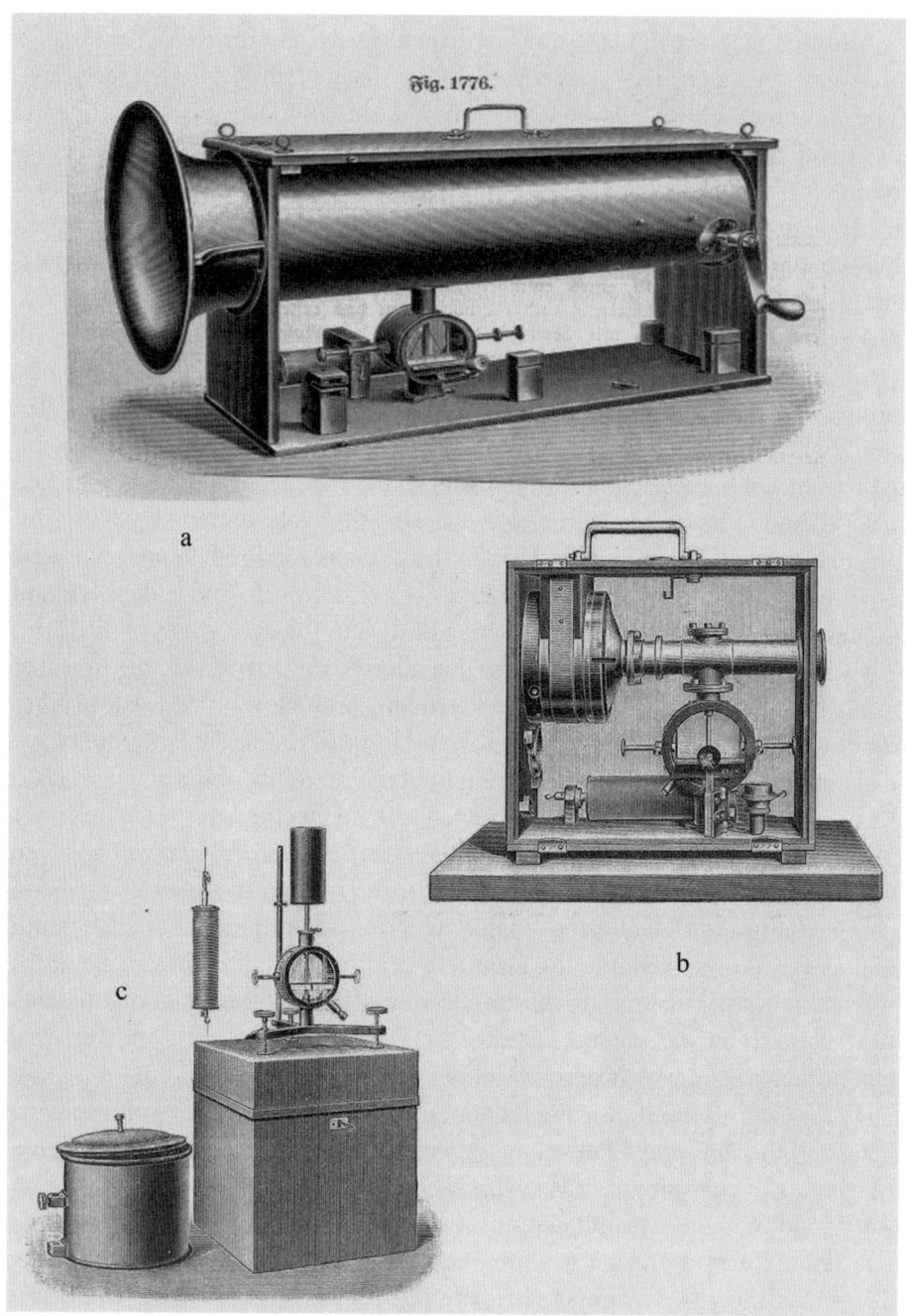

Figure 11.3. Various apparatus used at the beginning of the twentieth century: (a) Gredien's aspirator-condenser for measuring the electrical conductibility of the air; (b) Ebert's ion-counting apparatus; and (c) Elster and Geitel's apparatus for measuring the dispersion of electricity.

Thunderstorm and Lightning Electricity

Lightning phenomena are atmospheric, transient, high-current electric discharges that during thunderstorms (together with rain and corona discharges) tend to keep the Earth charged negatively and the atmosphere positively.[14] Approximately two thousand thunderstorms are in progress around the world at any given moment. A "typical" lightning strike, which can be several miles long, can have a voltage of about 10^8 volts with a current of 10^4 amperes. The charge carried by it can be 20–30 coulomb.

Certainly, these facts were not known by eighteenth century natural philosophers and electricians who, after the pioneering experiments of Franklin and Dalibard, were inspired to play with thunderstorm electricity in their "electrical cabinets." These experiments were made in small pavilions into which the bottom end of a lightning rod or a wire holding a kite penetrated (fig. 11.4).

The conductors were connected to spark gaps, electroscopes, electric chimes, and Leyden jars. When a thunderstorm approached, the spark gaps arched, the chimes began to ring, and the electroscopes went crazy. Certainly, eighteenth century electricians did not have the slightest idea of the huge amount of electricity developed by thunderstorms, and they largely underestimated the danger involved in these kinds of observations. In 1753 the Petersburg professor Georg Wilhelm Richmann, who unfortunately became popularly known as "the first martyr of electricity," was fatally hit by a side discharge that had struck the lightning rod of his cabinet. But these impressive and dangerous games did not greatly contribute to the knowledge of atmospheric electricity; they merely confirmed that its effects were identical or at least very similar to the electrostatic ones produced with friction machines. In fact, between the middle eighteenth century and the very end of the nineteenth century, measurements of thunderstorms and lightning did not make substantial progress. However, for most of the nineteenth century, scientists observed and described lightning, recorded damages due to lightning, and compiled statistics of death by lightning, but made no real measurements. Academy reports, scientific journals, and even newspapers regularly published observations of violent lightning strikes, of extraordinary and mysterious ball lightning, and of Saint Elmo's fires. But there was no real advance in the knowledge of these phenomena. The French physicist and astronomer François Arago and others tried to classify lightning from its appearance, but the "science of lightning" remained a purely descriptive and qualitative discipline. Conversely, there were still heated arguments about the best way to construct lightning rods. Since the fierce scientific (and political) controversies of the 1770s, which were represented by opposing partisans of the pointed rod and spherical conductors, scientists had tried to increase the efficiency of lightning protectors.[15] Table 11.1

Figure 11.4.   Cuthbertson's "electrical cabinet" of 1786.

provides an idea of the kinds of papers concerning atmospheric electricity and lightning protection devices presented to the Parisian Academy of Science from 1835 to 1895.

Most of these articles concerned lightning observations and lightning protectors (both for buildings, ships, and telegraphic equipment). During the nineteenth century, dozens of different types were proposed: long single rods, multiple conductors, and points of different shapes and sizes. Several materials such as platinum, gold, iron, carbon, and even straw were tested for their points or for rods and wires. Different types of ground "dispersers" were conceived and various geometries were proposed for the conductor.[16] Some investigators tried to determine the volume and the surface protected by lightning rods. Was it a cone whose height corresponded to the tip of the rod with a base radius double its height? Or was it a cone with the height of the rod and the radius equal to 7/4 of its height? In spite of the large number of proposed solutions, lightning often missed the conductor, which was supposed to direct it to the ground. Instead it killed people and destroyed clock towers, houses, and ships. Many theories were advanced, and generally the failures of lightning conductors were attributed to poor ground connections. But experiments were hardly possible; at the time no transformers or electrostatic machines could produce a discharge comparable to atmospheric ones. At the same time,

Table 11.1

Years	Description and Observations of Lightning	Lightning Protectors	Measurement and Telegraphic Protectors	Utilisation of Atmospheric Electricity
1835–1850	58	10	22	
1851–1865	44	35	8	
1866–1880	44	41	37	2
1881–1895	30	10	47	2

knowledge of the mechanisms and of the parameters of atmospheric discharges was rather simple, and a mathematical model or theoretical prediction appeared to be impossible. Finally, the events (e.g., lightning strikes on a precise conductor) were so rare that a systematic, statistical study of the quality of lightning protectors was quite difficult. Consequently, the devices were built with the wrong assumption that lightning currents behaved like normal direct current that obeyed Ohm's law. One could argue, therefore, that many lightning rods installed in the nineteenth century, instead of being a protection, represented an additional danger for the buildings they were supposed to preserve.

Experiments and Instruments

In the second half of the century, lightning caused a lot of problems for a new and rapidly developing technology: telegraphy. Static electricity and atmospheric discharges not only disturbed the telegraphic transmissions but, if lightning struck a line, it could also badly damage or even destroy the transmitting apparatus and the ancillary equipment in the telegraphic bureaus. Several telegraphic lightning protectors existed, but due to the lack of knowledge about lightning, they were not efficient. Engineers generally thought that the most important characteristic of a lightning protector should be a very low resistance, and they subsequently treated lightning discharges like a direct current.[17]

In 1888 the British physicist Oliver Lodge was asked to prepare a series of lectures concerning lightning protectors for the Royal Society of Arts. The topic carried a practical concern: hundreds of thousands of lighting protectors had been installed by the post office, yet their efficiency was far from proven. Lodge became quite involved with the subject and, after becoming convinced that lightning was similar to the discharge of a condenser, he began a series of experiments using an electrostatic machine, Leyden jars, and spark gaps connected in various ways (fig. 11.5a). He observed that his jars did not necessarily discharge across the gap connected to them with a circuit of lower ohmic resistance. Often sparks chose an "alternative path." Lodge, who was wrongly convinced that lightning resulted from oscillatory discharges, nevertheless correctly understood that they generally follow the path with minimal impedance and not of minimal resistance.[18] Lightning in fact is a series of very rapid unidirectional discharges, and the self-induction of conductors greatly influenced the behavior of such accelerating currents. Starting from these considerations, Lodge suggested several improvements in the construction of lightning rods for the protection of buildings and telegraphic lightning protectors. But in the era of the rapid evolution of electrical science, Lodge's experiments and advanced theoretical analysis were strongly criticized and aroused violent debate between physicists and engineers.[19] Lodge continued his experiments, however, with oscillating circuits, which led to the study of electromagnetic waves. He would become one of the pioneers of wireless telegraphy.

In 1888 Hertz's experimental discovery of electromagnetic waves opened the way to the wireless transmission of signals. The first electromagnetic wave detectors (the metallic powder of an imperfect contact coherer developed by Calzecchi-Onesti, Edouard Branly, Lodge, and others[20]) not only revealed the presence of manmade electromagnetic waves but also reacted to the influence of electromagnetic disturbances produced during a thunderstorm. In 1895 the Russian engineer Alexander Stepanovich Popov demonstrated his thunderstorm detecting and recording instrument to the members of the Russian Physical and Chemical Society.[21] Popov first connected a coherer to an aerial, which, connected to the ground, was inserted to the circuit of an electric bell (fig. 11.5b). The train of electromagnetic waves produced by a lightning strike (several miles away from the apparatus) was detected by the apparatus, which rang an electric bell. More sophisticated devices of this kind could record the number of strikes on a telegraphic receiver. Various types of lightning and thunderstorm detectors were proposed by instrument makers at the beginning of the twentieth century.

In spite of this theoretical and instrumental progress, nothing was done concerning lightning current and voltage measurements. Lightning energy and current could be estimated roughly from the damage caused by them. For example, strength could be assessed by the length and thickness of metal wire melted by a strike. But these estimations were far from accurate, and in the nineteenth century, the inertia of the available current meters was far too big for measuring the short current pulses of strikes.

The first "scientific" determination of lightning currents was due to Friedrich Pockels in the final years of the nineteenth century. Pockels discovered that the residual magnetism of pieces of basalt submitted to a magnetic field was only dependent on the maximum value of the field, or on the maximum value of the current generating this field. Starting from this consideration, Pockels analyzed various pieces of basalt found near trees hit by a lightning strike, or pieces placed

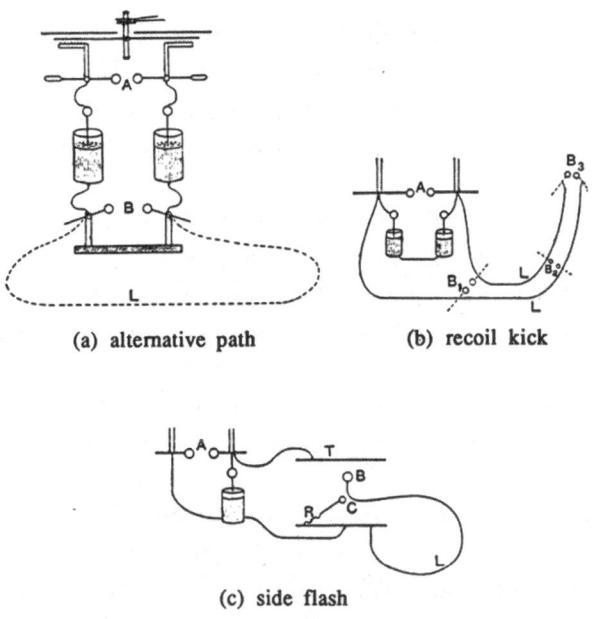

(a) alternative path

(b) recoil kick

(c) side flash

Figure 11.5. *Top*: Diagrams of some of Lodge's experiments. *Bottom*: Popov's thunderstorm recording instrument.

near the lightning rod of an observing tower in the Apennine Mountains. He calculated current on the order of 10^4 amps. The practical value of Pockels' method
was beyond question, and in the 1920s and '30s, these measurements were refined
and various types of test probes made of steel bars or wires with high remanence
were introduced for measuring peak currents of lightning strikes. Widely used after
World War II, a large number of them were installed near the protectors of towers,
chimney stacks, and buildings.[22] Another interesting current-measuring device,
using magnetizable materials, was the fulchronograph, introduced by Wagner and
McCann in 1940. The fulchronograph was a recording apparatus consisting of a
fast rotating aluminum wheel with magnetic links fixed along the periphery. These
links passed between coils through which the current to be measured flowed. The
principle of these instruments was identical to the magnetizable test probes, but
they also recorded a series of different measurements, whose maximum time resolution was on the order of 10^1 sec for a total duration of a few milliseconds. Other
magnetic instruments invented at the time were the magnetic surge front recorder
and the magnetic surge integrator. The former allowed one to measure the rate of
increase of lightning current while the latter made it possible to calculate the
charge carried by the strikes.

In the 1920s J. F. Peters proposed the klydonograph, an instrument for measuring lightning voltage on electric transmission lines, which exploited the well-
known phenomenon of Lichtenberg figures (fig. 11.6).[23] In the late 1770s, the
German natural philosopher, Georg Christoph Lichtenberg, studied the ar-
borescent figures generated by a high-tension pointed electrode (e.g., connected to the conductor of an electrostatic machine) set vertically near an
insulating surface covered by an electroscopic powder (generally sulfur and red
lead oxide). The ions generated by the corona effect of the electrode produced
circular and branched figures whose diameter depended on the potential, and
the structure on the polarity, of the discharge. In the nineteenth century, Lichtenberg's figures were obtained on a photographic plate. In Peters' klydonograph, the electrode was connected to an electric line through a shunt and
properly calibrated.[24] It then rested on a slow revolving photographic plate
moved by a clockwork mechanism. When lightning struck, the line generated
a figure on the plate. The analysis of the figures allowed one to measure voltages on the order of several million volts. These values were given with an approximation of about 20 percent to 30 percent.

Finally, in the 1910s and 1920s, Charles Thomson Rees Wilson, well known
for the invention of the cloud chamber, performed pioneering work to measure the electric field and its variations due to lightning. Wilson carried out his
measurements with an antenna that was alternatively shielded and unshielded
by a grounded, metallic, moveable plate. The latter was connected with a gold
leaf or with the capillary electrometer.

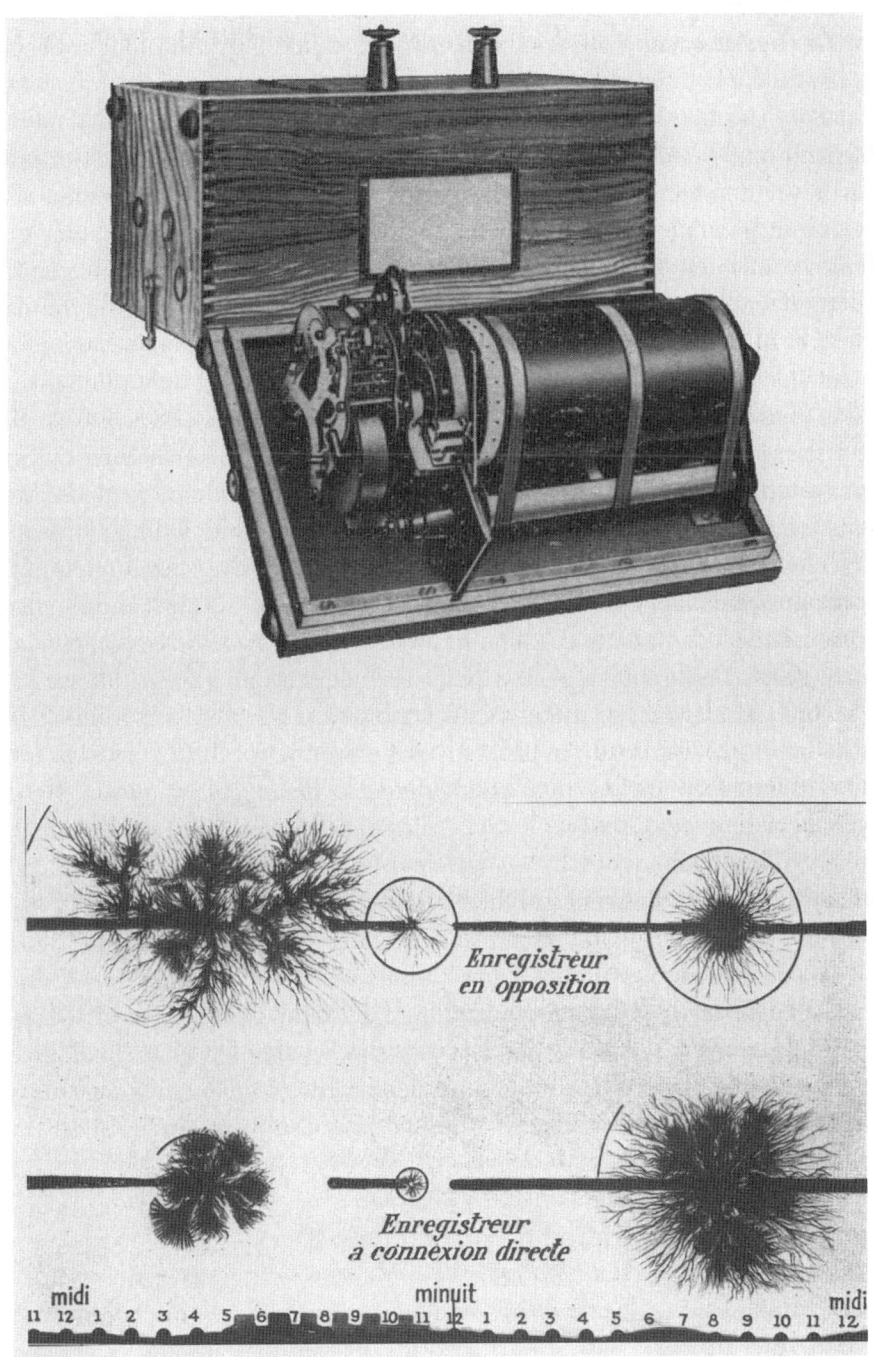

Figure 11.6. *Top*: The klydonograph. *Bottom*: Some recordings showing the traces left by voltage surges.

In the first decades of the twentieth century, research on atmospheric electricity acquired increased urgency with the extension of the electric network, whose protection against lightning was of paramount importance. Thus, several new instruments and techniques were developed, but certainly the most revolutionary was the oscillograph, derived from Braun's tube. In the late 1920s, with these early electronic apparatus (connected with an antenna) it was possible to trace on a fluorescent screen the rapidly varying currents and voltages of lightning strikes.

At the same time lightning photography and spectroscopy became standard practice. Already in the 1830s the British physicist Charles Wheatstone, during experiments to measure the speed of electricity, also tried to determine the duration of strikes with a simple stroboscopic wheel. In the late 1880s the first photographs appeared, which showed multiple-strike flashes. Some of these pictures were obtained by rapidly moving a normal camera in the horizontal plane. A few years later in Germany flash-resolved pictures were made with a clockwork-driven camera. However, the breakthrough came with the introduction of the cameras invented by Charles Vernon Boys in the 1920s. The first of these had a fast-revolving lens camera, while an improved version had stationary optics and moving film. These apparatus, and their subsequent improvements, allowed one to record clear images of strikes and contributed to the determination of their structure. Furthermore, in the first half of the twentieth century, it was possible to build transformers and high voltage generators capable of producing discharges of several million volts. These apparatus were used for testing the strength of insulators as well as for reproducing (even if on a reduced scale) lightning strikes to better understand their behavior and effects.

In the last fifty years, researchers concerning lightning and thunderstorms have made enormous progress. Knowing the mechanisms underlying lightning is essential for developing better protection systems in a world filled with delicate electronic computer networks and electric grids, where thousands of vulnerable airplanes are in the air at the same time, and where a blackout in a big area can be disastrous.[25]

Utilization of Atmospheric Electricity

Lightning has always been seen as a symbol of power.[26] Scientists, engineers, inventors, and dreamers have always seen the tremendous amount of energy in lightning as a promising and endless source of cheap energy. Who could watch a lightning show on a stormy night and not ask "Could we use this electricity for our industry and our homes?" Martin Uman, a world-renowned expert on lightning, once estimated this energy.[27] If we consider that a typical cloud-to-ground lightning (10^8–10^9 volts, and a charge of tens of coulombs) has energy of about 1 to 10 billion watt-seconds, and if we assume that there are about one hundred lightning flashes per second over the whole Earth, we have a maximum value for the total electrical power input to worldwide cloud-to-ground lighting of about 1,000 billion watts. That is quite a lot. But unfortunately there is now no practical method of harnessing it. Not only is a large part of this energy dispersed in electromagnetic radiation, heat, and acoustic energy, but it would be necessary to also build hundreds of thousands of one-thousand-foot-tall towers for a few strikes a year. Even neglecting the technical problems of such a project, these structures would represent a prohibitive investment and would have a disastrous ecologic impact. Conversely, Uman pointed out that "if its total energy were available, a single light flash would run an ordinary household light bulb for only a few months." So, as far as we know today using the electrical energy of the atmosphere is absolutely unrealistic. Nevertheless, it is interesting to mention some of the more significant projects that have been realized, or at least proposed, with the hope of practically exploiting atmospheric electricity.

One of the first attempts to practically use it was in the field of wireless telegraphy. Perhaps the most intriguing series of experiments in this direction were made by the American dentist and inventor Mahlon Loomis.[28] Loomis, who dreamed of seeing atmospheric electricity in the service of mankind and who proposed various devices for collecting it, was able to successfully realize a primitive form of wireless transmission as early as 1864. The basic principle of his system was simple. On two mountain peaks in Virginia, he launched two kites fourteen miles apart, each covered with copper gauze and held with six hundred feet of thin copper wire. Each wire ran through a galvanometer to the kite. When one of the wires was grounded or removed from the ground, there was a disturbance of the atmospheric electric field, which was detected by the galvanometer of the other kite.[29] Every time the action was repeated, the needle of the galvanometer deflected. Loomis tried for several years to improve this system and struggled to find support, but unfortunately his "aerial telegraph" died with him.

Patents, proposals, and inventions for using the energy accumulated in the sky flourished especially between the years 1890–1930. In this period the electrical industry boomed, large power plants were installed in Europe and in the United States, and electrical machinery and transmission lines became emblems of modernity. It is not surprising that in this era of electrical euphoria and engineering optimism even the most far-fetched technological dreams seemed to be possible. Most of these projects involved the use of kites and balloons for collecting static electricity, which had to be converted into a more useable form of energy. Just a few examples will give an idea of the equipment proposed by some of these dreamers.

In 1897 Heinrich Rudolph Hertz of Germany proposed to install a three-hundred-meter-long metallic net with about 3.6 million collecting needles. It was elevated by kites. Owing to a special commutator, the static electricity was supposed to charge a battery of twenty thousand accumulators. In 1900 Andrea

Palencsar of Budapest patented a complicated system consisting of a double-walled heated balloon covered by a movable net with pointed collectors. In this project, the electricity had to feed a rheostatic machine that charged a battery.[30] Similar technological monstrosities were conceived in 1912 by Heinrich Johannsen of Lübeck, Germany, and by Walter J. Pennock of Philadelphia.

There is no doubt that the most sophisticated and elaborate projects came from Estonian Hermann Plauson.[31] In the late 1910s and early '20s, Plauson lived in Germany and in Switzerland and was the director of the Otto Traun'schen Forschungslaboratorium in Hamburg.[32] He received several patents and wrote the book *Gewinnung und Verwertung der atmosphärischen Elektrizität* in which, after a short historical introduction illustrating and critically analyzing the machinery of the above-mentioned inventors, he presented his own apparatus and experiments. Plauson performed a great deal of research and spent several years in improving his system. Compared to his predecessors, Plauson had an ingenious and original idea: to convert static electricity in high-frequency oscillating currents. It is impossible to give here a detailed description of all of Plauson's devices, but his basic system was supposed to work in the following way: Atmospheric electricity was collected by metallized balloons covered with collectors and was used to charge a battery of condensers. These were periodically discharged with a spark gap connected to a Tesla transformer.[33] The rapid oscillatory sparks generated a high-voltage and high-frequency current in the secondary of the transformer. With this current it was possible to activate a special type of high-frequency motor or "condenser motor." Plauson proposed several types of balloons, collectors, circuits, transformers, spark gaps, and motors (fig. 11.7). He made some small-scale tests in Finland where, with two collector balloons and a large condenser, he was able to obtain about 82 kW of energy in twenty-four hours. But in spite of several experiments, attractive economic estimations, and optimistic claims, his system failed.

In the late 1920s there was another series of promethean experiments that tried to make practical use of atmospheric electricity. Arno Brasch, Friedrich Lange, and Kurt Urban of the Friedrich Wilhelm University of Berlin decided to exploit the high voltage (and not the power!) of the atmospheric field for their pioneering research in nuclear physics.[34] Before the advent of accelerators for splitting the atoms and for generating penetrating high-energy rays, it was necessary to have a very high tension capable of accelerating the particles as "projectiles." The three researchers tried to develop a vacuum tube capable of withstanding very high voltages. After having considered various locations where the atmospheric electrical activity was particularly strong, they chose a small valley near the top of Monte Generoso, a mountain overlooking Lugano Lake in Ticino, the southern part of Switzerland. A railroad easily transported the equipment to the top of the mountain (about 1,700 meters). In 1927 they installed the first electricity collector composed of a metallic net of four hundred square miles

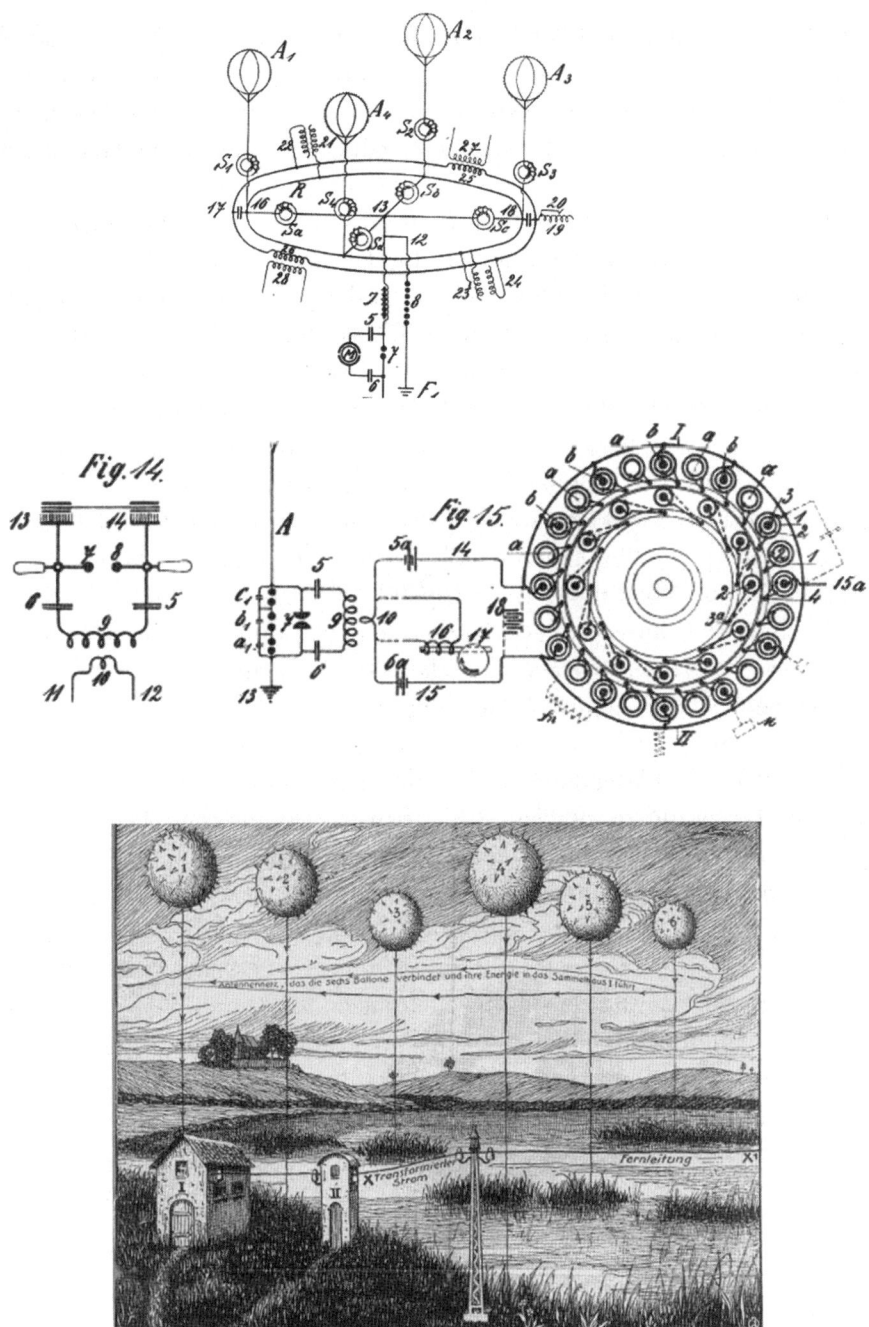

Figure 11.7.   *Top*: Schemes of some of Plauson's apparatus. *Bottom*: An idealized popular version of Plauson's system.

suspended by an eight-hundred-meter insulated cable at eighty meters from the bottom of the valley (fig. 11.8). A spark gap was supposed to measure the high voltages that fed the vacuum tubes, and the instrument located in a small metallic cabin acted as a protecting Faraday cage. But many problems appeared. The insulators of the conductor were not good enough, the spark gap was too small, and the cable risked potential breakage. Nevertheless, the system was tested in August. The following summer the three researchers returned to Switzerland. They installed a better aerial and a larger spark gap. More tests were made, and during thunderstorms, it was possible to produce sparks of eighteen meters arcing the gap. (A theodolite was used for measuring the length of the gap.) But the working conditions were extremely difficult, the cables broke a few times, and there was no vacuum tube ready to be tested. Furthermore, in a tragic accident, Kurt Urban fell to his death during exploration of the difficult terrain. Finally, after all that, Brasch and Lange abandoned Mount Generoso and went back to Berlin, where they continued their research in the high voltage laboratory of the Allgemeine Elektrizitäts-Gesellschaft (AEG).

In the first decades of the twentieth century the visionary projects of Plauson as well as the experiments of Brasch, Urban, and Lange, and other scientists and inventors were largely reported, often with some exaggeration, in the contemporary newspapers and in the popular science reviews. Readers often could not distinguish between the realistic descriptions of a high-tension industrial laboratory used for testing the dielectric strength of insulators, and the sometimes overly optimistic articles illustrating futuristic projects. Lightning was depicted in films and books as the ultimate source of energy. This excited the public imagination as well as authors of science fiction. In 1939 the successful German writer and engineer Hans Dominik wrote a book titled *Himmelskraft* (*The Force of the Sky*). In it, a powerful German electrical company (AEG?) fought against an American one (General Electric?) for supremacy in harnessing the atmospheric electrical energy. The key patents used by both companies concerned some "flame collectors" fixed on a huge metallic net supported in the air by helium balloons. The contraptions were supposed to transmit atmospheric electrostatic charges to a gigantic power plant. These apparatus sounded very similar to the ones described in Plauson's patents.

Conclusion

After more than 250 years of studies, measurements, and research of atmospheric electricity, an impressive amount of data has been collected and sophisticated models and theories have been developed. Nevertheless, if we read some contemporary works related to this topic, we can easily see how our knowledge of phenomena related to the "electricity of the sky" is still incomplete. A short

Figure 11.8.   The experiment of Monte Generoso, Switzerland, in 1927–28. *Top to bottom*: The installation of isolators, the antenna across the valley, and the spark gap before and during a thunderstorm (the length of the spark in this photograph is about four meters).

survey of this history shows us how research into "fair weather" atmospheric electricity developed at a constant pace since the second half of the eighteenth century. For most of the nineteenth century, measurements continued to follow the direction inaugurated around 1750. Certainly, instrumentation became more and more sophisticated: better electrometers and collectors allowed one to refine the measurements, and recording apparatus were introduced to continuously register the variation of atmospheric fields. But the words written in 1860 by William Thomson remain true. Only near the end of the nineteenth century with the discovery of new phenomena (such as ionization, radioactivity, cosmic rays, photoelectric effects) did the study of "fair weather" atmospheric electricity enter into a new and fruitful era. It should also be pointed out that most of the research in this field was carried on by physicists, meteorologists, and scientists from the academic world. The study of lightning also interested the community of engineers and technicians, who were mostly concerned with protection devices for buildings, telegraphic equipment, and later for electric transmission lines and machines. However, up to the turn of the nineteenth century, lightning research was essentially limited to observations and descriptions. Large-scale experiments and systematic measurements started only with the introduction of technical instruments developed by electric companies and industrial laboratories. Following these developments and due to the spectacular success of the electrical industry, the dream of harnessing electricity from the sky seemed realizable. Today, even if balloons carrying flame and radioactive collectors and "condenser motors" are buried in the cemetery of unsuccessful inventions and the field is not seen as profitable, scientists and inventors have not abandoned this fascinating field of research. Military researchers, for example, are still hoping to use artificial or natural lightning as weapons while projects such as the Franco-German "Teramobile" study the possibility of triggering and guiding lightning to the ground with a very powerful laser beam.[35] Even if atmospheric electricity can only run small electrostatic toy motors connected to balloons but not vacuum cleaners, light bulbs, and refrigerators, there are still web sites dedicated to this so-called free energy and alternative energy.[36] The projects of Plauson and other visionaries of the first half of the twentieth century are still quoted and are a matter of speculation and discussion. The spectacular effects of atmospheric electricity will never cease to fascinate us and excite our curiosity. And fascination and curiosity certainly are the very powerful driving forces that push men and women to investigate nature and its phenomena.

Notes

1. John Heilbron, *Electricity in the 17th and 18th Centuries: A Study of Early Modern Physics* (Mineola, N.Y.: Dover Publications, 1979), 339f; and Hans Prinz, *Gewitter-*

elektrizität, Deutsches Museum Abhandlungen und Berichte Heft 1 (München: R.Old-enbourg Verlag, 1979), 17f.

2. The primary literature related to these topics is enormous. Therefore in this essay I will mention only some of the most important secondary articles and texts in which it is possible to find a much more detailed bibliography.

3. William Thomson, *Reprint of Papers on Electrostatics and Magnetism* (London: Macmillan, 1872), 209.

4. For an introduction to the history of atmospheric electricity up to the 1910s, see B. Chauveau, *Electricité atmosphérique, Premier Fascicule, Introduction historique*, Librairie (Paris: Octave Dion, 1922).

5. See the essay of Paola Bertucci in the present volume.

6. For a detailed history of early electroscopes, see Willem Hackmann, "Eighteenth Century Electrostatic Measuring Devices," *Annali dell'Istituto e Museo di storia della Scienza* 3 (1978): Fasc. 2, 3–58. Nineteenth century electroscopes are described in detail in Eleuthère Mascart, *Traité d'électricité statique*, vol. 1 (Paris: Masson, 1876), 344–36.

7. A similar system was used several years later by Antoine César Becquerel, who confirmed an increase of positive electricity with height. With a bow he launched into the sky an arrow connected to an electroscope with a very thin metallized silk wire.

8. For a detailed description of Thomson's instruments, see William Thomson, *Reprint of Papers on Electrostatics and Magnetism* (London: Macmillan, 1872), 260–309; George Green and John T. Lloyd, *Kelvin's Instruments and the Kelvin Museum* (Glasgow: University of Glasgow Press, 1970); and Mascart, *Traité d'électricité statique* (cit. n. 6).

9. Auguste De la Rive, *Traité d'électricité théorique et appliqué*, vol. 3 (Paris: J. B. Baillère et fils, 1858), 89.

10. A detailed article about atmospheric electricity studies, theories, and instruments at the beginning of the twentieth century is H. Gredien, "Die atmosphärische Elektrizität," in *Handbuch der Physik*, ed. A. Winkelmann, vol. 4: *Elektrizität und Magnetismus*, 2nd ed. (Leipzig: J. Ambrosius Barth, 1905), 687–729. See also: O. D. Chwolson, *Traité de physique*, vol. 4, Premier fascicule, *Champ électrique constant* (Paris: Librairie A. Hermann et fils, 1910).

11. The most important instruments used between the end of the nineteenth century and about 1930 are described in the chapter dedicated to atmospheric electricity in E. Kleinschmidt (ed.), *Handbuch der meteorologischen Instrumente und ihrer Auswertung* (Berlin: Julius Springer Verlag, 1935). These volumes include a very extensive bibliography.

12. Chauveau, *Electricité atmosphérique* (cit. n. 4).

13. Franz Exner, "Sur les recherches récentes relatives à l'électricité atmosphérique," *Rapports présentés au Congrès international de physique réuni à Paris en 1900*, vol. 3 (Paris: Gauthier Villars, 1900), 415–37.

14. Some fundamental facts concerning lightning and thunderstorm can be found in Martin A. Uman, *Lightning* (New York: Dover Publications, 1969); Martin A. Uman, *All about Lightning* (New York: Dover Publications, 1986); Claude Gary, *La foudre des mythologies antiques à la recherche moderne* (Paris: Masson, 1995); Johannes Wiesinger, *Blitzforschung und Blitzschutz*, Deutsches Museum Abhandlungen und Berichte, Heft 1–2 (München: R. Oldenburg Verlag, 1972); and Philip E. Krider, "75 Years of Research

on the Physics of a Lightning Discharge," in *Historical Essays on Meteorology 1919–1995: The Diamond Anniversary History Volume of the American Meteorological Society*, ed. J. R. Fleming (Boston, Mass.: American Meteorological Society, 1996), 321–50.

15. See the essays of Willem Hackmann and Roderick Home in this volume.

16. See, for example, *Instruction sur les paratonnerres adoptée par l'Académie des Sciences. Instructions ou Rapports de 1784, 1823, 1854, 1867 et 1903* (Paris: Gauthier Villars, 1904).

17. See the essay of Elizabeth Cavicchi in this volume.

18. When a changing current (alternating, pulsating, etc.) passes through a conductor, it produces a changing magnetic flux, and this in turn generates an electromotive force that is proportional to the self-inductance of the conductor itself. Because of Lenz's law, this induced emf acts against the primary current. Therefore, in this case we have not only a simple ohmic resistance but also an opposition to the current caused by the self-inductance. The total resistance of a circuit (which can also include a capacity) to a changing current is called impedance. Depending on the geometry of the circuit, the inductive (or capacitative) part of the impedance can be much greater than the simple ohmic resistance.

19. See Ido Yavetz, "Oliver Heaviside and the Significance of the British Electrical Debate," *Annals of Science* 50 (1993): 135–73; and Ido Yavetz, "Between High Science and Practical Engineering: Two Studies of Lightning by Simulation," *Physis*, Nuova Serie 33 (1996): 221–58. I would like to thank Anna Guagnini, who gave me these references.

20. This type of coherer is composed of a glass or ebonite tube containing some metallic filings (silver, nickel, or other metals) between two electrodes. Normally its electric resistance is of the order of the megaohm, but a train of electric waves (e.g., produced with an oscillating discharge of a Leyden jar) reduces the resistance to $10^1 - 10^2$ ohms. In an electric circuit that includes a battery and an electromagnetic bell, the coherer acts as on−off switch. A little mechanical shock on the tube (produced, for example, by the hammer of the bell itself) reestablishes the high resistance condition of the device.

21. The Popov detector was at the center of a well-known controversy concerning Marconi and the priority of the invention of wireless.

22. Normally minium (lead oxide) and sulfur were used.

23. See, for example, R. Villiers, "La foudre et les figures de Lichtenberg: Le klydonographe," *La Nature*, August 1927: 97–100.

24. The shunt, generally formed by a series of insulators that worked as resistances, was necessary for reducing by a known fraction the surge tension created by lightning.

25. See the essay of Moore, Aulich, and Rison in this volume.

26. See the article of Fuhrmeister in this volume.

27. See Uman, *Lightning* (cit. n. 14).

28. Several web sites are dedicated to Loomis. See, for example, http://www.smecc.org/mhlon_loomis.htm; http://www.loc.gov/exhibits/treasures/trr083.html; and http://www.acmi.net.au/AIC/LOOMIS_BIO.html. See also Mary Texanna Loomis, *Radio Theory and Operating* (Washington D.C.: Loomis Publishing Company, 1928).

29. It has to be pointed out that as far as I could see, Loomis's system did not have anything to do with the early "classical" wireless activities of Lodge, Marconi, Branly, and

others. Unlike the latter group, Loomis did not use electromagnetic waves, in spite of the fact that in his articles he mentioned waves and vibrations. He simply was able to provoke and to detect perturbations of the atmospheric electric field.

30. In the nineteenth century the French physicist Gaston Planté invented the rheostatic machine. It was composed of a series of condensers that, owing to a rotating switcher, could be charged in series and discharged in parallel. Planté used his machine for his high-voltage experiments.

31. Hermann Plauson, *Gewinnung und Verwertung der atmosphärischen Elektrizität. Beitrag zur Kenntnis ihrer Sammlung, Umwandlung und Verwendung*, 2nd ed. (Hamburg: Boysen & Maasch, 1920). See also "Power from the Air," *Science and Invention for Air*, March 1922, available at Alternative Energy Research Archive, http://www.nuenergy .org/alt/PlausonMarch1922.htm.

32. As far as we know, Otto Traun was the owner of the first German company (the Harburger Gummi-Kamm Co., founded in 1856 by Christian Justus Traun) producing rubber, ebonite, and gutta-percha. Traun probably sponsored the research of Plauson.

33. The Serbian-born scientist and inventor Nikola Tesla was one of the pioneers of the polyphase alternating current system. At the end of the nineteenth century he conceived of a special coreless high-frequency transformer (Tesla coil) that could produce voltages of several million volts. In 1899–1900 in his laboratory in Colorado Springs, Tesla was able to generate huge artificial sparks. For these spectacular experiments and for his visionary (and often excessive) claims, Tesla is today a mythical figure for many "alternative science enthusiasts."

34. This story was reconstructed in detail by Burghard Weiss in "Blitze für Kernphysik und Strahlentherapie. Die Stossspannungsexperimente von Brasch und Lange am Monte Generoso und bei der AEG in Berlin 1925–1935," *Technikgeschichte* 66, no. 2 (1999): 173–203.

35. See, for example, "'Teramobile Hurls Thunderbolts," CNRS, http://www2.cnrs.fr/ en/195.htm; "Teramobile, First Laser with Terawatt Mobile Power, Controls Thunderstorm," Machine-History.com, available at http://www.machine-history.com/node/553; and Winn, William, et al., "Teramobile Laser and Lightning," Air Force Research Laboratory, available at http://www.dtic.mil/cgi-bin/GetTRDoc?AD=ADA432059&Location =U2&doc=GetTRDoc.pdf.

36. See Nu Energy Research Institute, http://www.nuenergy.org/. In these often nebulous writings, free energy can refer to many things—atmospheric electricity, vibrational electricity of ions in the atmosphere, or even an omnipervasive radiation. In short, "free energy" is there and available in unlimited quantities to be captured.

A Modern Assessment of Benjamin Franklin's Lightning Rods

C. B. Moore, G. D. Aulich, and William Rison

Introduction

DURING SOME PARLOR DEMONSTRATIONS with electrical apparatus imported from England around 1746, Benjamin Franklin and his associates Thomas Hopkinson, Ebenezer Kinnersley, and Philip Syng discovered the point-discharge phenomenon in which the air around a sharp-tipped conductor becomes electrically conductive when exposed to strong electrical charges. They found that they could discharge an isolated but highly charged metal ball silently, without a spark, by approaching it holding a sharp-tipped metal needle in one of their hands. This discovery led Franklin to speculate in 1750:

> May not the knowledge of this power of points be of use to mankind in preserving houses, churches, ships, etc., by directing us to fix on the highest parts of these edifices, upright rods of iron made as sharp as a needle, and gilt to prevent rusting, and from the foot of these rods a wire down the outside of the building into the ground, or down one of the shrouds of a ship and down her side until it reaches the water? Would not these pointed rods probably draw the electric fire silently out of a cloud before it came nigh enough to strike, and thereby secure us from that most sudden and terrible mischief?[1]

After 1751 Franklin broadened his view of the function of lightning rods. While still holding to the idea that the primary use of a rod was to discharge the cloud silently, he also wrote that an alternative function was to dispose safely any stroke that came to it.[2] The essence of Franklin's final views on the functioning of his rods as stated in a letter from Paris in 1767 was that either his sharp-tipped lightning rods would discharge thundercloud electricity or, if they did not, that a sharp-tipped lightning rod would be the favored strike receptor and could conduct the lightning charge harmlessly to Earth.[3] While both of these ideas can be questioned, there is no denying the success of Franklin's lightning-protection system.

Early Demonstrations of Protection against Lightning Using Franklin Rods

Following Franklin's announcement of the lightning rod in *Poor Richard's Almanac* for 1753, protective, sharp- tipped rods were soon installed on tall structures in the American colonies and in Europe where their utility in connecting to lightning and conducting it to ground was almost immediately demonstrated. After a survey of lightning damage, Schonland recorded in his 1950 book *The Flight of Thunderbolts*:

> The record of damage to churches, whose elevated steeples attract lightning is voluminous. . . . Perhaps the most famous of these structures is the Campanile of St. Mark in Venice. This has had a very bad lightning history. It stands over 340 feet high in an area which, as already mentioned, experiences many thunderstorms. It was severely damaged by a stroke in 1388, at which time it was a wooden structure. In 1417 it was set on fire by lightning and destroyed. In 1489 it was again reduced to ashes. In 1548, 1565, and 1653 it was damaged more or less severely, and in 1745 a stroke of lightning practically ruined the whole tower, Repairs cost 8,000 ducats (3,000 pounds sterling in those days), but in 1761 and 1762 it was again severely damaged. In 1766 a Franklin rod was installed on it and no further trouble from lightning has occurred since.[4]

A report of the successful use of Franklin's rods in Siena has been provided by Krider.[5] Anderson reports on the same event and notes that the protection of the 102-m high tower by a Franklin rod on the city hall when it was struck by lightning on April 18, 1777 prompted the Senate of Venice to issue "a decree ordering the erection of lightning rods through the republic. It was the first recognition of the value of conductors by any government."[6]

The early installations of Franklin's lightning rods were on tall structures, high above other parts of the buildings, and most of these installations were quite successful as strike receptors. In later years, as a need arose for the protection against lightning damage on smaller structures, the standards for the installation were relaxed; they ceased to require, as Franklin had recommended in 1767, that the "rod extend six to eight feet above the highest part of the building." When we look at modern standards for lightning protection, we note that Franklin's recommendation for the height of the rod is not followed. The current American standard for the installation of lightning protection systems, NFPA 780, merely requires that the tip of a lightning rod used as a strike receptor "shall be no less than 10 inches (254 mm) above the object or area it is to protect."[7]

Moreover, during the years after Franklin announced his invention, there have been reports that sharp-tipped Franklin rods are not always the preferred strike receptors.[8] Numerous anecdotal reports have been circulated about lightning strikes to the corners and parapets of buildings equipped with sharp-tipped Franklin rods, but it has not been clear as to whether the installations of the protective systems were faulty or if the protective actions of the rods were limited. Recently, a report of strikes to the edges of building roofs at distances of about 0.5 to 1.0 meter from the installed sharp-tipped Franklin rods was provided by Hartono and Robiah.[9]

A Reexamination of Franklin's Invention

The story of Benjamin Franklin's invention of the lightning rod contains a lesson on the hazards of using preconceived notions to provide explanations for unexpected discoveries. Franklin and his associates had found that bringing a sharp needle near an electrified object caused the object to lose its electrical charge. This led Franklin to suggest that thunderclouds might similarly be discharged by erecting sharp-tipped rods beneath them, thus preventing "the mischief caused by lightning." However, when he tried this idea, sometime around 1752, with the erection of a grounded iron rod in the upper tip of which a small brass wire was inserted, instead of preventing lightning, his rod was struck by lightning. Even though the outcome of his first experiment with a lightning rod was exactly opposite to what Franklin apparently expected, he was so enamored with the "power of a point" in discharging the electricity carried on electrified objects in his demonstrations that he attributed the strike to his rod as being caused by its sharp tip. He also continued to hold that charge neutralization was an important function for his rods even though he had no evidence that they did so and he had some clear evidence that they did not. This is not surprising, given the limited knowledge of physics at the time. From his writings, it appears that he did not appreciate the scale of a thundercloud; even if he had understood that, he had no way to calculate the time that would have been required for the emissions from his rod to have neutralized one.

More significantly, he ignored the long-known observations dating from Roman and Greek antiquity (e.g., Lucretius, 55 BCE) that lightning preferentially strikes elevated structures. His writings contain no hint that he ever considered that his first rod was struck because its upper end was elevated and that its sharp tip may have had little to do with the initiation of the strike. Instead, he published an account of his experiment in which he recommended the use of sharp-tipped lightning rods, a practice that has continued into modern times, even though the desirability of making the tip sharp for improved strike reception has never been established. In the years since Franklin published his ideas, his rods have generally been effective in providing protection against lightning because they have intercepted strikes and have conducted them to ground without damage to the structures on which they were mounted, largely because of their elevated exposure above competing salients on those structures.

The modern view of Franklin's experiment is that the sole function of a grounded lightning rod is to be a strike receptor. It is now recognized that the emissions from sharp-tipped rods on the Earth do not neutralize the charges in thunderclouds that produce lightning. Most of the lightning strikes to Earth occur over the continents where all well-exposed, earth-connected objects emit charge to some degree when thunderclouds pass overhead. Clearly, these clouds are not being neutralized because they continue to produce lightning as they pass. In addition, our recent studies have shown that the high curvature of sharp-tipped rods, far from enhancing their protective abilities, can actually interfere with their strike reception.[10]

Modern Observations Versus Franklin's Ideas on Lightning Protection

Questions about the functioning of sharp-tipped rods arose in our minds after a long series of measurements of the point discharge currents that flowed into the air from elevated, small-diameter wires (see fig. 12.1) and, later, from sharp-tipped rods exposed to the ambient electric fields under thunderclouds.[11]

It was noteworthy that none of our exposed wires or the sharp-tipped rods was ever struck by lightning over a period of more than thirty years although cloud-to-ground strikes often occurred in their vicinity and the wire supporting balloons were usually burst by discharges nearby. We were puzzled by the observations that the long vertical wires were not struck by lightning. One explanation that occurred to us was that their emissions created space charges around their upper ends, which weakened the local electric fields below the strengths necessary for spark formation. This explanation then raised the question "What would happen if we made the tips of the wires move upward faster than the drift velocities of the ionic space charges?" To test this idea, we used the 5-million volt Van de Graaff generator at the Museum of Science in Boston and placed a length of 0.25-mm diameter wire into the strong electric field region beneath the generator's high-voltage electrode (See fig. 12.2).

When the wire was injected rapidly into the region with strong electric fields, it always initiated a strong spark discharge, but when a stationary wire was exposed in the strong field region, it glowed in the dark with St. Elmo's fire without producing a spark.

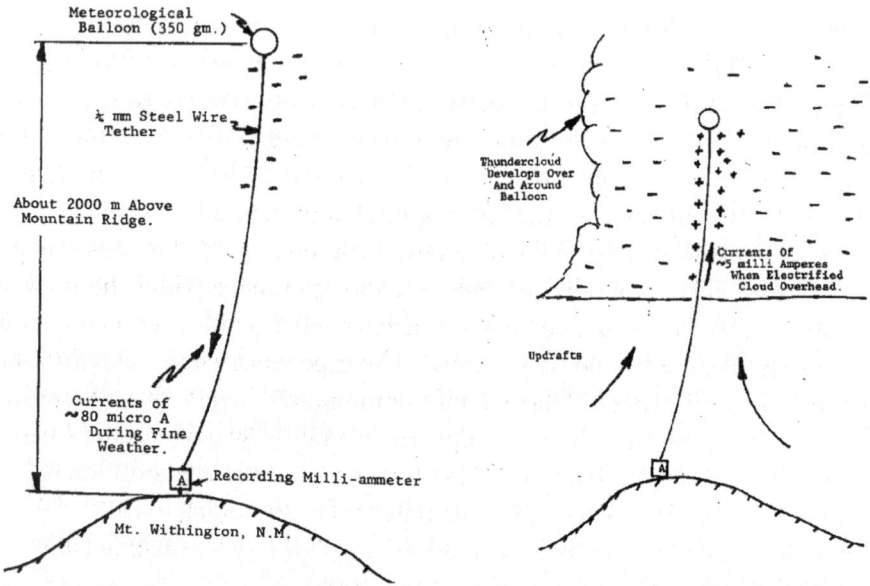

Figure 12.1.   Sketches illustrating an experiment carried out in 1956 aimed at initiating electrification in convective clouds by priming them with corona discharges from the tips of long, 0.25-mm-diameter wires that cut across large potential differences in the fine weather atmospheric electric fields. Although the convective clouds that formed over these vertical wires became electrified and produced lightning that burst the supporting balloons in more than twenty experiments, none of the vertical wires was ever struck by lightning, apparently because of the protection provided by their corona discharge emissions. Moore et al., *Observations of Thunderstorms* (cit. n. 11)

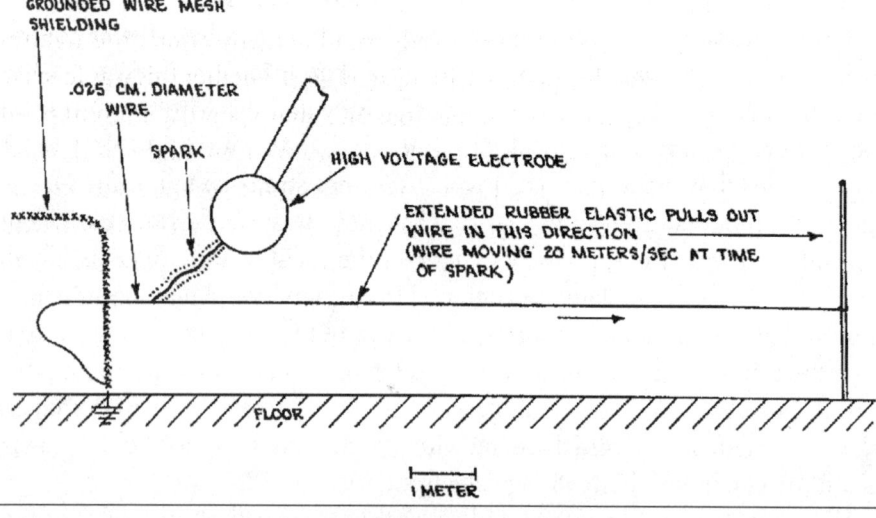

Figure 12.2.   Sketch illustrating experiment with the 5-million-volt generator at the Boston Museum of Science in which a fine wire was injected into strong electric fields in an attempt to initiate spark discharges. Brook et al., "Artificial Initiation" (cit. n. 12).

260

Figure 12.3. Photograph of lightning initiated by use of a wire injected into the air beneath a thundercloud with a small rocket. This lightning was initiated when the rocket and the upper end of the grounded wire reached the altitude at the top of the "bar" of light. At this level, a leader discharge formed and raced upward from the top of the wire.

It is interesting to note that this same effect can easily be demonstrated with a small Van de Graaff high-voltage generator and a sharp-tipped metal object in the hand. To get a spark to the sharp object, it must be moved quite rapidly toward the generator's high-voltage electrode. If the object is simply held in a near-stationary fashion near the electrode, the electric current that the sharp point emits will neutralize some of the charges produced by the high-voltage electrode, thus reducing the electric field applied to the object and no sparks will occur.

The results of the Museum of Science experiment confirmed our hypothesis that the emissions from the wires exposed to strong electric fields created a space charge that opposed the formation of spark-like discharges and caused us to question Franklin's argument for the use of sharp-tipped lightning rods. Soon after we published these results, Morris Newman and his associates initiated lightning by injecting the upper end of a grounded wire towed aloft by a rocket into the air beneath a thundercloud over the ocean.[12] Newman's success has led to the use of this technique for initiating lightning strikes at many different locations. A photograph of lightning initiated by use of a rocket towing aloft the upper end of a grounded wire is shown in figure 12.3.

When we began these studies in 1956, it had long been accepted that Franklin's first idea about the functioning of his rods (their charge emissions neutralize electrified clouds thereby preventing lightning) was not valid. Modern determinations indicate that the charges carried to Earth by a lightning strike are of the order of tens of coulombs while the currents emitted from sharp-tipped rods exposed beneath thunderclouds are of the order of tens of micro coulombs per second.

These measurements indicate that the flow of such point discharge currents for periods of as long as one week would be required to neutralize the charge in a single strike. Under vigorous thunderstorms, cloud-to-ground lightning often occurs at the rate of several strikes per minute, clearly unhampered by the point discharge currents emitted from many exposed objects on the Earth. In 1956, however, there was no consensus of scientific opinion whether sharp-tipped objects offer preferential sites for strike reception or for the formation of sparks. Franklin's ideas were not even questioned.

Current Ideas on the Initiation of Lightning

Modern measurements of the initiation of cloud-to-ground lightning discharges show that they start high above the Earth, at altitudes ranging from 5 to 7 km, where they are initially unaffected by objects and actions on the ground. The electrical breakdown of the air aloft that starts a discharge apparently begins with the liberation of free electrons from air molecules, possibly by collisions with cosmic rays. Under the influence of the very strong electric fields in electrified clouds, these free electrons accelerate and acquire sufficient kinetic energy to liberate other electrons from the air molecules with which they collide. These new electrons in turn may be accelerated, acquiring kinetic energy and liberating more electrons on collisions with other air molecules. The resulting "avalanches" of electrons can become organized into a "streamer" that culminates in the formation of a negatively charged "leader" that descends toward the Earth. Schonland and Collens discovered that these leaders descend from thunderclouds in a stepwise fashion, advancing ten to fifty meters, then pausing for some tens of microseconds before making another step.[13] As one of these "stepped leaders" approaches the Earth, the negative charge carried on its tip intensifies the electric fields acting on objects exposed on the surface of the Earth, accelerating downward any electrons that may be present and initiating an upwardly propagating "return stroke" leader that connects the stepped leader to Earth.

The Electric Fields and Positive Ion Formation
over Sharp-tipped Lightning Rods

Around the highly curved tips of exposed conducting objects, the electric fields are locally intensified manyfold over the ambient strengths.[14] This intensification, however, decreases rapidly with distance from the tip so that most of the action occurs in the immediate vicinity of the tip. Under very strong electric fields, a nearby accelerated electron can acquire sufficient kinetic energy between collisions with air molecules that, on the next collision, causes the liberation of another electron from the impacted air molecule, giving rise to an avalanche of electrons falling into the tip of an exposed conductor.

An electron avalanche leaves behind a trail of positively charged ions created by the loss of electrons from air molecules impacted by the accelerated electrons. Each of these ions is about fifty thousand times more massive than an electron and therefore moves, under the influence of the local electric fields, far more slowly than the avalanching electrons. These electrons "fall" into the tip while the positive ions move slowly away creating a positive ion "space charge" in the air just above the tip, which acts to weaken the electric field acting on it.

The results of this field weakening can be readily seen in time-resolved measurements of the electric currents that flow into the air from sharp-tipped electrodes that are exposed to strong electric fields. The currents are not continuous but consist of repetitive bursts. An example of this pulsating current is shown in figure 12.4.

Under the influence of the strong local electric field, a free electron, possibly liberated from a negative ion in the vicinity, is accelerated in toward the electrode tip, liberating other electrons that avalanche into the tip. The charges on the positive ions, created by the air molecules losing electrons to the avalanche, weaken the local electric field, reducing the electron accelerations so that the avalanches soon cease. No further ionization occurs until the "cloud" of positive ion space charge is cleared away from the tip by its migration under the influence of the external electric field or by the local wind, whereupon the field above the tip again becomes strong, another free electron may appear, and the process is repeated.

Warburg and Chapman found that the current flowing from a sharp-tipped electrode under the influence of a strong electric field varies with the square of the field strength in excess of a threshold value that depends on the tip radius of curvature.[15] We now know that this dependence arises because (a) the strength of the field determines the number of positive ions that may be formed before the cessation of electron avalanching, and (b) the local electric field at the tip is the dominant driving force for the migration of positive ions away from the tip; it controls the time required for clearing of the ions before another series of electron avalanches can begin. As a result, because the current flowing from a sharp-tipped electrode depends on the square of the applied field, any increase in the strength of the local field is counteracted quickly by the creation of more positive ions.

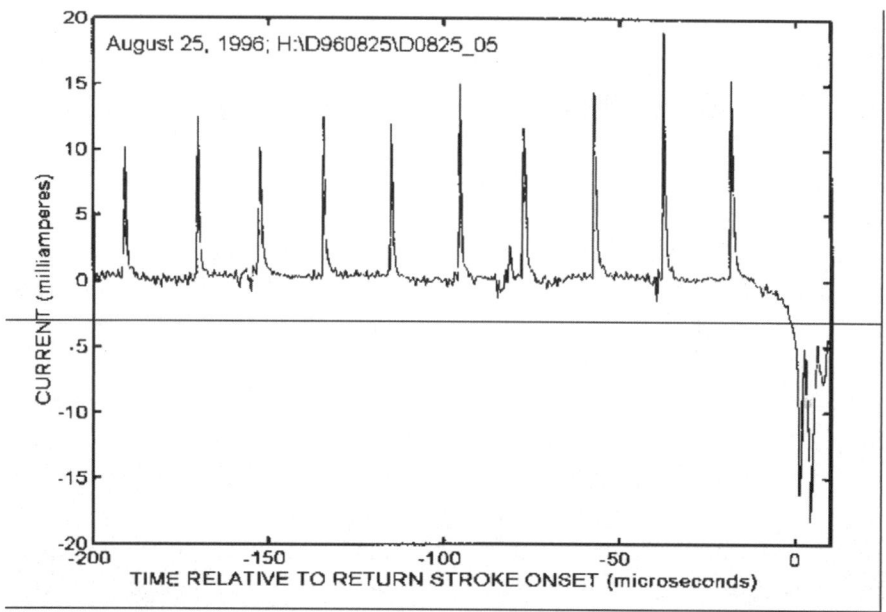

Figure 12.4.   Plot of the current that flowed from a sharp-tipped Franklin rod mounted on top of a six-meter-high mast on South Baldy Peak at the start of a nearby lightning strike on September 7, 1996, showing the repetitive nature of the charge emission bursts from the sharp tip.

Consequently, sharp-tipped grounded electrodes exposed to very strong, nearly constant electric fields do not produce "sparks"; rather, they emit significant "point discharge" currents. For a "spark" to form, the electric field ahead of the cloud of positive ions must become sufficiently strong that electron avalanches occur in that volume, creating more ions at greater distances from the electrode. This can only happen if the local electric fields intensify faster than they are weakened by the formation and transport of positive ions around the tip. The rate of this weakening—that is to say, the rate at which positive ion space charge is cleared from around the tip—has been shown in recent studies to be dependent on the tip radius of curvature.[16] The larger the radius of curvature, the more slowly the ions clear the tip, which means that a lower rate of field intensification is required for a spark to form. This implies that a lightning rod with a blunt tip would be a better receptor than one with a sharp tip.

Competitions for Strike Reception between Sharp-tipped and Blunt Lightning Rods

Tests of these findings have been made by arranging competitions for strike reception between UL-listed, sharp-tipped Franklin rods and variously-curved

Figure 12.5.   Photograph of three competing lightning rods installed above an underground laboratory, the Iron Kiva, on top of South Baldy Peak in the Magdalena Mountains of west-central New Mexico.

blunt rods constructed from round cylinders of aluminum, the upper ends of which were made into polished hemispheres having the same diameters as those of the rods from which they were formed.[17] In our experiments, three competing rods were installed on six-meter-high masts near the summit of South Baldy Peak in the Magdalena Mountains of west-central New Mexico (see figure 12.5). The masts were separated by distances approximately equal to their height so that the calculated electrostatic perturbation of the electric field at the tip of one lightning rod caused by the presence of an adjacent rod is an enhancement of less than 1 percent of the ambient strength. This indicates that, although the three rods were exposed to essentially the same fields under thunderclouds, they responded individually with no significant initial perturbations produced by their neighbors.

As of July 2001, the results of this contest are that thirteen blunt-tipped rods have received lightning strikes while none of their nearby, sharp-tipped competitors have been struck during an eight-year period (see figure 12.6). The scars and discolorations on the rods in the figure were caused by the lightning strikes to the rods.

Figure 12.6.   Photograph of six blunt-tipped lightning rods that have been struck by lightning while in competition for strike reception with nearby sharp-tipped Franklin rod, none of which were struck. The diameter of the two rods on the left was 12.7 mm; that of the rods on the right was 25.4 mm and the diameter of the other rods was 19 mm.

An Assessment of Franklin's Ideas on Lightning Protection

With his fascination for the point-discharge phenomenon and for the "power of points" Franklin never recognized his rediscovery of the long-known phenomenon that is the basis of modern lightning protection practices. Franklin's rediscovery, which he never separated from his views about sharp-tipped rods, is simply that cloud-to-ground lightning preferentially strikes well-exposed, tall, conducting objects that are connected to the earth. In a later letter on how to protect buildings from lightning, Franklin wrote: "Thus the pointed rod either prevents a stroke from the cloud or, if a stroke is made, conducts it with safety to the earth with safety to the building."[18]

Both of Franklin's ideas about the functioning of his sharp-tipped rods have now been shown to be untenable; sharp-tipped Franklin rods neither prevent lightning nor are they the preferred strike receptors when conductors with less curvature in their tips are nearby. Nevertheless, both of these misconceptions are firmly embedded in modern folklore. Fortunately, despite the misconceptions, Franklin rods can continue to provide protection against lightning as long as they have no nearby competitors for strike reception. Our field experiments and our analyses just indicate that the strike reception probabilities of Benjamin Franklin's lightning rods are greatly increased when their tips are made moderately blunt.

At the present time, the major unsolved problem associated with lightning strikes to exposed conductors has to with understanding the "striking distance," the distance from the tip of a descending stepped leader to the object that is to receive the strike at the instant that the successfully connecting upward-going return stroke leader is launched. It has recently been found that the descending stepped leader propagates by ionizing the air ahead of the leader tip.[19] This advanced ionization may be caused by the emission of x-rays or other energetic radiation by the leader but the effect of the stepped-leaders' emissions of energetic radiation on the formation of an upward-going return stroke leader from the object that is to receive the strike has not yet been established.[20]

Notes

1. B. F. J. Schonland, *The Flight of Thunderbolts* (Oxford: Clarendon Press, 1950), 9.

2. B. F. J. Schonland, "Franklin's Work on Thunderstorms," *Journal of the Franklin Institute* 253 (1952): 384.

3. I. B. Cohen, *Benjamin Franklin's Experiments* (Cambridge, Mass.: Harvard University Press, 1941), 129.

4. Schonland, *Flight of Thunderbolts* (cit. n. 1), 9.

5. E. P. Krider, "Lightning Rods in the 18th Century," Second International Conference on Lightning and Mountains, Chamonix Mont Blanc, France 1997.

6. Richard Anderson, *Lightning Conductors: Their History, Nature, and Mode of Application* (London: E.& F. Spohn, 1879), 48.

7. National Fire Protection Association, *NFPA 780, Standard for the Installation of Lightning Protection Systems* (Quincy, Mass.: NFPA, 1997), 7.

8. C. B. Moore, "Improved Configurations of Lightning Rods and Air Terminals," *Journal of the Franklin Institute* 315 (1983): 61–85.

9. Z. A. Hartono and I. Robiah, "A Long Term Study on the Performance of Early Streamer Emission Terminals in a High Keraunic Region," in *Proceedings of the IEEE Asia-Pacific Conference on Applied Electromagnetics*, Shah Alam, Malaysia, 2003: 146–50.

10. C. B. Moore, G. D. Aulich, and William Rison, "The Case for Using Blunt-Tipped Lightning Rods as Strike Receptors," *Journal of Applied Meteorology* 42 (2003): 984–93.

11. C. B. Moore, B. Vonnegut, and A. G. Emslie, *Observations of Thunderstorms in New Mexico*, Report to the Office of Naval Research, 1959, Contract Nonr 1684, 44.

12. M. Brook, G. Armstrong, R. P. H. Winder, B. Vonnegut, and C. B. Moore, "Artificial Initiation of Lightning Discharges," *Journal of Geophysical Research* 66 (1961): 3967–69.

13. B. F. J. Schonland and H. Collens, "Progressive Lightning," *Proceedings of the Royal Society London* A 114 (1934): 654–74.

14. C. B. Moore, W. Rison, J. Mathis, and G. D. Aulich, "Lightning Rod Improvement Studies," *Journal of Applied Meteorology* 39 (2000): 593–609.

15. E. Warburg, "Ueber die Spitzenentladung (On Point Discharge)," *Wiedemanns Annalen d. Phys. u. Chem.* 67 (1899): 67–83; S. Chapman, "Corona Discharge from an Isolated Point," Cornell Aeronautical Report Series, CAL 61 (Buffalo, N.Y.: Cornell Aeronautical Laboratory, Inc., 1967), 59.

16. Moore et al., "Lightning Rod Improvement Studies" (cit. n. 14).

17. C. B. Moore, G. D. Aulich, and W. Rison, "Measurements of Lightning Rod Responses to Nearby Strikes," *Geophysical Research Letters* 27 (2000): 1487–90.

18. Letter XXIV from Paris in 1767, reproduced in Cohen, *Franklin's Experiments* (cit. n. 3), 129.

19. E. M. Bazelyan and Y. P. Raiser, *Lightning Physics and Lightning Protection* (Bristol: Institute of Physics Publishing, 2000), 83–88.

20. C. B. Moore, K. B. Eack, G. D. Aulich, and W. Rison, "Energetic Radiation Associated with Lightning Stepped Leaders," *Geophysical Research Letters* 28 (2001): 2141–44.

Epilogue
An Invisible Technology—What Remains to Be Seen

Peter Heering and Oliver Hochadel

This book began with the claim that the lightning rod, despite its apparent self-evidence and technical simplicity, has a history that has been widely neglected. The contributions in this volume aimed at making visible several facets of the lightning rod's historical complexity. Yet no complete image emerges. Several aspects have been omitted or merely touched upon. In this epilogue, we will suggest some of these aspects that, in our view, merit further investigation and highlight the rich potential of the lightning rod as an object of interdisciplinary research.

Alternative Protection Devices

Our focus on the lightning rod omitted almost completely "alternative" practices of protecting against lightning. In some of the essays, the ringing of church bells is mentioned, but only as something that was about to be abolished. Yet why not look at this well-established practice from a different perspective? When lightning became an accepted electrical phenomenon and other practices were criticized, attempts were made to develop a rational explanation for these protection procedures. "Thunder can be disrupted and diverted by the sounds of several bells or the firing of a cannon; thereby great agitation is excited in the air which disperses the parts of the thunder" claimed no lesser mind than Pieter van Musschenbroek, discoverer of the Leyden jar in the flagship publication of the Enlightenment itself, the *Encyclopédie*.[1] Similarly, French natural philosopher Jean-François Pilâtre de Rozier attempted to scientifically justify the ringing of church bells by theorizing that lightning was the result of an accumulation of inflammable gases that were meant to be dissipated through the sound of the bells.[2]

There were, of course, countless other practices of protection against lightning, some dating back to antiquity. To name but a few, in the Mediterranean people put laurel leaves or branches of olive trees on top of their houses, around their fields, or on their beds (Bertucci); the bell tower of St. Stephen in Vienna had been protected by antlers since the sixteenth century; and individuals protected themselves by wearing amulets made of oak wood.[3]

One might also wonder what means of protection were practiced in non-European cultures. What kind of understanding did or do they have of phenomena such as lightning? To give an example from Southern Africa, "Among most of the Bantu tribes . . . the belief is general that lightning is produced by a magical thunderbird, Umpundulo, which dives from the clouds to earth and whose vivid plumage and beating wings give rise to the flash and to the thunder."[4] In Japan people used specially made bamboo instruments that produced frightening sounds when waved to keep rice seedlings safe from lightning. This kind of magic is called "kandachioi," which means "escorting the heavenly power to another place." Personal protection was sought through the use of mosquito nets and from special antilightning pills, which Japanese noblemen carried in small boxes.[5]

All these practices deserve study in their own right. They point to different systems of meaning and consequently different ways of communicating with nature or God. So far this has been the domain of anthropologists. But as far as we can see, there are only a few scattered articles but no systematic investigation of these alternative practices, let alone comparative approaches.

Admittedly, this volume has been fairly Eurocentric. Except for Clark's essay on Spanish Mexico, we did not deal with questions of how the lightning rod was introduced in South America, Africa, or Asia. Yet this might be a very fruitful approach. Was it similar to or entirely different from Europe and North America? Which groups "pushed" the new technology in India, Egypt, and China, and what were their motivations? How were lightning rods adapted by these countries in terms of both their design and their cultural meaning? In this context, the "contest" between Western technology and indigenous knowledge should receive more attention. Did the lightning rod lead to conflicts with older practices of protecting house and home against lightning? As recent studies in colonial science clearly show, it would be misleading to assume a mere "transfer" of Western knowledge. The interactions between colonizers and the colonized were far more complex.[6]

A Mighty Metaphor: Repercussions of the Lightning Rod in the Arts

Christian Fuhrmeister offers a first look at the meaning of lightning and lightning rods in the realm of political propaganda. Yet this is the only essay in which the reception of the lightning rod is discussed explicitly beyond the debates about its utility and design. However, other transfers and metaphorical uses could be equally telling. Not only the fine arts but also music and literature are sources of interest for research on the perception of lightning and its change through the acceptance of the lightning rod.[7]

A broader, more interdisciplinary analysis might produce an even more differentiated and thus telling image of the cultural impact of lightning rods. Here it would be of particular interest if different cultures were compared because this may also help to develop insights into the transfer mechanism mentioned earlier. Again, despite some pioneering works, there seems much left to discover with respect to the appropriation of technological inventions such as the lightning rod. What does a piece of music, literature, or art dealing with lightning or the lightning rod tell us about changing mentalities? A piece such as Johann Strauss' polka "Unter Blitz und Donner" ("Lightning and Thunder," composed in 1867–68) cannot be composed if one is afraid of thunderstorms, as Roland Schmenner pointed out. A more complicated case is Beethoven's Sixth Symphony (the "Pastorale") composed in 1807–8, in which the fourth movement emulates a thunderstorm. Already by that time, it seems, audiences were able to appreciate such dangerous spectacles of nature.[8]

New Technologies and the Military

Elizabeth Cavicchi showed in her essay how in the nineteenth century the new technology of telegraphy led to problems with regard to protection against lightning. What technologies developed in the twentieth century—e.g., aviation and microelectronics—that may have posed new challenges for lightning protection? In August 2006 the Space Shuttle, for example, was struck by lightning while stationed on the launch pad, much to the distress of NASA. Again, we might ask how the required protection could be ensured and what approaches were used.

Such a study could also be related to the essays of R. W. Home and Peter Heering. They make clear that lightning rods were a relevant topic for the military. The protection of powder magazines as well as warships was a major issue at least up to 1850. As technologies such as airplanes and microelectronics were adopted by the military, their protection also became a question of military relevance. Lightning rods might thus be an intriguing topic in the analysis of the connection between the demands of the military and scientific research. This appears to be of particular interest because the problems posed in this case are different from those in standard applications. Microelectronic components, for instance, can be damaged by induced currents, which common lightning protection techniques may not be able to prevent.

With respect to modern research on protective technologies against lightning, computer simulations have become an increasingly important research tool.[9]

These tools raise questions comparable to those discussed by Willem Hackmann with respect to model experiments. How can we be sure that the tools of analysis are adequate with respect to large-scale phenomena?

Archeology of Technology

Nowadays historical lightning rods are hard to find. In Philadelphia's Independence Hall, where one of the first lightning rods was installed in the fall of 1752 (when that building served as the Pennsylvania State House), fragments of the down conductor were found under the plaster. Philip Krider has shown that the technical details of the initial design of the lightning rod were changed very quickly.[10] It only took a few years, however, for the design to "stabilize," and Krider and many others correctly claim that the basic principles of the lightning rod are still the ones formulated by Franklin. Yet this claim tends to level past and present, turning the lightning rod into an ahistorical and therefore invisible object. These basic principles are not identical with the technological object and its scientific understanding. The necessity for this distinction becomes obvious in the face of the continuing debates on the proper technical design and controversies about how to improve the lightning rod. Even though the basic principles of the lightning rod remained unchanged, the historical actors were still able to interpret these principles in various different technical or conceptual ways. One example would be the question how far the protection of the rod extended and how high the rod should be elevated above the roof. Therefore, the analysis of historical lightning rods should not be limited to the very first examples. It should include later artifacts in order to understand the development of the lightning rod.

The Weather

In the late eighteenth century, French émigré Charles Volney was stunned by the ferocity of thunderstorms in Philadelphia: the difference between Europe and North America with respect to the electricity of the atmosphere "may be made perceptible to the senses at any time, without any complicated apparatus."[11] And indeed, the number and intensity of thunderstorms vary widely from region to region. Perhaps this is part of the explanation as to why the lightning rod spread much faster in British America than it did in Europe. However, taking Clark's contribution into consideration, it becomes obvious that one has to be very cautious in establishing causal relationships: José Antonio Alzate y Ramírez claims that, due to its natural conditions, Spanish Mexico is particularly in need of protection from lightning; nonetheless, in Mexico the lightning rod was established only very slowly.

However, it is not only the climate (*longue durée*) but also the weather (short-term conditions, single events) that deserves closer attention. Despite some substantial pioneering efforts, there is still an enormous potential for interdisciplinary work.[12] To give but one example, Hochadel's essay is an attempt to link climatology to history of science and technology, claiming that the dry fog of 1783 triggered the erection of lightning rods on a large scale in the German Empire. This is of particular interest because the dry fog of 1783 was a single incident. Yet the question of to what extent climatic conditions or individual weather events influence the acceptance of protective technologies such as the lightning rod could be discussed in an even broader context of how people historically experienced catastrophes and the economic consequences of extreme weather.[13]

Commodification, Standardization, and Risk Management

This volume focused a great deal on the lightning rod in the eighteenth century but said less about developments in the nineteenth century and even less about the twentieth. We very much suspect that there are more histories hidden in this later period. For example, the lightning rod did change substantially over time as a physical object and technological artifact. Many kinds of material were used to construct the rod. The rod itself diversified; the "simple" (singular) rod has been succeeded by larger systems of multiple rods. Scientific theories of lightning have changed (and are still changing). Methods of production and distribution have varied, and industrial standards have evolved as well. As mentioned in the introduction, from very early on economic considerations played a role in whether lightning rods were erected. Therefore, the transition from individual manufacturing to industrial mass production, a topic addressed by Mohun, deserves more attention. This is closely tied to the question of standardization. When were standards developed for lightning rods, on what basis, by whom, and who was in charge of controlling them? Insurance companies seemed to have played a major role here.

The late eighteenth century not only witnessed the spread of the lightning rod but also the birth of the insurance industry (with considerable local variations), in particular fire insurance. Risk was commodified, as Arwen Mohun demonstrated, for the United States in the antebellum period. To follow that lead with respect to modern technologies might yield new insights into the business around risk and havoc caused by nature. Every year, lightning caused — and still causes — enormous damage. Calculations of the harm done are notoriously difficult, but it is estimated that in a small country such as Austria it amounts to 100 million U.S. dollars per year.[14] This includes not only the destruction of property but also the loss of profits suffered by enterprises due to interruptions of electrical power caused by lightning. The fact that electronics play such an important role in business and in our everyday life makes reliable protection devices more important than ever. In other words, there is a huge market for lightning protection systems, which brings us to our final point.

Current Debates

The essay of Charles Moore, Graydon Aulich, and William Rison provides insights into current questions of lightning rod research. One major bone of contention since the 1970s, particularly in recent years, has been new, allegedly "high-tech" lightning protection devices such as the so-called charge transfer systems (CTS) and, more importantly, early streamer emission (ESE) technology. The idea behind CTS is nothing less than to eliminate lightning by producing "copious amounts of charge that will neutralize the cloud or at least form a cloud of space charge that will neutralize the downward leader." The predecessors of the ESE were the radioactive rods mentioned in the introduction: "An ESE device releases a charge at its tip earlier than done by a Franklin rod. The charge is claimed to form a giant upward streamer which acts as an extension to the rod." This is supposed to vastly increase "the protective range of the rods."[15]

CTS and ESE have been aggressively (and successfully) marketed by their makers. The established scientific community thinks little if anything of these new devices, holding that they provide at best the same measure of protection but at a much higher price.[16] The companies have responded by threatening to sue scientists or organizations who publicly question or denounce the utility of their devices. In at least one case, a private body, the National Fire Protection Association, was brought to court. The vendors of CTS and ESE systems have also "retaliated" by questioning the scientific basis of conventional protection lightning systems.[17]

These recent disputes about alternative lightning protection devices provide all the ingredients for an intriguing case study in the sociology of scientific knowledge: they raise questions about authority in science and in the courts; the battle over standards among insurance companies, vendors and experts; and the use and abuse of science in the marketing of technological products. In this respect, it might be telling to compare the strategies for establishing authority in this modern controversy with the ones used during the eighteenth and nineteenth centuries.

We have come a long way. More than 250 years after its invention, the lightning rod, a seemingly simple artifact, lends itself to a host of fruitful interdisciplinary approaches. The intriguing question of non-Western appropriation of the lightning rod, the attempt to situate history of technology within environmental history by taking a closer look at the role of climate and the weather, and the continuing discussions on the proper design of the protection mechanism in which technological disputes are closely interwoven with economic and legal wrangling are but three promising avenues for further research. It has been the aim of this collection of essays to historicize the lightning rod and to make an invisible object visible again. Still, much more remains to be seen.

Notes

1. I. Bernard Cohen, "Prejudice against the Introduction of Lightning Rods," *Journal of the Franklin Institute* 253 (1952): 393–440, at 396n9.

2. See, e.g., Jean-François Pilâtre de Rozier, *La Vie et les Mémoires de Pilâtre de Rozier* (Paris, 1786), 103–9. Looking at the rationale of the proponents of the lightning rods reveals that things were not simply right or wrong. For example, Langenbucher argued (correctly from our modern perspective) that the ringing of church bells has no protective value but had rather the opposite effect. Yet he assumed (incorrectly from our modern perspective) that due to the movement in the air that is produced by the sound, the air will be more electrified, and instead of weakening the thundercloud, it will be strengthened. See Jakob Langenbucher, *Richtige Begriffe vom Blitz und von Blitzableitern* (Augsburg, 1783), 33f.

3. Joachim Knuf, "Traditioneller Blitzschutz als Kommunikationsproblem: Anmerkungen zu materiellen und nichtmateriellen Zeichen," *Volkskunst. Zeitschrift für volkstümliche Sachkultur* 8 (1985): 18–22; Viktor Flieder, "Die Hirschgeweihe von St. Stephan in Wien," *Österreichische Zeitschrift für Volkskunde* 69 (1966): 261–66. With respect to the German-speaking states, see also Karl-Heinz Hentschel, *Kleine Kulturgeschichte des Gewitters*, http://www.karl-heinz-hentschel.net/Gewitter2.html, February 10, 2007.

4. Basil F. J. Schonland, *The Flight of Thunderbolts* (Oxford: Clarendon, 1950), 1. Schonland also discussed the understanding of lightning in ancient Egypt as well as in Greek and Roman culture. For another brief discussion of the understanding of lightning in other cultures, see Hans Prinz, "Gewitterelektrizität (Nach dem nachgelassenen Manuskript bearbeitet durch Heinz Steinbigler)," *Deutsches Museum: Berichte und Abhandlungen* 47, no. 1 (1979): 5–74, esp. 9–17.

5. We owe this information to Hans-Joachim Knaup (Keio University, Yokohama), who presented a paper on "Lightning as an Object of Adoration and Fear in Japan" at the 2002 conference at The Bakken.

6. To give but two recent examples, James Delbourgo and Nicholas Dew, eds., *Science and Empire in the Atlantic World* (London: Routledge Chapman & Hall, 2007); and Benedikt Stuchtey, ed., *Science across the European Empires 1800–1950* (Oxford: Oxford University Press, 2005).

7. To give but two examples from German romantic literature, Maria M. Tatar, *Spellbound: Studies on Mesmerism and Literature* (Princeton, N.J.: Princeton University Press, 1978), especially the chapter "Thunder, Lightning, and Electricity: Moments of Recognition in Heinrich von Kleist's Dramas," 82–120; and Maximilian Rankl, *Jean Paul und die Naturwissenschaft* (Frankfurt a.M./Bern: Lang, 1987), 269–72.

8. Roland Schmenner, *Die Pastorale: Beethoven, das Gewitter und der Blitzableiter* (Kassel: Bärenreiter, 1998), 3.

9. See, e.g., Edward R. Mansell, Donald R. MacGorman, Conrad L. Ziegler, and Jerry M. Straka, "Simulated Three-Dimensional Branched Lightning in a Numerical Thunderstorm Model," *Journal of Geophysical Research* 107 (2002): ACL 2–1—ACL 2–12; and H. Lammer, T. Tokano, G. Fischer, W. Stumptner, G. J. Molina-Cuberos, K. Schwingenschuh, and H. O. Rucker, "Lightning Activity on Titan: Can Cassini Detect It?" *Planet: Space Science* 49 (2001): 561–74.

10. E. Philip Krider, "Benjamin Franklin and Lightning Rods," *Physics Today*, 2006: 42–48; for eighteenth century lightning rods in southwest Germany that still exist, see Rainer Laun, "Historische Blitzableiter. Eine unbeachtete Gattung technischer Kulturdenkmale," *Denkmalpflege in Baden-Württemberg* 15 (1986): 85–92.

11. Quoted in James Delbourgo, *A Most Amazing Scene of Wonders: Electricity and Enlightenment in Early America* (Cambridge, Mass.: Harvard University Press, 2006), 50.

12. Emmanuel Le Roy Ladurie, *Histoire du climat depuis l'an mil* (Paris: Flammarion, 1967); Hubert H. Lamb, *Climate, History and the Modern World*, 2nd. ed. (London: Routledge, 1995); and Christian Pfister, *Climatic Variability in Sixteenth-Century Europe and Its Social Dimension* (Dordrecht: Kluwer, 1999).

13. The perception of extreme weather conditions is discussed, for example, by Michael Kempe and Christian Rohr, eds., *Coping with the Unexpected—Natural Disasters and Their Perception, Environment & History*, vol. 9, no. 2, Special Issue (Isle of Harris: The White Horse Press, 2003). Studies in this field may also become exemplary with respect to Chang's concept of "complementary science"; see Hasok Chang, *Inventing Temperature: Measurement and Scientific Progress* (Oxford: Oxford University Press, 2004).

14. Interview with Gerhard Diendorfer, director of the Austrian Lightning Locating System (ALDIS), Vienna, June 2002.

15. Abdul M. Mousa, "War of the Lightning Rods," *Electricity Today*, 2004: 45–47, at 46.

16. See ibid. and the references he gives as well at www.lightningsafety.com, in particular William Rison, "There is no Magic to Lightning Protection: Charge Transfer Systems do not Prevent Lightning Strikes," http://www.lightningsafety.com/nlsi_lhm/magic.pdf, July 29, 2007, 1–4.

17. Mousa, "War of the Lightning Rods" (cit. n. 15), 45.

Contributors

GRAYDON AULICH earned a B.S. in physics and a Ph.D. in meteorology. Currently he is an atmospheric research scientist at the Langmuir Laboratory for Atmospheric Research, New Mexico Institute of Mining and Technology, Socorro, New Mexico. Among his recent publications are "Lightning Rod Improvement Studies" (*Journal of Applied Meteorology* 39 [2000], with C. B. Moore, W. Rison, and J. Mathis); "Measurements of Lightning Rod Responses to Nearby Strikes" (*Geophysical Research Letters* 27 [2000], with C. B. Moore and W. Rison).

PAOLA BERTUCCI is an Assistant Professor of History of science at Yale University. She has published extensively on eighteenth century electricity, and she is the author of a book on the Italian journey of Abbé Nollet: *Viaggio nel paese delle meraviglie: Scienza e curiosità nell'Italia del Settecento* (Turin, 2007) and co-editor (with Giuliano Pancaldi) of *Electric Bodies: Episodes in the History of Medical Electricity* (Bologna, 2001).

PAOLO BRENNI is researcher for the Italian CNR (National Research Council) and works between Florence (Istituto e Museo di Storia della Scienza and Fondazione Scienza e Tecnica) and Paris (associate researcher to the Centre de Recherche en Histoire des Sciences et des Techniques). In 2000 he was given the Sarton Chair at the University of Ghent for the academic year 2000–2001, and in 2002 he was awarded the Bunge Preis of the H. Jenemann Foundation for his work in the history of instruments. Since the end of 2002 he has been president of the Scientific Instrument Commission of the International Union of History and Philosophy of Science.

ELIZABETH CAVICCHI is an instructor at MIT's Edgerton Center. Her teaching involves students in their own explorations of science phenomena and historical experiments. She researches nineteenth century electromagnetism by re-creating apparatus and experiments. She received the Ed.D. in 1999 from Harvard University; master's degrees at Harvard, Boston University, and MIT; and bachelor's degree at MIT. She was a postdoctoral fellow at the former Dibner

Institute, MIT and has taught at University of Massachusetts in Boston and Lowell, MIT, and Harvard Graduate School of Education. Her papers include "A Witness Account of Solar Microscope Projections: Collective Acts Integrating across Personal and Historical Memory" (*British Journal for the History of Science*, 2008) and "Nineteenth Century Developments in Coiled Instruments and Experiences with Electromagnetic Induction" (*Annals of Science*, 2006).

FIONA CLARK (Ph.D. Queen's University Belfast, 2003) is lecturer in Colonial Latin American Studies at Queen's University Belfast. Until December 2006 she was Irish Council for the Humanities and Social Sciences Postdoctoral Fellow at St Patrick's College, Dublin City University, working on medical networks between Ireland, Europe, and Mexico, with a particular focus on the late eighteenth century Irish physician Daniel O'Sullivan. Her previous research focused on the Mexican literary-scientific periodical press and the use of transatlantic networks in the early modern period.

CHRISTIAN FUHRMEISTER is an art historian who was trained as a stonemason before studying art and English at the universities of Oldenburg, Germany, and Towson State University, Maryland. After completing his studies, he became a member of the DFG Research Group "Political Iconography" from 1994 to 1997 and received his Ph.D. in art history from the University of Hamburg in 1998. From 2000–2002 he held a position at the Sprengel Museum Hannover, from 2002–2003 he was executive secretary of the Department Kunstwissenschaften at the Ludwig-Maximilians-University Munich. Since 2003, he is project manager at the Zentralinstitut für Kunstgeschischte in Munich.

WILLEM D. HACKMANN retired as Senior Assistant Keeper of the Museum of the History of Science, University of Oxford. He is Emeritus Reader in the History of Science at Oxford, an Emeritus Fellow of Linacre College at Oxford, and for a short time the Honorary Archivist of the Clarendon Laboratory at Oxford. He has written papers on aspects of eighteenth century experimental philosophy, in particular in relation to electrical instrumentation and on visualization in science, and is the author of *Catalogue of the Pneumatic, Magnetic, Electrostatic, and Electromagnetic Instruments in the Museo di Storia della Scienza* (Florence: Giunti, 1995); *Seek & Strike: Sonar, Anti-submarine Warfare and the Royal Navy, 1914-1954* (London: HMSO, 1984); and "Electricity from Glass: the history of the frictional electrical machine" (Alphen aan den Rijn: Sijthoff & Noordhoff, 1978), and joint editor (with Anthony Turner) of *Learning, Language and Invention* (Aldershot: Variorum, and Paris: Société Internationale de l'Astrolabe, 1994).

PETER HEERING is Professor of Physics and Physics Didactics at the Universität Flensburg. His current research in history of physics focusses on the analysis of experiment practices and on the history of science education. He published *Das Grundgesetz der Elektrostatik: Experimentelle Replikation und wissenschaftshistorische Analyse*, Wiesbaden 1998 (DUV) and edited *Welt erforschen—Welten konstruieren: Physikalische Experimentierkultur vom 16. bis zum 19. Jahrhundert*, Oldenburg 1998 (Isensee), (together with Falk Rieß and Christian Sichau) *Im Labor der Physikgeschichte: Zur Untersuchung historischer Experimentalpraxis*, Oldenburg 2000 (BIS), and (together with Daniel Osewold) *Constructing Scientific Understanding through Contextual Teaching*, Berlin 2007 (Frank & Timme).

OLIVER HOCHADEL is a historian of science and currently a Postdoctoral Fellow at the Centre d'Estudis d'Història de la Ciència (CEHIC) at the Universitat Autònoma de Barcelona working on the history of human-origins-research in the late twentieth century. He is the editor (with Ursula Kocher) of a volume on lying and fraud: *Lügen und Betrügen: Das Falsche in der Geschichte von der Antike bis zur Moderne*, Cologne 2000 (Böhlau), and the author of a book on electricity as a public science: *Öffentliche Wissenschaft. Elektrizität in der deutschen Aufklärung*, Göttingen 2003 (Wallstein). He also works as a science journalist, edits the Austrian science magazine *heureka!* and is the cofounder of a postgraduate program in science communications in Vienna.

R. W. HOME was Professor of History and Philosophy of Science at the University of Melbourne, 1975–2003, and is now Emeritus Professor. He has published widely on the history of physics, especially in the eighteenth century, and on the history of Australian science.

ARWEN MOHUN is an Associate Professor of History at the University of Delaware. She is the author of *Steam Laundries: Gender, Technology, and Work in Great Britain and the United States, 1880–1940* (Johns Hopkins University Press, 1999), and co-editor of two collections of essays: *His and Hers: Gender, Technology, and Consumption* (University of Virginia Press, 1998); and *Gender and Technology: A Reader* (Johns Hopkins University Press, 2003). She is currently completing a book-length study of risk and technology in American history.

CHARLES B. MOORE is retired as Professor of Atmospheric Physics at the New Mexico Institute of Mining and Technology, Socorro, New Mexico and a former chairman of the Langmuir Laboratory. He has been involved in research into thunderstorms for fifty years and is the author or co-author of more than fifty works in the scientific literature.

DAVID J. RHEES is Executive Director of the Bakken Museum in Minneapolis and adjunct Assistant Professor in the Program in History of Medicine at the University of Minnesota. He has curated numerous exhibitions, including *Benjamin Franklin and the Lighting Rod* (2002). His research and writing have focused on the history of chemistry, the popularization of science, and the evolution of medical technology.

WILLIAM RISON is a Professor of Electrical Engineering at the New Mexico Institute of Mining and Technology, where he has been involved in lightning research for the past twenty-five years. His research focuses on the design and use of instrumentation for the study of lightning.

Index